Sebastian A. Gerlach

Marine Pollution

Diagnosis and Therapy

With 91 Figures

Springer-Verlag
Berlin Heidelberg New York 1981

Professor Dr. Sebastian A. Gerlach
Institut für Meereskunde
Düsternbrooker Weg 20
2300 Kiel, FRG

Translators

Dr. R. Youngblood
German Department
Washington and Lee University
Lexington, VA 24450, USA

Mrs. S. Messele-Wieser
Frielinger Straße 44
2800 Bremen, FRG

Title of the German Edition
S. A. Gerlach, Meeresverschmutzung
© by Springer-Verlag Berlin Heidelberg 1976

ISBN 3-540-10940-4 Springer-Verlag Berlin Heidelberg New York
ISBN 0-387-10940-4 Springer-Verlag New York Heidelberg Berlin

Library of Congress Cataloging in Publication Data. Gerlach, Sebastian A. Marine pollution. Rev. translation of: Meeresverschmutzung. Bibliography: p. Includes index. 1. Marine pollution. I. Title. GC1085.G4713 1981 628.1′686162 81-9267 AACR2

This work is subject to copyright. All rights are reserved, whether the whole or part of the material is concerned, specifically those of translation, reprinting, reuse of illustrations, broadcasting, reproduction by photocopying machine or similar means, and storage in data banks. Under § 54 of the German Copyright Law, where copies are made for other than private use, a fee is payable to 'Verwertungsgesellschaft Wort', Munich.

© by Springer-Verlag Berlin Heidelberg 1981.
Printed in Germany.

The use of registered names, trademarks, etc. in this publication does not imply, even in the absence of a specific statement, that such names are exempt from the relevant protective laws and regulations and therefore free for general use.

Offsetprinting and bookbinding: Konrad Triltsch, Graphischer Betrieb, Würzburg.
2132/3130-543210

Preface

When, in 1966, the German Research Society directed the attention of oceanographers in the Federal Republic of Germany to problems of marine pollution, I was not enthusiastic. Emphasis on this problem area meant that other important research plans had to be postponed. But the lectures at the Third International Oceanographic Congress, September 1970, in Tokyo, and at the FAO Conference on Marine Pollution and its Effects on Living Resources and Fishing, December 1970, in Rome, convinced me that research on problems of marine pollution is a social obligation, and that the oceanographer has to take a stand. I issued public warnings about the continuing use of pesticides and had to defend myself against protests by the fishing industry and many colleagues who were, in November 1970, unaware of the extent of the threat. Thus, I was required by my profession to acquire an overview of the problems of ocean pollution.

In 1971 I only needed to familiarize myself with some one hundred bibliographical items. In the interim, the flood of data has risen dramatically, and in the year 1975, no fewer than 868 publications under the heading of "Marine Pollution" were reported (Table 1). It is, therefore, more and more difficult to distinguish new results of scientific research from the many repetitions and variations, and I fear that from year to year my efforts to illustrate the actual status of the problem at a given moment will be subject to more gaps.

It will be apparent in my book that I was not a commission member dealing with the questions of effluents and did not serve as an advisor on legislation. My main concern was with those long-term effects and dangerous materials which could produce a global threat. I also witnessed the progress in the field of marine pollution research, particularly associated with the numerous insights which are presently revolutionizing knowledge of marine chemistry and biology.

In recent years, I have given lectures on the problems of ocean pollution at the Universities of Hamburg, Bremen, Copenhagen and Buenos Aires. While doing so, I felt the lack of a detailed textbook as keenly as did my audiences, who would have liked to pursue controversial questions in detail. Consequently, the first edition of this book in German came into being during my two years as professor of marine biology at the University of Copenhagen and at the Marine Biology Laboratory in Helsingør (Gerlach 1976).

This book provides a broad view of fundamental problems of marine pollution which are evident in all parts of the world oceans, on all shores. The reader will realize, however, that examples are included from the German coasts and from research activities in the Federal Republic of Germany, whenever there was a choice. To those wishing to inform themselves in more detail about North Sea problems seen from the German angle, I recommend the *North Sea Opinions* (Rat von Sachverständigen, 1980) which appeared while I was finishing this book. Valuable information not completely included in the text of this book can be found in the Proceedings of the Symposium *Protection of Life in the Sea*, Helgoland 1979 (Kinne and Bulnheim 1980). To the reader who wants more detailed information about the general problems, I recommend the book on marine pollution by R. Johnston (1976a).

For this second, English language edition, the text has been revised commensurate with the rapid progress of research in the field of ocean pollution. A particularly large number of changes will be found in the chapter on oil pollution; new chapters on waste heat and radioactivity in the ocean have been added. My thanks are due to the scientists at Institute of Marine Research in Bremerhaven and to many other colleagues for corrections and stimulating discussions, to the librarian Miss Heineking for unstinting bibliographical assistance, and to Mr. Hagen Westphal for the illustrations.

The first edition of this book was dedicated to the people of Minamata who suffered from mercury poisoning [the book of W.E. and A.M. Smith (1975) keeps their memory alive], and to the terns, seagulls and auks, to the petrels, pelicanes and other fish-eating feathered creatures who will be condemned to extinction if marine pollution is not stopped.

This second edition is dedicated to the memory of Klaus Grasshoff who died, much to early, on March 11, 1981.

Bremerhaven, June 1981   Sebastian A. Gerlach

Contents

1 Introduction

1.1 General 1
1.2 The Definition of Pollution 4
1.3 Units of Measurement and Seawater Composition . 4

2 Domestic Effluents

2.1 Biodegradable Organic Substances 6
2.2 Infections 13
2.3 Eutrophication 16
2.4 The Situation in the Baltic 20
2.5 Global Eutrophication? 27
2.6 Sewage Treatment Plants in Coastal Regions? 27
2.7 Detergents 28
2.8 Residual Heat 29
2.9 Gradual Coastal Changes 33

3 Industrial Effluents

3.1 General 37
3.2 Effluents Containing Mercury 37
3.3 Effluents from Pesticide Plants 47

4 Pollution of the Sea by Ships

4.1 Discharging of Wastes on the High Seas 53
4.2 Waste Acids and Ferrous Sulfate of the Titanium Pigment Industry and the Problem of the Indicator Communities 57
4.3 Discharging of Dredge Spoil and Sludge 64
4.4 Toxic Substances in Antifouling Paints 68
4.5 Garbage from Ships 69

5 Oil Pollution

5.1 The Composition of Petroleum 71
5.2 Fate of Oil on the Surface of the Ocean 72

5.3	Sources of Marine Pollution by Petroleum	80
5.4	Effect of Oil on Marine Life	87
5.5	Combating Oil Pollution	96

6	**Radioactivity**	
6.1	Natural Background Radioactivity and Fall-out	104
6.2	Reprocessing Plants, Plutonium Plants, and Nuclear Reactors	109
6.3	Radioactive Waste	112
6.4	Effects of Radioactivity	116

7	**General Problems of Harmful Substances in the Sea**	
7.1	Toxicity	120
7.2	Accumulation	126
7.3	Geochemical Processes	136
7.4	Global Considerations	143

8	**Global Contamination of the Oceans by Heavy Metals**	
8.1	How Much Mercury May Be Tolerated in Fish for Human Consumption?	146
8.2	Mercury in Large Fish, Seal, and Open Sea Marine Birds	149
8.3	Mercury in the Water of the World's Oceans	151
8.4	Sources of Mercury in the Sea	152
8.5	Has Man Increased the Concentration of Mercury in the World's Oceans?	154
8.6	Contamination of the Oceans with Cadmium	158
8.7	Contamination of the Oceans with Lead	160

9	**Global Pollution of the Oceans with Chlorinated Hydrocarbons**	
9.1	General	164
9.2	What Quantity of Chlorinated Hydrocarbons May Be Tolerated in Marine Food for Human Consumption?	169
9.3	Ways of Transport, Transformations, and Concentrations	174
9.4	Effects of Chlorinated Hydrocarbons	179

10	**Laws Against the Pollution of the Oceans**	184
11	**Diagnosis and Therapy**	191
References		196
Subject Index		213

1 Introduction

1.1 General

In the years 1968–1972, public concern was greatly attracted to the problem of the pollution of the ocean: press and television reported on it and marine scientists were pressured by politicians research the questions of marine pollution. In the meantime, public interest has turned elsewhere to the energy question and economic crises, and interest in things relating to marine pollution has decreased. In 1980, however, dumping of sewage sludge and of wastes from titanium dioxide fabrication caused considerable public concern in the Netherlands and in the Federal Republic of Germany. There is a general feeling that the sea, the last paradise on earth, must be kept clean at any price. However, according to my information, there are good reasons for a decreasing interest in matters of marine pollution:

First, measures were taken to lessen or even to prevent pollution of the sea in coastal regions or by ships. New industrial construction in coastal areas is being strictly monitored by agencies and by public interest groups. Treatment plants are under construction or in the planning stage. Laws have been passed, and regulatory agencies are beginning to achieve positive results. By virtue of this, it has been possible to prevent new catastrophes and to stop, or at least slow, insidiously damaging processes.

Second, people have come to terms with some threats to marine life because effective measures to combat them would be so costly that no one would be willing to pay the price. This can be demonstrated, using the examples of oil and chlorinated hydrocarbons.

Pollution by oil must be combated in the most decisive manner. Everybody knows, however, that there will be accidents; they could only be eliminated if ocean surface transportation and subsurface piping were eliminated,

Every year, probably around 80,000 t of chlorinated hydrocarbons, including significant amounts of DDT, continue to reach the world's oceans via the atmosphere. As long as the underdeveloped countries of the world can point with convincing arguments to the fact that, on economic and technological grounds, DDT is irreplaceable for combating mosquitoes and cotton-damaging insects, nothing will change. Nothing will change, that is, unless science advances significant new arguments.

These are just two examples in which oceanographic research finds itself in conflict between the quality of life and the ecosystem on the one hand and obvious economic interests and realities on the other.

Third, there have been no new sensational news items about ocean catastrophes in the years 1971 to 1980, except oil accidents, It has been possible to revise some gloomy findings from the previous years, findings which initially seemed even more threatening. Mercury is obviously a poison and harmful to life processes, even in those small amounts in which it naturally occurs in seawater. Where it accumulates locally, the consequences are catastrophic. But the view is held that the quantity of mercury released by man is insufficient to increase its concentration worldwide in seawater significantly.

There are, however, still various other substances on the list of those that are potentially dangerous. Systematic research on the ecological behavior of synthetic organo-compounds has just begun. One must not forget that ocean pollution is a very new area of research.

Marine pollution research started 30 years ago with studies on radioactive wastes dumped into the sea, and the first international congress on marine pollution took place in 1959. But the true magnitude of the problem was not and could not be appreciated at the time. It has only been 15 years since the German Research Society put priority to studies on marine pollution (Caspers 1975). Numerous publications describing research by the maritime nations on marine pollution are only found after 1966. To report on developments without the passage of time and the resulting objectivity may seem superficial. It is, however, necessary to do so in order to account for the funds consumed by research and to inform a public which is still very much interested in the problems of marine pollution.

In the future, it will be increasingly difficult to present the entire sphere of marine pollution in the form of a critical report (Table 1). An ever-increasing specialization in research is becoming evident from year to year. From basic problems, which also interest the layman and which he can attempt to evaluate, the development moves on to various geophysical, chemical, biochemical, toxicological, geochemical and biological topics which only the specialist can evaluate.

Both research on problems of marine pollution and research into the basis of life in the sea are inextricably linked. Determination of the damaging effects of marine pollution must be based on knowledge of the ocean and marine organisms. On the other

Table 1. In 1972 it was possible to gain an overview of the problems of ocean pollution after reading approximately a hundred publications; this is certainly no longer the case. The 1975 issue of the periodical *Aquatic Sciences and Fisheries Abstracts* reports 868 publications on the questions of marine pollution which are divided into the following categories:

General surveys:	30
Regional studies, domestic effluents, eutrophication:	70
Waste heat:	20
Pathogens:	28
Detergents:	11
Oil, hydrocarbons:	246
(of these, 21 discuss degradation of oil in the ocean, 61 discuss measures to combat oil)	
Control of harmful substances by bio-indicators:	23
Trace elements:	204
(of these, 43 discuss mercury, 61 discuss radioactive substances)	
Chlorinated hydrocarbons and other synthetic organic compounds:	94
Industrial wastes, dumping, and incineration:	53
Model concepts on the dispersal of harmful substances and on toxicity:	55
Various other topics:	34

hand, programmed research on problems of ocean pollution have also given a tremendous stimulus to marine chemistry and biology.

We should remember that 20 years ago dissolved organic substances that occur naturally in the ocean were practically unknown. Since that time harmful organic substances in ocean water have had to be analyzed and the means as well as deployable methods to implement them have been available. Not only have harmful substances been found, but natural substances as well, and in a few years the material for a chapter on "Dissolved Organic Substances" was at hand for textbooks on oceanography.

Ten years ago, very little was known about the bio-accumulation of trace elements. Today, we are familiar with a whole catalog of harmful substances that are absorbed from seawater and accumulated without harmful effects. They are, in part, important for vitamin syntheses and enzyme maintenance. A new area of research has developed: to investigate the mechanisms involved in the uptake, storage, and elimination of trace elements.

Today a discipline "Marine Aerial Chemistry" exists because it was discovered ten years ago that the atmosphere is a transporting agent for harmful substances. Geochemistry is gaining insight into the fate of harmful substances and in doing so is expanding our basic knowledge about the conditions for sedimentation in the ocean. The measurement of the radioactivity of seawater has opened up new avenues for the investigation of ocean water movements. Toxic gasses like carbon monoxide that are relevant to the environment are also being produced by ocean vegetation (Junge et al. 1972). Unexpected results often occur during research on questions of ocean pollution: these are often more exciting than those expected when the tests are planned (Goldberg 1974).

We can hope that the obvious dangers to marine life and the health problems for man eating seafood and swimming in the sea have been identified. These problems could be solved with existing technology and at expense which could be paid. Science, however, has to go one step further, and identify and evaluate small-scale pollution impacts. Very often it is a difficult task to separate man-made effects from the natural background, and a much better knowledge of the natural variability of factors in the sea, of physical and chemical interactions at the air-sea interface and at the border between water and sediment is essential for the progress of science. In fact, about half of this book devoted to marine pollution questions does not deal with man-made contaminations, but with stress of natural origin. More knowledge of trends in atmospheric pollution and of the fate of pollutants in the atmosphere is necessary for a better evaluation of the continuing threat of chlorinated hydrocarbons to the marine environment.

What will the next ten years of research on problems of ocean pollution bring? Probably no one is willing to advance a prognosis in view of the numerous still unanswered questions. We can only hope that the problems of ocean pollution will continue to present themselves in approximately the way we are able to view them now. Then they would be largely open to solution — not only locally, but world wide as well. And the future of life on Earth would not have to be questioned as a product of this threat. My prognosis is, however, only valid under the condition that reasonable decisions are made and reasonable measures taken. I am aware of the fact that such conclusions do not fit with political intentions of people who want to demonstrate how miserably our present way of life is organized. And I do not agree with those who react against any waste disposal in the sea, even of relatively harmless wastes, and accept instead risks of economic damage and ecological problems on land. Certainly I am not free from intentions of my own, how could I when writing a book on marine pollution? I have tried to present the scientific evidence in a way that can be checked and criticized.

1.2 The Definition of Pollution

The Intergovernmental Oceanographic Commission (IOC) describes marine pollution precisely: "Marine pollution is the introduction by man, directly or indirectly, of substances or energy into the marine environment (including estuaries), resulting in such deleterious effects as: harm to living resources; hazards to human health; hindrance to marine activities including fishing; impairing the quality for use of seawater and reduction of amenities".

1.3 Units of Measurement and Seawater Composition

Indications of concentration are obviously those which relate the amount of a harmful substance to the amount of a substance under analysis. Since 1 l of ocean water with 35‰ salt content weighs approximately 1.028 kg, it does not make much difference in a rough calculation whether the concentrations refer to liters or kilograms.

On the other hand, the differences are large if wet weight, dry weight, extractable fat (lipid), or carbon are selected as a basis for measurements. Fish, mussels, and crabs contain 75%–80% water. The figure for concentration by wet weight must be multiplied by 4 or 5 to derive concentration by dry weight. If the fat content of an organism amounts to 10%, the figure for concentration to wet weight must be multiplied by 10 to attain those concentrations in relation to extractable fat. However, the water and fat content of organisms and tissues varies widely. One must, therefore, be very careful in the application of rough measurements.

An additional element that contributes to the confusion is that concentrations are often labeled "ppm", "ppb", or "ppt":

ppm (parts per million, 10^6) means μg/g or mg/kg as well as about mg/l
ppb (parts per billion, American use, 10^9) means ng/g or μg/kg, as well as about μg/l or mg/m^3
ppt (parts per trillion, American use, 10^{12}) means pg/g or ng/kg or μg/t, as well as about ng/l or μg/m^3.

If plant nutrient elements are involved, marine biologist and marine chemist often indicate concentrations in μg-at/l or, the more modern term, in μmol/l. To come to this figure, the weight in μg/l is divided by the atomic weight of the element or compound in question.

Table 2 summarizes analysis of the elementary composition of seawater.

Table 2. The concentration of chemical elements in seawater (according to data of Brewer 1975). The classical textbook figures have been modified in recent years by new methods of analysis and are also still undergoing corrections; for example, according to recent analyses, mercury concentration only stands at 7 ng/l, lead concentration at 2 ng/l

	mg/l		µg/l		ng/l		ng/l
Chlorine	18,800	Zinc	4.9	Xenon	50	Lanthanum	3.0
Sodium	10,770	Argon	4.3	Cobalt	50	Neodymium	3.0
Magnesium	1,290	Arsenic	3.7	Germanium	50	Tantalum	2.0
Sulfur	905	Uranium	3.2	Silver	40	Yttrium	1.3
Calcium	412	Vanadium	2.5	Gallium	30	Cerium	1.0
Potassium	399	Aluminum	2.0	Zirconium	30	Dysprosium	0.9
Bromine	67	Iron	2.0	Mercury	(30)	Erbium	0.8
Carbon	28	Nickel	1.7	Lead	(30)	Ytterbium	0.8
Strontium	7.9	Titanium	1.0	Bismuth	20	Gadolinium	0.7
Boron	4.5	Copper	0.5	Niobium	10	Praseodymium	0.6
Silicium	2	Caesium	0.4	Thallium	10	Scandium	0.6
Fluorine	1.3	Chromium	0.3	Tin	10	Holmium	0.2
Lithium	0.18	Antimony	0.2	Thorium	10	Thulium	0.2
Nitrogen	0.15	Manganese	0.2	Helium	7	Lutetium	0.2
Rubidium	0.12	Selenium	0.2	Hafnium	7	Indium	0.1
Phosphorus	0.06	Krypton	0.2	Beryllium	6	Terbium	0.1
Iodine	0.06	Cadmium	0.1	Rhenium	4	Samarium	0.05
Barium	0.02	Tungsten	0.1	Gold	4	Europium	0.01
Molybdenium	0.01	Neon	0.1				

2 Domestic Effluents

2.1 Biodegradable Organic Substances

Man, like all animals, consumes organic foodstuff and leaves undigested organic remains behind in his feces. In the preparation of his food organic substances are also left over as garbage, whether in the commercial processing of foodstuffs or as household waste. In natural environments, decomposers exist: organisms which specialize in the breaking-down of a dead organic substance. More precisely, they are organisms which satisfy their energy needs from a dead organic substance. For the most part, they are bacteria and micro-fungi.

Under ideal conditions, the cycle of carbon and oxygen is balanced in nature (Fig. 1). When effluents from a city are introduced into a body of water, this means an additional supply of dead organic substances, and the question is whether or not it can be absorbed by nature. The number of bacteria in bodies of water can respond to the amount of available organic substance (to their food). However, bacteria consume oxygen in their respiration process. To serve as a rule of thumb, the Population-Equivalent Unit has been created; one unit represents the amount of oxygen that is consumed to break down during a 5-days' experiment the readily degradable fraction of the feces and garbage produced by 1 person per day. This figure is called Biochemical Oxygen Demand (BOD_5), and it differs according to life styles: in the Federal Republic of Germany a couple of years ago it was 54 g, and now it is 60 g of oxygen per day; in the U.S.A. it is 75 g of oxygen per day. Such a quantity of oxygen is contained in 6–10 m^3 of seawater saturated with oxygen (Table 3).

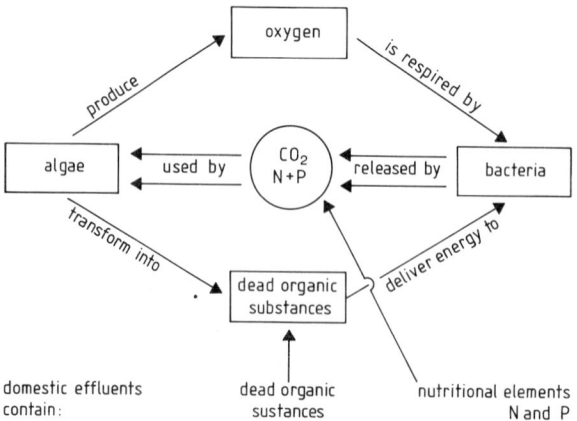

Fig. 1. Greatly simplified schematic of the circulation of oxygen and of nutrients. The role of animals in the circulation of substance has been omitted. Attention is focused on the contribution of dead organic material and nutrients which domestic effluents may support

Table 3. Saturation concentration of oxygen (in mg O_2/l) at different temperature and different salinity

Temperature (°C)	Salinity (‰)				
	0	9.0	17.9	26.6	35.3
5	12.80	12.09	11.39	10.70	10.01
10	11.33	10.73	10.13	9.55	8.98
20	9.17	8.73	8.30	7.86	7.42
30	7.63	7.25	6.86	6.49	6.13

The term BOD describes either (in kg O_2/day) a quantity of organic wastes, or (in mg O_2/l) it refers to water quality. To identify the BOD of a water sample, first the oxygen concentration is analyzed, then the sample is maintained for 5 days at a temperature of 20°C in a closed bottle, and the oxygen concentration is again measured. From the difference between the two measurements one calculates the amount of free dissolved oxygen which has disappeared, the BOD.

Within 5 days a reasonable fraction of the degradable organic matter in the sample is degraded by water bacteria. What is left over are organic substances which are difficult to degrade; they could be analyzed by chemical methods (COD, Chemical Oxygen Demand).

It is, then, important to know first whether enough oxygen is available to accomplish the decomposition of organic materials through exchange with the atmosphere, through photosynthesis, or from both sources. For if the available oxygen is depleted, the subsequent decomposition of organic materials takes place anaerobically, that is, without oxygen. Anaerobic bacteria, however, work more slowly, and as the final product of the transformation they produce various organic compounds which smell like feces. For this reason, it is, as a rule, desirable to biodegrade effluents aerobically and not leave the job to anaerobic bacteria, unless this is done in sealed containers.

The situation is the same in marine as in freshwater, even though different kinds of bacteria are involved. In high concentrations of organic effluents, decomposition in seawater does, to be sure, begin somewhat later than in freshwater (Fig. 2). This time lag, however, is due to experimental conditions. There is no reason to assume, in general, that marine bacteria are less efficient in degrading organic substances, compared with freshwater environments. Experiences with biodegradation of domestic effluents in freshwater can be applied to conditions on the coast (Wachs 1972). However, one must not lose sight of the fact that oxygen dissolves at a somewhat lower rate in seawater than in freshwater (Table 3).

On the sea bottom, there are all types of sediments from sand to mud, from well-oxygenized interstitial pores to situations where only the upper millimeter of the sediment contains oxygen and the deeper layers are anaerobic. The different types of sediment are inhabited by different communities of animals and bacteria. If sewage is introduced into a marine region without strong currents, the sediment close to the inlet may become enriched with organic particles, and the tendency for the sediment is to change to anaerobic conditions. In consequence, a fauna develops which is specialized to such conditions (Figs. 3 and 4) or the region becomes devoid of animal life.

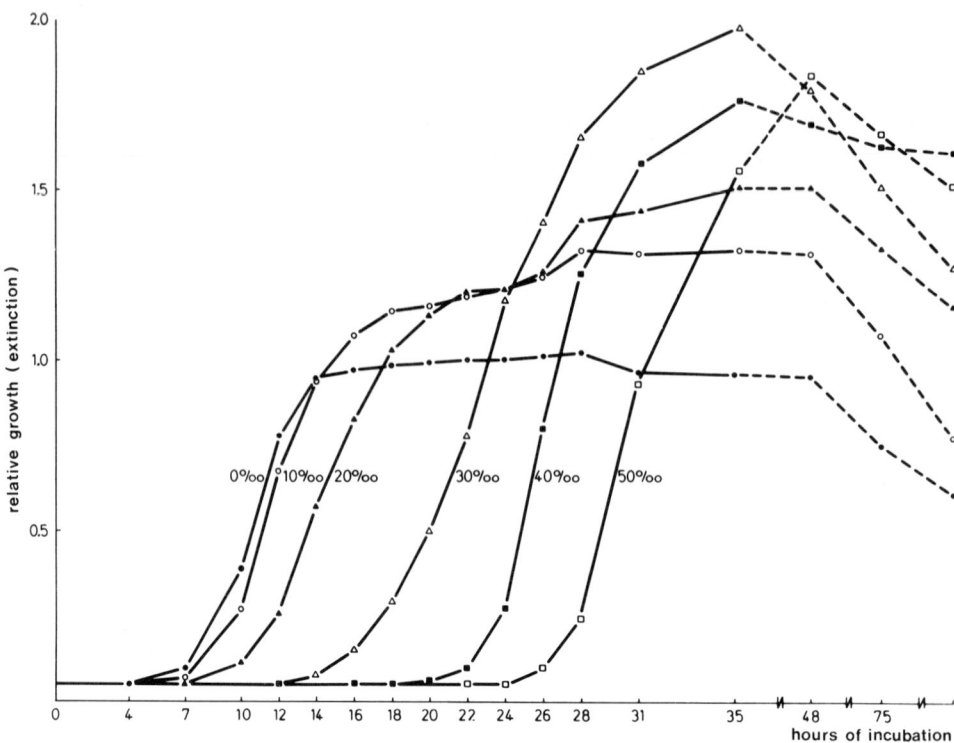

Fig. 2. The biodegrading of domestic wastes is extended in salt water, because salt-tolerating bacteria must first replace the freshwater bacteria. Under good culture conditions, a population of effluent bacteria is quickly produced in freshwater after a 7-h incubation period. However, if table salt is added to the cultures, then the beginning of the growth of the bacteria is delayed, corresponding to the amount of table salt (by up to 20 h at a concentration of 50 ‰ sodium chloride). In the diagram, the relative growth of bacteria cultures is represented by the extinction of light in the culture medium (Gocke 1975)

Downstream from the West Germany city of Bremen, approximately 1.4 million Population Equivalent Units of organic substances are introduced daily into the Weser River (Table 4). As a consequence, the oxygen content in the water of the Weser sinks to 2–4 mg/l or 20%–40% of saturation in the summer months. That is barely sufficient to sustain fish life. At present, treatment plants are being built everywhere along the Lower Weser, and these will take care a considerable portion of the organic load in the future (Fig. 9). One will then see what part man's domestic effluents has contributed to the bad oxygen balance in the Lower Weser and what part is to be attributed to natural causes. The natural strain on the oxygen balance is already high in the brackish water zone of an estuary. The marine plankton organisms and the freshwater plankton both die out here. Through the effect of the "turbidity trap", a 200-fold concentration of suspended matter results (Fig. 5): because of the salt stratification, the carrying power of the tidal current at the bottom of the Weser River is stronger in upstream than in downstream direction. Therefore, suspended sediment and other materials in the vicinity of the bottom are transported upstream.

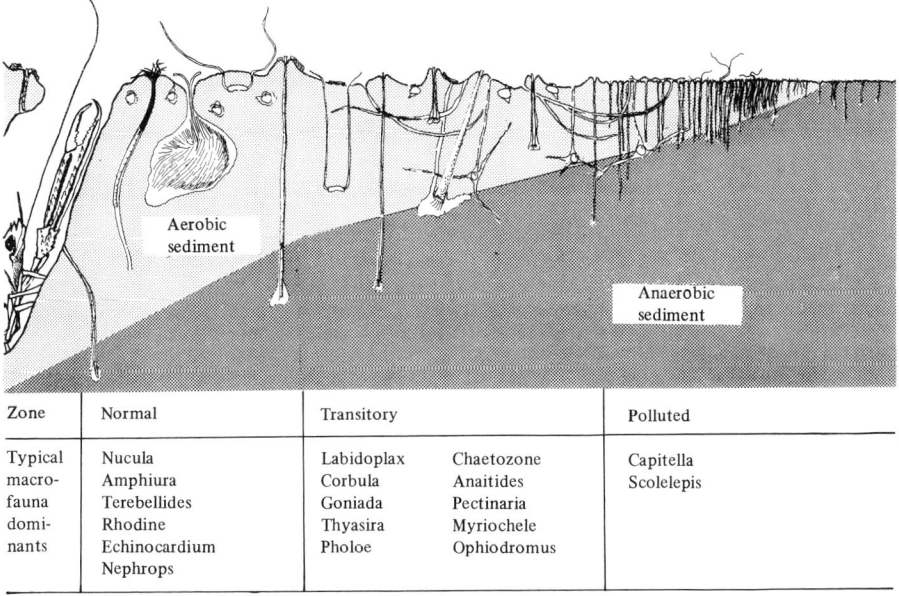

Fig. 3. Waste products containing lignin are found in the waste water of cellulose factories and paper mills that use the sulfite procedure. These wastes are only with difficulty biodegradable. Where they cover the sea bottom the fauna die out (grossly polluted). In regions where the lignin particles accumulate in the sediment the anaerobic layer extends close to the surface of the sediment (polluted), and the composition of the macrofauna species is corresponding: organisms are found that under natural, undisturbed conditions live in such areas of marine coasts where from natural causes the oxygen supply to the sediment is poor. Schematically presented are the faunistic changes which occur in North Sea environments (Pearson and Rosenberg 1976)

In the Thames, organic remnants and waste products of various sorts are present which, it is known, come from the North Sea. Marine bryozoan remnants were observed up to 20 km below London Bridge. The quantity of these suspended particles is so large that it sometimes clogs the cooling water filters of the power station at Tilbury (Board 1973). Detritus, conveyed to the estuaries from sea areas, thus contributes to the bad oxygen balance.

It is to be hoped that the Weser will be spared the fate of many an English estuary. Oxygen-less zones regularly occurred at ebb tide in the summer along a 10 km-long stretch of the Tyne Estuary because domestic effluents from 1 million inhabitants and a corresponding amount of industrial waste material was introduced into it (Ratasuk 1972). Since 1920, a worsening of the situation in the Thames has been observed. From 1947 on, annually from July to September, an oxygen-less (and from 1950 on a hydrogen sulfide rich) region has been observed along a 20 km-long stretch of the river 50 km downstream from Teddington Weir. In the meantime, treatment facilities have been put into operation, and since 1960, the situation has definitely improved (Fig. 6). After 1966, anaerobic conditions were no longer observed; in 1969, approximately 50 different species of organisms were noted in those regions

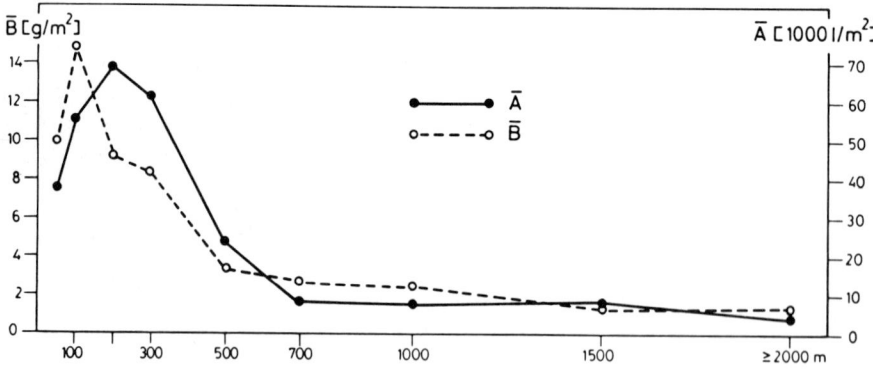

Fig. 4. In the composition of the bottom fauna the effects of sewage disposal are reflected. Until April 1972, the city of Kiel, Germany, discharged 50,000 m^3 of untreated sewage daily into the Kiel Bight, containing 18 t particulate organic matter, 2.4 t phosphorus and 1.4 t nitrogen. The outlet was located 200 m from the beach at a water depth of 2.5 m. About 15 animal species comprise the bottom fauna in the outlet region. At a distance of 200–700 m away from the outlet sensitive sand-inhabiting species like the amphipod *Bathyporeia* disappear; less demanding forms such as the polychaete *Pygospio* get the upper hand. Still closer to the outlet the sediment is well enriched with organic matter. Oligochaets and the polychaet *Capitella* dominate at a distance of 50–100 m from the outlet. In this transect the number of species hardly varies, but the biomass of the bottom fauna near the outlet is 15, and the number of individuals 50 times higher than further distant. Therefore the calculation of diversity results in lower values. The graph represents a transect north of the outlet, at a water depth of 3 m. Indicated are biomass ($\bar{B}$ in g/m^2 wet weight without molluscs and echinoderms) and abundance ($\bar{A}$ in 1000 individuals/m^2) (Anger 1975)

Table 4. Pollution stress on the Lower Weser River, Germany, by domestic effluents and organic industrial effluents in 1978. Treatment plants are under construction or being planned so that the strain will be alleviated in the coming years. (Data according to Wasserwirtschaftsamt Bremen and Brake, and Wasserbehörde Bremerhaven)

km below Bremen Weir	Contributors	BOD$_5$ kg O$_2$/day
8.5	Waste treatment plant (mechanical) Bremen-Seehausen (construction of biological plant started 1980)	33,000
11.0	City of Delmenhorst and wool spinning-mill (treatment plant to be ready in 1979)	10,000
11.2	Klöckner Steel Factory	6,000
16–23	Various villages and industry	100
21.5	Bremen Wool Factory	3,500
25.2	Waste water trreatment plant Bremen-Farge	300
32.5	Waste water treatment plant Elsfleth and industry	90
40–42	Waste water treatment plants Brake	820
42.5	Fat refinery Brake (treatment plant to be ready in 1980)	2,000
48.1	Treatment plant Rodenkirchen	150
59.1	Treatment plant Nordenham	110
57–66	Various industry of Nordenham	1,000
65–67	City of Bremerhaven including fish industry and harbors (treatment plant to be ready in 1982)	21,300
Total (corresponding to about 1.4 million Population Equivalent Units)		78,000

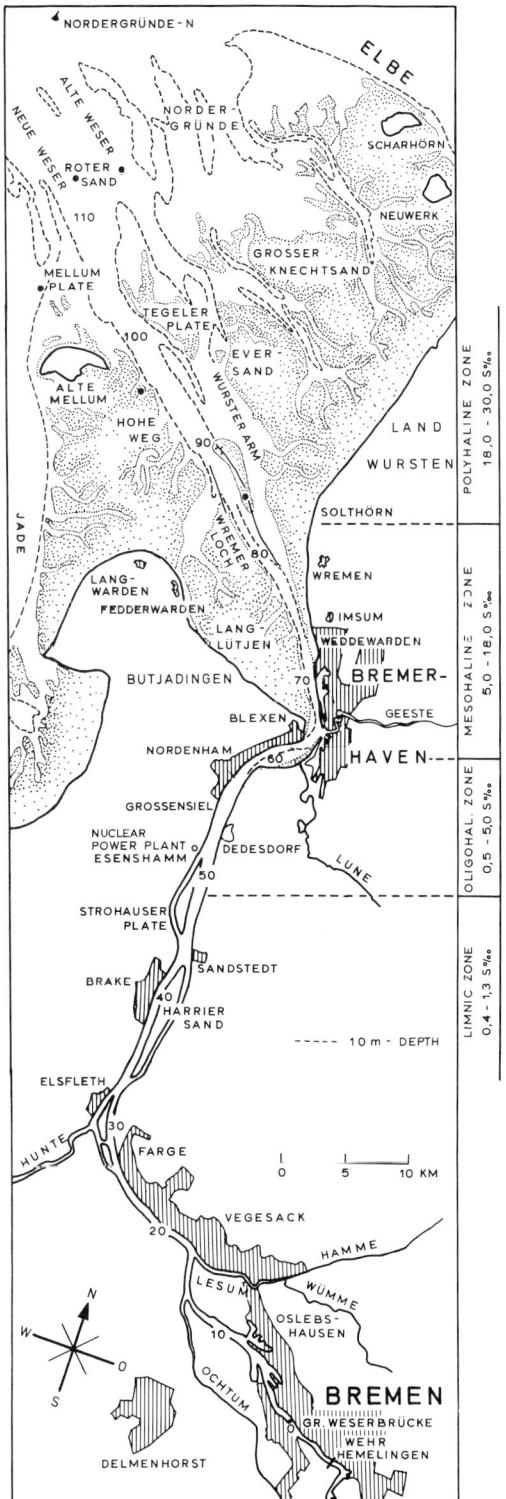

Fig. 5. At Bremerhaven, Germany, the Lower Weser, which is influenced by tides, expands into the delta of the Outer Weser with its Wadden-Sea tidal flats and sandbars. The brackish water border lies between the freshwater zone and the oligohaline zone with from 0.5‰ to 5 ‰ salinity; it moves between Brake and Bremerhaven, depending on the amount of freshwater which is streaming into the estuary (mean value 250 m^3/s, or between 11–19 million m^3 per tide). The oligohaline zone is characterized by the turbidity cloud with high concentration of suspended particles in the brownish water. Downstream from Bremerhaven large quantities of North Sea water are mixed with the Weser water. At Robbenplate (km 85) about 563 million m^3, at Mellum (km 100) about 909 million m^3 of water flow seaward during each tide (according to measurements of Wasser- und Schiffahrtsamt Bremerhaven). *Numbers* along the channel of the Weser are km from Bremen (Lüneburg et al. 1975)

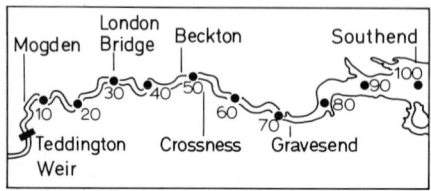

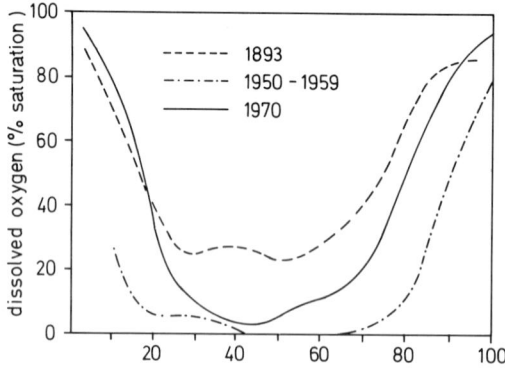

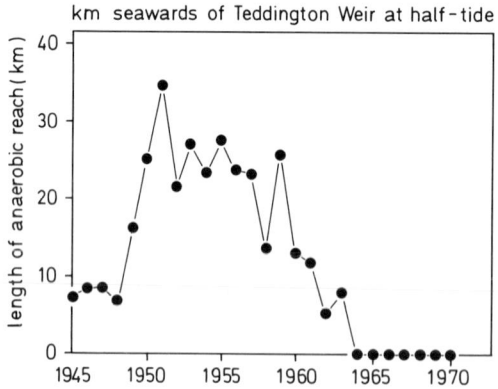

Fig. 6. Water quality in the Thames. *Above:* map of Thames Estuary and situation of treatment plants Mogden (full biological treatment), Beckton (introduction of 206 t of BOD_5 daily in the year 1950), and Crossness (79 t). In the years 1950–1953, the total amount of effluents introduced between Teddington Weir and the confluence with the North Sea was 427 t of BOD_5 daily. *Numbers* represent the distance in km from Teddington Weir. *Middle:* long-term vacillations of oxygen saturation in the water of the Thames Estuary calculated on the basis of a summer water flow of 13 m^3/s of river water. *Bottom:* dimension of the stretch of river where anaerobic conditions prevailed in the summer. The high increase of 1949–1950 can be traced back to the fact that synthetic detergents were introduced. The improvement in the conditions can be traced back to the fact that the biological treatment plant at Beckton was put into operation in 1960, the one at Crossness in 1964, the latter with a daily capacity of 50,000 m^3 (Gameson et al. 1973)

in which they had previously died out, and since 1972, flocks of wintering birds on the Thamesmead tidal flats indicate that enough worms and shellfish are available there (Hattison and Grant 1976).

In 1970, a public service strike paralyzed the Thames waste water treatment plants for a month. They did not operate at capacity, and the sewage sludge, which normally is taken out to the North Sea by barge, was introduced directly into the Thames. The consequences of this strike were seen in the reduced water quality during the entire winter of 1970–1971; they also had an effect on the stocks of fish and shrimp. The Thames example shows that the expenditures for effluent purification are worthwhile.

2.2 Infections

At this point we cannot discuss whether treatment plants, from an economic point of view, are effective in destroying infectious organisms, or whether other means of sterilization might be applied. Since it is desirable to keep the sea coasts hygienically clean as recreational beaches, the removal of communicable organisms from effluents in coastal areas is often more important than the reduction of the organic effluent load.

Particularly strict bacteriological requirements must be met in those areas where there are shellfish cultures in the vicinity of effluent drain pipes. Mussels and oysters filter bacteria out of seawater and store them up without the vitality of the bacteria suffering. For this reason, oyster culture areas are rightly subject to very strict hygiene controls.

Of the 230 000 km^2 of shellfish culture area along the coasts of the U.S.A., a quarter is closed because the water quality is sub-standard. In these waters more than 700 coliform bacteria per 100 ml were found to be present (Wood 1972). There is a law that at 70–700 coliforms per 100 ml of water, shellfish must be held in clean water so that they will give off the organisms they have absorbed.

Sometimes shellfish are disinfected in various ways before they are sold. Boiled shellfish are only then completely safe when the temperature has had enough time to sterilize the inner area of the shellfish. At the beginning of this century, guests at a banquet given by the English royal family came down with typhus. Raw oysters were on the menu. This occurrance guaranteed that at this early date in England throughgoing hygienic measures were taken in the oyster beds and the oyster trade (Korringa 1968; Fig. 7).

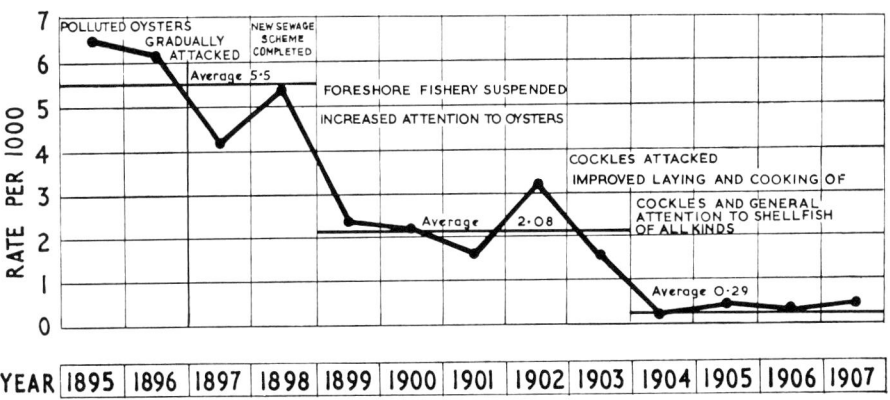

Fig. 7. In 1895, in the English seaside resort Brighton, it was observed that typhoid fever and the consumption of oysters were connected. Oysters out of sewage-influenced areas were thenceforth avoided. In 1898, all oyster fishery directly on the coasts was given up, and 1902 cockles (Cerastoderma = Cardium) were included in the surveillance program. Within 10 years the danger of typhoid fever due to the consumption of seafood was considerably lowered. Indicated are the number of cases per 1000 inhabitants of Brighton (Mosley 1975)

We would very much like to know more exactly how cholera bacteria, typhus, and paratyphus germs, polyomyelitis viruses, jaundice viruses, and certain other pathogenic organisms really behave in water, how long they are capable of living, and how they react to seawater. We know that seawater has an adverse effect on some soil and intestinal bacteria which does not stem from salt content alone (Fig. 8). Compared with their living conditions in freshwater, the conditions for life of nonmarine bacteria in seawater are probably worse and more complex (Mitchell and Chamberlin 1975). This may be due to their being in competition with marine bacteria, to antibiotica released by marine bacteria, or that they are attacked by bacterial parasites, or consumed by protozoa. Bacteria-eating unicellular organisms are in all likelihood hardly specialized in intestinal bacteria. However, they contribute among other influences to the elimination of *Escherichia coli* because, under normal conditions, this intestinal bacterium cannot multiply in marine environment (Enzinger and Cooper 1976). When, however, seawater contains more than 100 mg/l of organic substance, *Escherichia coli* grows and holds its own against marine bacteria. Some research assumes that heavy metals which have a poisonous effect in pure seawater are bound by organic substances (Jones and Cobet 1975). *Salmonella* can multiply in water outside the host animal if the protein concentration in the water is high enough.

Some disease-causing viruses survive a long time in seawater. Even amoeba of the genus *Acanthamoeba* that cause diseases in the eyes of humans have been isolated from the sediment in the Bay of New York — and not only in spots where sewage sludge is dumped (Sawyer et al. 1977).

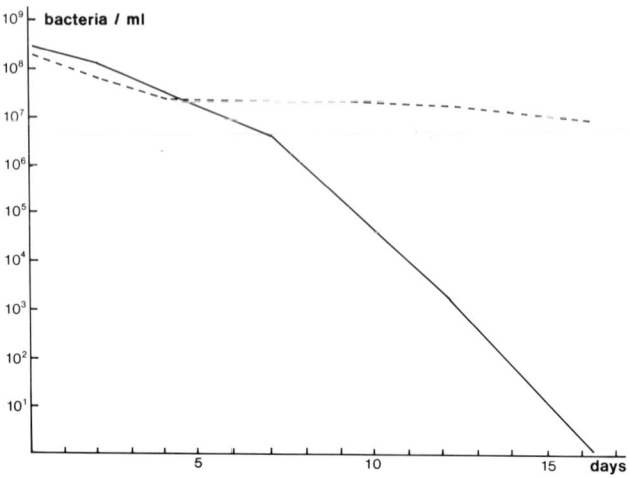

Fig. 8. The bacteria-killing effect of seawater can be shown by introducing *Escherichia coli* cultures into fresh seawater *(solid line)*: the bacteria count in the culture diminishes quickly. The bacteria-killing effect remains constant if the seawater is diluted by 99 parts of freshwater or if it is heated up to 42°C for 10 min. On the other hand, *Escherichia coli* survives well *(dotted line)* in cultures with seawater sterilized at 105°C, or if the seawater entered through a filter with 0.45 μm pores, and in culture of artifical seawater. From these tests, the conclusion can be drawn that the bacteria-killing effect stems from living organisms in seawater (Guelin 1974)

Disease-causing organisms can only be found in bodies of water if the disease is prevalent in the human population. Basically then, it does not make a lot of sense to look specifically for dangerous pathogenic agents in bodies of water if one is looking for a yard stick for water quality measurement. For this reason the amount of *Escherichia coli,* is monitored. *Escherichia coli* is under normal circumstances a non-pathogenic bacterium living in the intestine of man and of warm-blooded animals, occurring in the feces and therefore used as an index organism, along with some other fecal microorganisms. According to different methods of detection, strictly fecal coliform germs are distinguished from more numerous total coliform germs, and both numbers are used to monitor sewage pollution and to asses health risks from drinking water, from food stuff, and from bathing.

Escherichia coli is always present in excrement and fecal effluents. Naturally, the presence of *Escherichia coli* in a body of water does not indicate anything beyond the fact that fecal effluents are present in a diluted form.

Civil health authorities in many countries of the world consider a ban on swimming if concentrations of *Escherichia coli* higher than 100–1000 per 100 ml are found in freshwater. Health authorities permit more leeway with regard to the bacteria-killing effect of seawater in coastal areas. In Great Britain in particular, there are adherents of the view that the problem does not need to be worried about seriously because the danger of infection is almost nonexistent (Moore 1975) and a vacationer swallows scarcely more than 10 ml of water per day when swimming.

Other researchers believe they have proven the connection between various infections and swimming in polluted seawater, especially in the Mediterranean. Where the borderline of a presumably dangerous health hazard lies is open to question (Regnier and Park 1972).

The swimming facilities on the Weser River just outside Bremerhaven, Germany, were not rebuilt after their destruction by seastorm in 1962 because levels of up to 52,000 *Escherichia coli* per 100 ml were found to be present there. Conditions have worsened since then: from 1964 to 1968, only a few tests showed less than 10,000 *Escherichia coli* per 100 ml; in 1969 and later, some tests showed over 100,000 *Escherichia coli* per 100 ml.

It is not hard to appreciate the fact that the situation has worsened. More and more people flock to resort areas and camping grounds at beaches, and sewage treatment plants there have only been built in the last few years. Cities like Bremen and Bremerhaven have expanded their sewage systems considerably in recent decades and have thereby attached more and more households to the main drainage channel of the Weser (Fig. 9). The mandatory treatment plants are only now following this development (Table 4). Let us hope that construction of sewage treatment plants will eliminate the problem and the *Escherichia coli* figures will reduce again.

A detailed investigation of the German part of the Baltic Sea coast in the area of Kiel Bay between Flensburg and Fehmarn showed heavy sewage pollution in certain areas in the vicinity of coastal cities. That 13% of all samples also contained *Salmonella* seemed serious. The *Escherichia coli* counts in all swimming areas were, to be sure, within the norms required by the German government. But the results of a 1974 study by the Institute for Hygiene of the University of Kiel indicated that only two-thirds of all the resort areas satisfied the more recent higher requirements as well.

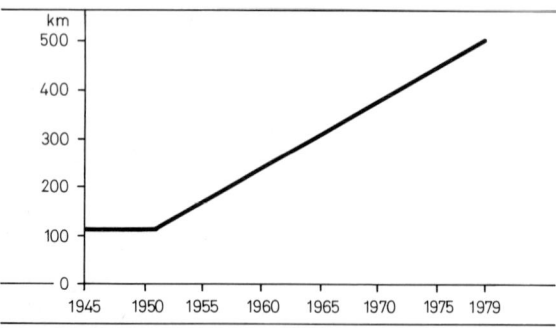

Fig. 9. In 1945 the homes of only half of the 80,000 Bremerhaven inhabitants were connected via drainage channels with the Weser River, the other half used septic tanks or other techniques of sewage disposal which are now regarded as health risks. From 1950 to 1979 400 km of drainage channels have been added to the original about 100 km, and in 1979 the homes of 134,000 Bremerhaven inhabitants (97% of 137,000) were connected with the Weser River. This explains the poor water quality in the Weser. Improvement may be expected when the sewage treatment plant now under construction starts to operate in 1982. The figure represents the length of drainage channels in Bremerhaven, Germany. (From newspaper information)

Since then, numerous treatment plants have been put into operation and conditions have improved.

Originally, in the Federal Republic of Germany no more than 20% of the samples were permitted to have *Escherichia coli* counts higher than 10,000 per 100 ml if a coastal water area was to pass the normal swimming standard. In 1975 the European Community issued guidelines for the quality of bathing waters which have been binding for member countries since 1978 (see Chap. 10). According to these, at least 95% of all analyzed samples must contain less than 2,000 fecal-coliform bacteria, and less than 10,000 total-coliform bacteria per 100 ml. Amounts under 100 fecal-coliform germs per 100 ml in at least 80% of the samples are desirable. The samples should normally be taken every two weeks during the swimming season.

These rules apply to Great Britain as well. The British are calculating 100 million £ for the construction of new treatment plants which they will need in order to come up to the European water standards for swimming. This has given rise to the question whether funds of this magnitude might not be better applied more effectively in other areas of public health (Moore 1977). By the way, chlorination of untreated domestic sewage reduces *Escherichia coli* in coastal areas to European Community swimming standard at minimal costs compared with waste water treatment plants (Jenkins 1981).

2.3 Eutrophication

When domestic effluents are biodegraded by bacteria, not only carbon dioxide and water, but also nitrogen and phosphorus are released as inorganic compounds which were originally contained in the protein component of plants and animals. Nitrates, phosphates, and other salts are nutrients essential for the growth of plants. The circulation of matter in nature functions only when these nutrient salts are continually released. Without them, no plant growth could take place (Fig. 1). However, it is too much of a good thing when more nutrients are introduced into bodies of water than are used for good plant growth. As a result, these bodies of water are genuinely over-fertilized, not eutrophized but hypertrophized, with the result that too lush a plant growth results. If a farmer fertilized without farming, nothing but weeds would be the result. In the sea, weeds cannot be pulled up. As a result of over-fertilization, a flora results which is not always desirable because it displaces the normal algae flora.

The North Sea receives nutrient imports from three sources: from the atmosphere, from the Atlantic Ocean, and from rivers and coastal towns. By far the most important source are the currents which bring water from the Atlantic Ocean into the North Sea: about 85% of all phosphorus input comes from the Atlantic. Rivers bring about 15%, 800,000 t of nitrogen and 70,000 t of phosphorus per year. Nutrient input from rivers has increased very much over the past decades.

The River Rhine, in 1932, brought approximately 3000 t of phosphorus to the estuarine region of the Netherlands. In 1955, the figure was 7000 t, in 1970, 30,000 t (Postma 1978). Most of this amount is from human activities. Evaluations for the Federal Republic of Germany show that 40% of the phosphorus in rivers is from detergents in washing ingredients, 27% from feces, 17% from agriculture (specially from liquified feces and urine produced by modern cattle-raising operations), and 13% from industrial sources. Evaluations from the W. German State Schleswig-Holstein support the view that agricultural land use contributes only little compared with domestic sewage: only 5% of the phosphorus and 32% of the nitrogen that go to the Western Baltic are from agriculture plus natural groundwater, that is 100 g P/ha agricultural area. All the rest comes with domestic sewage or from small industry, with a rate of 3 g P and 10 g N, for example, from each inhabitant of Kiel city per day (Hoffmann 1979).

The purification of effluents by waste water treatment plants has only a limited effect on this transportation of nutrients. Biological treatment plants hold back only around a third of the phosphorus, because waste water bacteria set phosphorus compounds free in dissolved form. Only by special chemical methods of waste water purification, by precipitation with ferrous or aluminum sulfate, can nutrients effectively be removed from effluents, and such measures are not applied with effluents to the North Sea.

In the coastal areas of the southwestern North Sea and the German Bight, the concentrations of nutrients are high for this reason (Figs. 10 and 11). The question then is, whether the increasing input of nutrients into the North Sea has an effect on phytoplankton productivity, herbivore production, and fisheries yield. Unfortunately, there are but few extensive long-term studies of nutrient concentrations and plankton, so that the effects of eutrophication can only be studied by a few examples. There are calculations that primary production in the Southern Bight of the North Sea and in the Dutch Wadden-Sea did increase (Postma 1978).

In the area around Helgoland, which is affected by the Elbe and Weser Rivers, the amounts of phosphorus in the seawater have apparently increased (Fig. 12). If one compares characteristic hydrographic situations during summer 1954 and summer

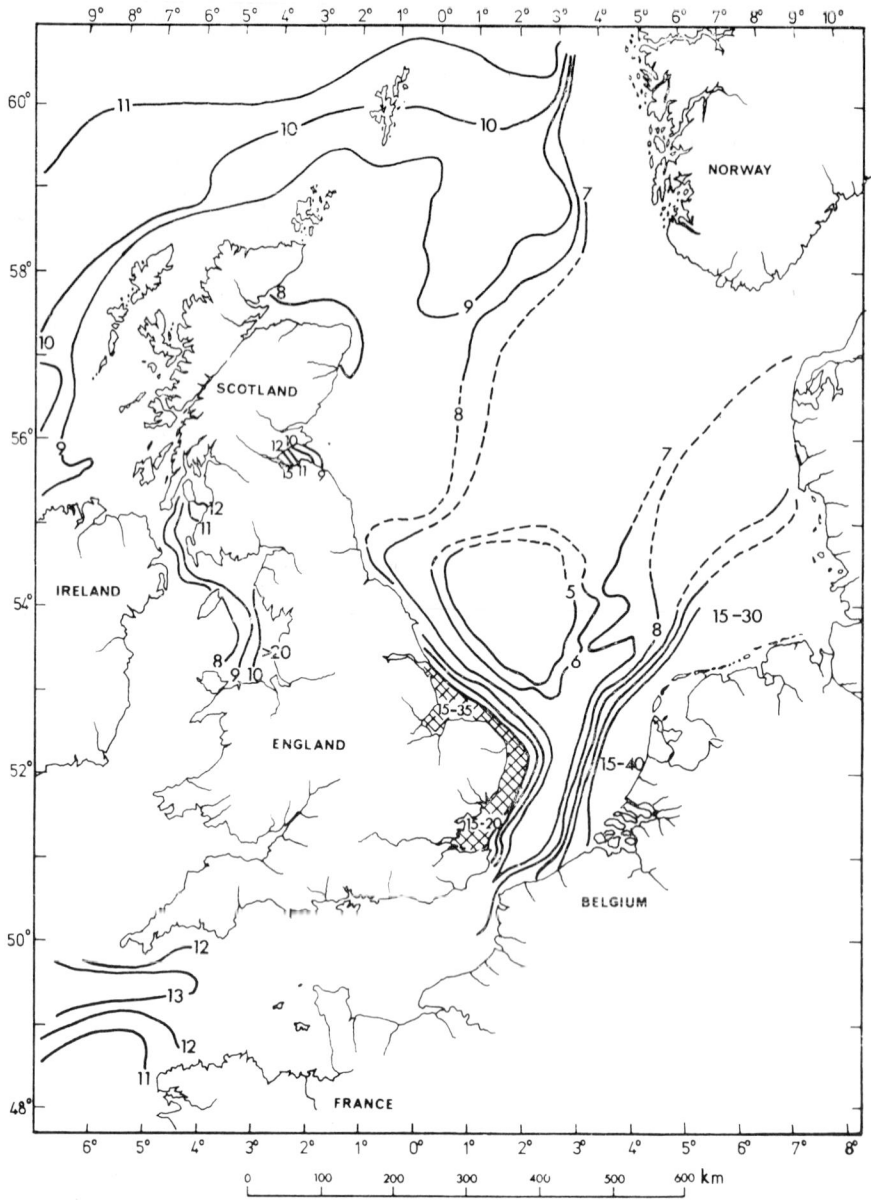

Fig. 10. Nitrate concentration in North Sea water during the winter. The figures refer to nitrogen reported as nitrate in $\mu mol/l$; they must be multiplied by fourteen if the values in $\mu g/l$ are wanted (McIntyre and Johnson 1975)

1977, one finds about 5 times higher PO_4-P concentrations, 1–5 μmol P/l in recent samples. One can conclude that silicium now is the limiting nutrient for diatom plankton growth, and one can correlate this increase of phosphorus with the introduction of phosphate-containing detergents in the early sixties (Hickel et al. 1980).

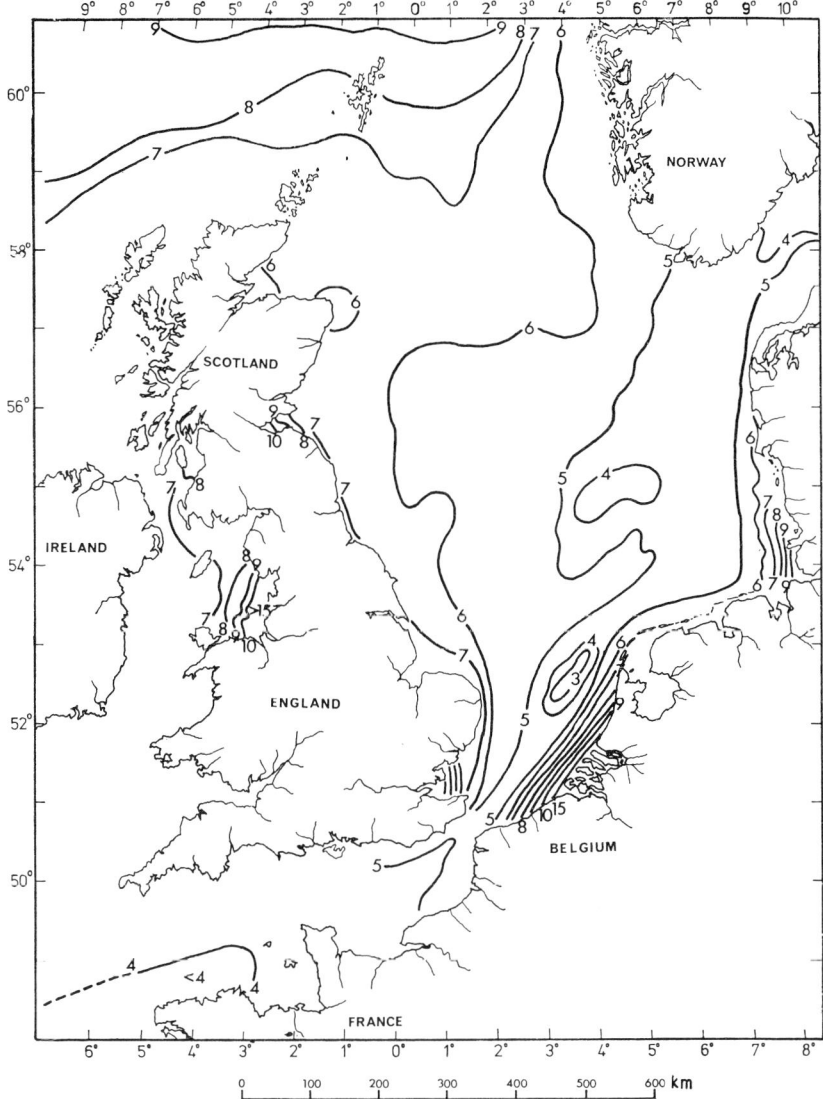

Fig. 11. Phosphate concentration in North Sea water during the winter. The figures refer to 10^{-7} mol/l of phosphorus reported as phosphate; they are to be multiplied by 3.1 if values in $\mu g/l$ are desired (McIntyre and Johnson 1975)

It will still be a number of years before scientists can determine whether an increase in phytoplankton has also resulted, because marked differences from year to year make it difficult to define trends statistically. It is an open question, then, whether observed summer situations of oxygen deficiency in the nearbottom water are the consequence of the eutrophication trend, and whether there is any effect on the fisheries yield. Around 1.3 million t of fish were caught annually in the North Sea in the 1930's, approximately 2.4 million t in the 1960's.

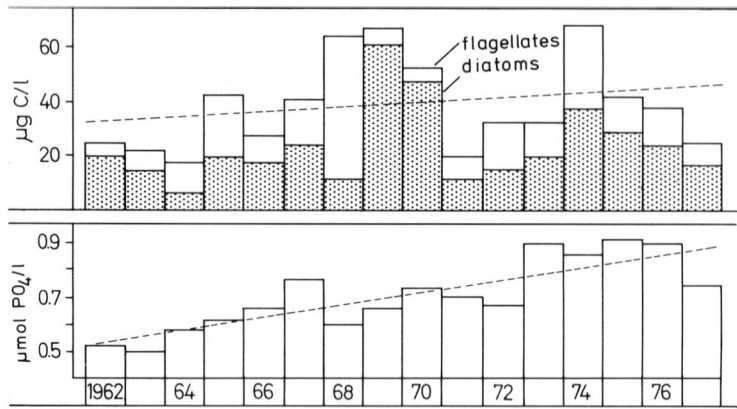

Fig. 12. In the Bay of Helgoland (German North Sea) the trend to eutrophication seems to be predominant. Regular measurements of the phosphorus in the seawater near Helgoland show an increase in the yearly average from approximately 0.5 μmol/l in 1962 to approximately 0.9 μmol/l in 1976 *(lower diagram)*. Note that the amounts vary greatly from year to year and depend very strongly on the amounts of freshwater that are transported into the German Bight by the Elbe and Weser Rivers. In years of heavy rainfall, the average salinity is lower and the average phosphate concentration higher. Even stronger are the annual variations of the average phytoplankton-biomass (in μg carbon/l, *upper diagram*). The regression line is not significant; it is therefore doubtful whether an increase of the phytoplankton has actually occurred in the German Bight (Hagmeier 1978; according to a personal communication)

There is definite hope that eutrophication of coastal waters by phosphates will decrease in the coming years. At present, 40% of the phosphate load in rivers of the Federal Republic of Germany originates from phosphate in washing powder, the component that softens the water by binding calcium. For a number of years it has been well known that zeolites have the same softening effect in washing powder, and they are nontoxic sodium-aluminum-silicates which do not cause eutrophication. Zeolites are minerals with a peculiar cristal structure; they work like ion exchange devices and take up calcium, heavy metals, and other substances from solutions. When incorporated in the zeolites, these substances are particulate and settle down in the sediment. Zeolites occur naturally as a product of vulcanism, and they can be produced by the chemical industry. At present, a plant for the production of 65,000 t of zeolites per year is under construction close to Cologne, and the government of the Federal Republic of Germany issued a regulation which reduces the permissable amount of phosphates in washing powder to 50% before 1984 (Anon 1980a).

2.4 The Situation in the Baltic

Mixed with some seawater, freshwater from the numerous rivers that flow into the Baltic moves on the surface of the Baltic to the Skagerak. This brackish water covers the saltier deep water like a blanket (Figs. 13 and 14). Because of this, all effective contact between the deep water and the atmosphere is prevented. Only a certain small degree of transportation of oxygen to the deep water takes place by means of diffusion and downwelling.

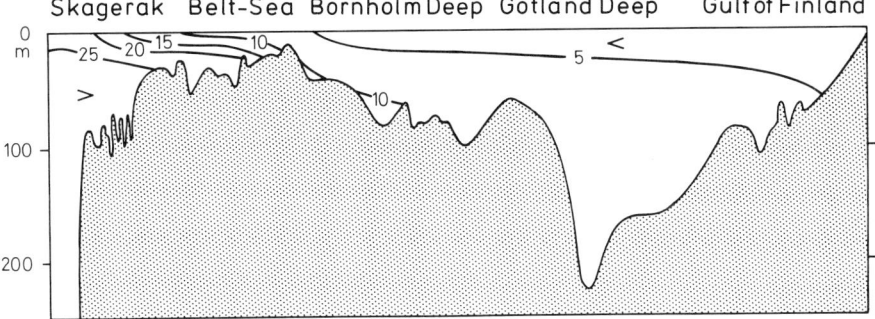

Fig. 13. In the Baltic Sea there is a thin surface layer above the deep water, and the gradient in density hinders the exchange of surface and deep water. The figures refer to σ_t as an expression of the temperature dependent density σ (sigma) = (specific gravity x 0.001) + 1 (Siedler and Hatje 1974)

Because of the brackish layer, the deep water of the Baltic basin is low in oxygen. The fact that oxygen is not completely absent there is due to the inflow of water from the Skagerak which is rich in oxygen and which is forced across the shallow sills of the Belt Sea under corresponding weather conditions.

Conditions have obviously worsened in recent years. Around 1900, 2.5 ml/l of oxygen were being measured in the nearbottom water of the Landsort Deep; in 1950, it was only around 1.5 ml/l. Since that time, there have more than once been situations in which no oxygen was present; the same is true for the other deep basins of the Baltic (Figs. 15 and 16). Under anaerobic conditions, except for some nematodes, the entire bottom fauna dies out. When in the following winter a regeneration of the bottom water takes place, bottom life is again present, however with a different species composition. Opportunistic species dominate (Table 5).

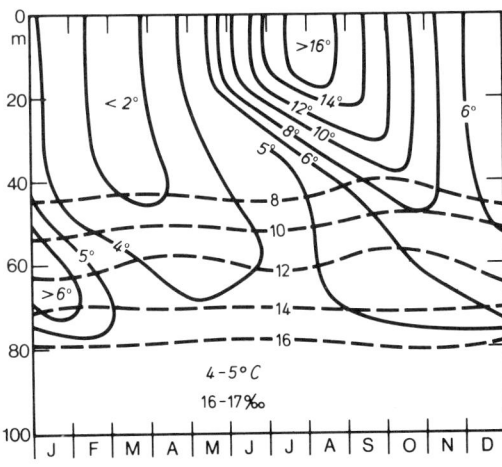

Fig. 14. In the Bornholm Deep (Baltic Sea), the salinity increases from 8‰ to 16 ‰ between 40 and 80 m depth. A salinity gradient (halocline) arises. It remains stable in all seasons *(broken lines)*. Even when the surface water layer cools and becomes denser in the winter months *(unbroken lines)*, this layering remains present. In the deep water, temperature throughout the entire year is 4°–5°C, the salinity 16‰–17‰ (Siedler and Hatje 1974)

21

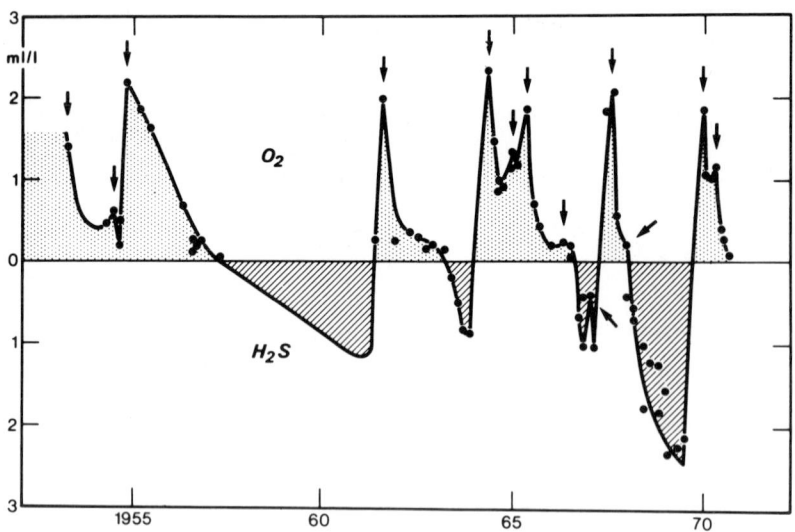

Fig. 15. Between 1957 and 1970, there were four periods with oxygen deficiency and with hydrogen sulfide in the nearbottom water of the Gotland Deep (Baltic Sea). Reproduced are the conditions at a 240 m deep station. *Upper half* oxygen concentration; *lower half* concentration of hydrogen sulfide. (According to data of Fonselius from Grasshoff 1974)

Table 5. In the Bornholm Deep (Baltic Sea) the clam *Macoma calcarea* was the dominent bottom fauna. As a result of the lack of oxygen, in 1968, the entire macrobenthos died out in the area deeper than 75 m. After water rich in oxygen streamed in during the winter of 1969–1970, bottom fauna was again observed. But molluscs were scarcely present. Some opportunistic species were dominant; the most common is the polychaete *Scoloplos armiger*. The table lists the names of the fauna found in 42 samples taken in April and September 1970 (Leppäkoski 1975)

Species	Frequency (%)	Abundance (individuals/m^2) Mean	Maximum	Biomass (g/m^2) Mean	Maximum
Scoloplos armiger	100	1028	2600	7.50	16.3
Harmothoe sarsi	95	79	260	0.95	3.4
Trochochaeta multisetosa	62	24	110	0.15	0.7
Heteromastus filiformis	45	19	87	0.15	1.4
Macoma sp. juv.	33	9	43	0.02	0.1
Capitella capitata	26	4	13	0.02	–

In addition: *Diastylis rathkei, Priapulus caudatus, Nephtys ciliata, Pontoporeia femorata, Pholoe minuta, Halicryptus spinulosus,* Nemerteans, *Terebellides stroemi*

At present, oceanographers in all Baltic countries are trying to find out whether the deterioration of the oxygen situation in the deep water of the Baltic is due to the fact that the large-scale climatic condition over the North Atlantic has changed. Weather conditions of the westwind type, which brought rain, storms, and high water have been occurring more and more seldom over the British Isles. The period 1965–1970

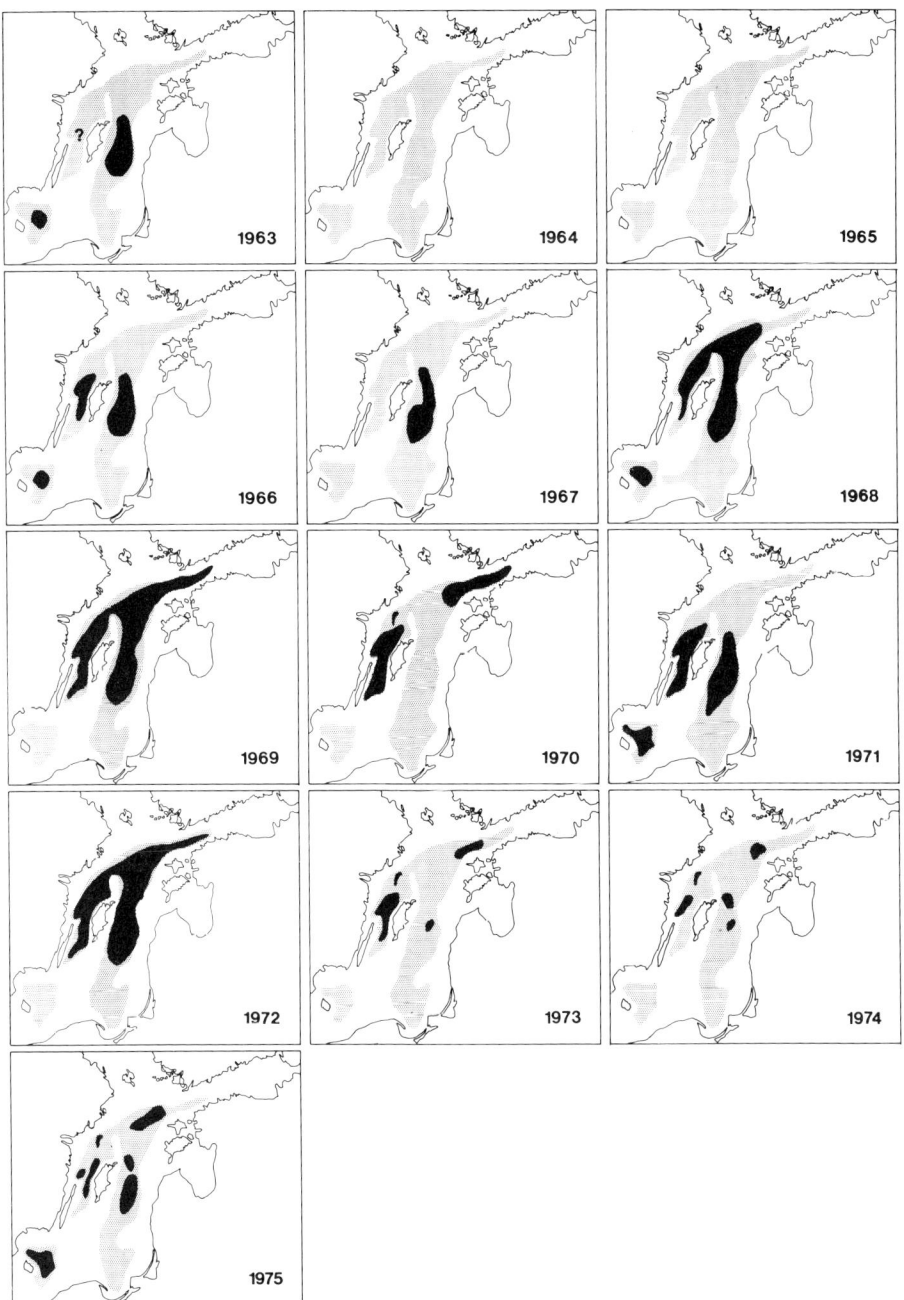

Fig. 16. The extension of zones with oxygen deficiency in the deep areas of the Baltic Sea, 1963 until 1975. *Hatched:* zones with less than 2 ml/l oxygen. *Black:* zones with hydrogen sulfide (Andersin et al. 1978)

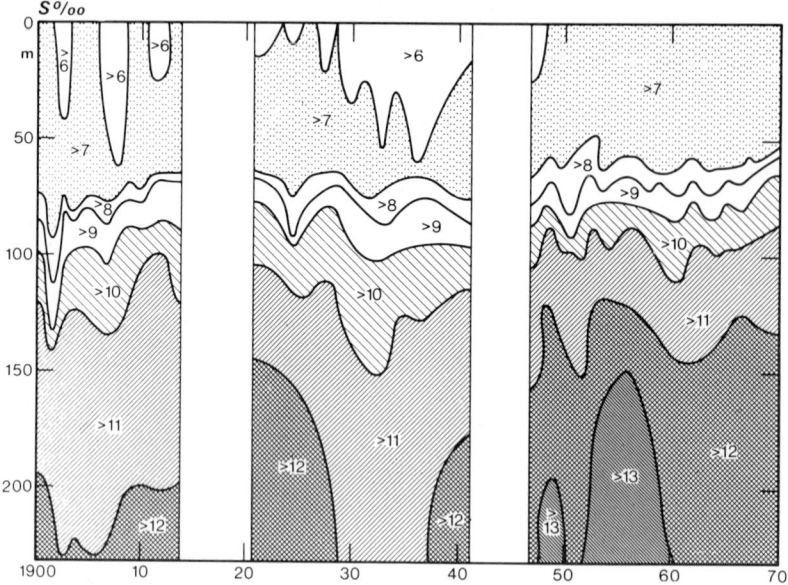

Fig. 17. Since 1900, salinity in the deep water of the Gotland Deep (Baltic Sea) has gradually increased, and the 8‰ salinity isohaline has risen from 80 m water depth to 65 m. This means that the volume of deep water below the halokline increased. The graph shows annual mean salinity in ‰. (According to data of Fonselius from Grasshoff 1974)

was the one with the least frequency of westwind weather conditions in 110 years. From 1970–1975, such conditions were observed again more often.

Climatologists are not in agreement on whether a reversal of this trend is starting to take place, because a reversal of the trend has only been observed in the high northern latitudes since 1970. In the central and equatorial part of the northern hemisphere, the trend to lower ocean water temperatures has continued with temperature declines of 0.1°–0.2°C per decade (Kukla et al. 1977).

Changes of sea level may be responsible for the fact that seawater infusions into the Baltic from the Skagerak are occurring less and less often and that the volume of water of high salinity in the depths of the Baltic has increased (Fig. 17). In the Baltic Sea region, post ice age changes in the coastal contour are still taking place. After all, as a sea area the Baltic is only 7000 years old.

Naturally, the question of what influence the human population of neighboring countries has on oxygen changes in the Baltic Sea is also being studied. In 1960, 1,5 g of phosphorus per inhabitant were introduced into the Baltic via effluents. Through phosphorus-containing detergents and increased agricultural and industrial production, the amount had risen to approximately 4 g of phosphorus per inhabitant by 1970. Like the tropical oceans, however, the surface water of the Baltic Sea is low in nutrients because nutrients are constantly lost with dead plankton and plankton fecal pellets that sink to the deeper levels.

If the effluents of cities contribute fairly large amounts of nutrients to the surface water of the Baltic, an increase in the production of phytoplankton and intensification of the nutrient cycle result. More dead organic material sinks down from the surface layer and reaches the deep water. There, the amount and activity of marine bacteria increases. The oxygen is consumed by the bacteria until it is no longer available. Then, due to anaerobic activity, hydrogen sulfide results. The benthos, the fauna on the ocean bottom, dies off.

One gram of phosphorus can result in a plant production equivalent to 50 g of organic carbon. But dead organic substances with 50 g of carbon need 150 g of oxygen for the process of bacterial decomposition.

Baltic marine scientists have no shadow of a doubt that eutrophication is occurring in the surface waters of the Baltic. There is good evidence that blooms of bluegreen algae that seem to appear more and more frequently in surface waters are the result of this eutrophication. Recent hydrographic observations and more modern calculations indicate that the increasing quantities of dead organic substances introduced via eutrophication or directly into the Baltic are the main reason for the trend to poorer oxygen concentrations in the deep basins of the Baltic. It seems that hydrographic trends are smaller than supposed a few years ago (Jansson 1980). Detailed research programs that are devoted to explaining the processes of water exchanges in the Baltic region and the oxygen and nutrient cycles will provide the answer to this question (Figs. 18 and 19).

If it turns out that human activity does indeed decisively influence the oxygen balance of the Baltic, then it is a frightening example of the power civilization has, not only over rivers and lakes, including America's Great Lakes, but also over a large inland sea like the Baltic. But the trend could be stopped by adequate measures.

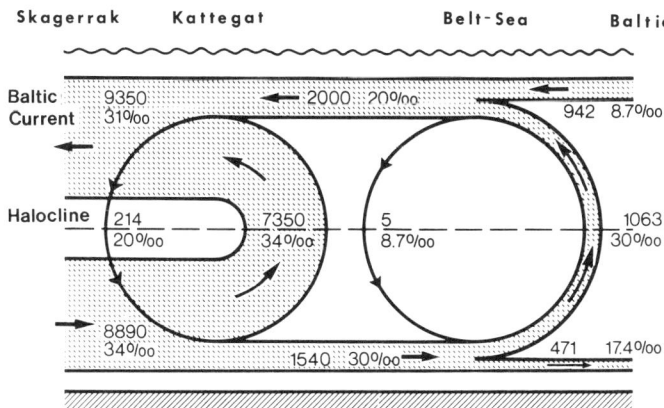

Fig. 18. Presently it is an open question whether marine pollution is responsible for the phenomenon in which hydrogen sulfide occurs in the deep water of the Baltic Sea. To answer this question one has to know exactly the water exchange between Skagerak and Baltic. The above scheme was introduced by Steeman Nielsen in 1941. Figures are according to the present status of knowledge. However, these figures will probably be outdated when the results of the present international hydrographic program on water exchange of the Baltic are available. Figures of water exchanges are in km^3/year (Grasshoff 1974)

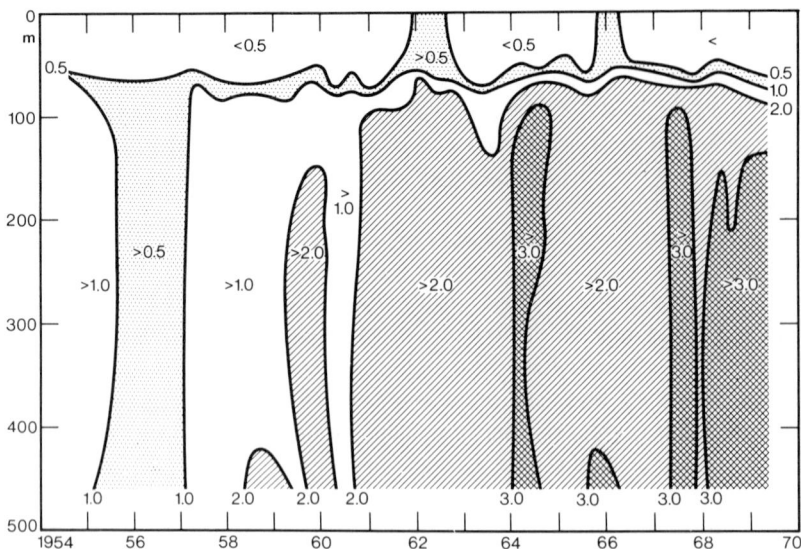

Fig. 19. The concentration of phosphorus increased in the deep water of the Landsort-Deep (Baltic Sea) between 1954 and 1969. Especially high values have been recorded in the years that follow periods of oxygen depletion (see Fig. 13). It seems probable that large amounts of phosphates are set free from the sediment whenever reductive conditions prevail at the water-sediment-interface. Figures refer to phosphate-phosphorus in μmol/l. (According to data of Fonselius from Grasshoff 1974)

If the Baltic Sea is naturally changing so significantly that eutrophication by humans plays only a subordinate role in the process, many measures currently being taken for the purification of effluents by Baltic countries would not be necessary. Then one would only need to set up purification plants for reasons of local hygiene and in order to maintain the self-purifying capacity of the surface water.

The Baltic could finally be approaching the hydrographic condition which has been prevailing in the Black Sea for millenia. There, a brackish surface layer of light water seals off the deep water and prevents contact with the atmosphere. The inflow of water from the Dardanelles is not sufficient to supply the oxygen required for a renewal of the deep water of the Black Sea. The result is that hydrogen sulfide is continually present below approximately the 175 m mark, and no animal life is possible in the deep water. On the water's surface and the beaches nobody at all notices this.

The Baltic Sea is smaller and more placid than the Black Sea and has a different coast line configuration. Under corresponding wind conditions, upwelling takes place here and there and salt-rich deep water rises to the surface. If fairly large areas of the Baltic's deep water contained hydrogen sulfide, people in certain regions would have to accept fish occasionally dying. In large areas of Baltic surface water, life would go on, however, eutrophication would continue, more plankton and conceivably more pelagic fish would live in the Baltic. It is true the cod *(Gadus morhua)* would lose their spawning grounds if a lack of oxygen in the deep water prevented their reproduction, and the character of the Baltic as a oligotrophic sea area with highly transparent water would be lost.

2.5 Global Eutrophication?

Apart from polar and upwelling regions, oceans on the whole are deficient in nutrients, oligotrophic, if one only observes the productive surface layer, which is permeated by light (Fig. 75). The water of the surface layer is warm and thus lighter than the cold deep water. Across broad expanses of the warm oceans in the tropics, only little exchange of water takes place between the surface and the deep water layers. Plankton which are in the process of dying and fecal pellets from plankton animals sink slowly and conduct phosphorus and nitrogen away from the surface. Therefore surface primary production is poor. If one could compensate for the loss of nutrients through fertilization, it would then probably be possible to increase the amount of commercial fish in tropical seas. Technically and economically, however, this idea is an illusion.

Looked at from the perspective of marine pollution, eutrophication is not a problem which is likely to assume ocean-wide proportions. The outlook for the supply of raw materials is more important. The chemical industry can produce nitrous salts in unlimited amounts as long as sufficient energy is available for the synthesis of nitrogen compounds from the atmosphere. The deposits of phosphate ore on the other hand cannot be replenished. At the present rate of consumption, the known deposits of ore would be sufficient for another 500 years. But if one takes a global view, thinks of the future, and keeps the well-being of future generations in mind, then the overuse of irreplaceable phosphates in fertilizers and detergents cannot be justified. The amounts of phosphorus that are present in the world's oceans contribute little to the fertility of the world; they are mainly in the light-less depths of the world's oceans and cannot be utilized.

2.6 Sewage Treatment Plants in Coastal Regions?

Should coastal city sewage processing plants be built which also remove phosphorus and nitrogen from waste water so that the seas are protected from the dangers of eutrophication? For freshwater, the answer is in many cases obvious: eutrophication must be avoided if rivers and lakes are to maintain their capacity for purifying themselves and supplying drinking water. For the Baltic, the answer depends on the evaluation scientists will provide regarding natural versus man-made causes of oxygen depletion in the deep water (see Chap. 2.4). For coastal areas of other seas, like the North Sea and Kattegat, there are, at present, not many arguments in favor of a chemical elimination of nutrients from sewage. One can, by the way, go one step further and question whether waste water treatment is at all necessary in coastal situations.

A biological sewage treatment plant reduces the nutrient load in the effluent by about one third and the amount of biodegradable organic substances, expressed as BOD, by about 90%. To protect narrow fjords, bays, and estuaries, such treatment plants are necessary, otherwise anaerobic conditions could spread. To a certain degree a reduction of pathogenic bacteria and viruses is achieved, too, by the treatment of waste water. The question, however, has to be discussed whether a waste water treatment plant, which is a rather complicated, costly kind of a factory, is the cheapest and most efficient technical means if no other effect is wanted but to kill pathogens.

If an open coastline is concerned with favorable geographic situations and a good water exchange pattern, one could as well argue that no sewage treatment plant is necessary at all. As regards pathogens, one could rely on the bactericidal properties of seawater and the effects of time and dilution. As regards oxygen demand, if enough seawater rich in oxygen flows by, then the degradation of organic wastes could as well be done in the sea as within a biological sewage treatment plant. Calculations have been made that the cost of a biological waste water treatment plant for about 4000 m^3 of effluents per day is comparable to the expenses for a piping system that could bring untreated sewage 4 km away from the shore into the sea; for a 38,000 m^3 plant, expenses are equal to a 16-km-long piping system (Pearson 1975).

Under conditions of good oxygen supply and good seawater transparency, there is little difference between the effects of biologically treated and untreated sewage when emitted into the sea. Nutrients which leave the treatment plant in soluble form cause an increased development of plankton in the seawater that receives the effluents. When the plankton dies, demand for oxygen results and the effect on the oxygen situation may then be almost the same as when the effluents are released untreated into the sea (Officer and Ryther 1977). Moreover, nutrient-rich effluents from treatment plants seem to especially enhance the growth of *Ulva* and *Enteromorpha*, green algae which cover tidal flats and cause anaerobic conditions in the sediment beneath.

There is one important difference between emitting either untreated or biologically treated domestic sewage into the sea: biologically treated sewage effluents contain only a fraction of the original content of heavy metals and persistent organic compounds. The rest has settled down with the sediment and can be found in the sewage sludge. From a number of coastal cities sewage sludge is dumped into the sea (see Chap. 4.2).

2.7 Detergents

Detergents, or tensides are defined as substances which reduce the surface tension of water. They are effective in wash powders. For several decades they have played a large role in households and industry. As a result of bad experiences with persistent compounds, laws now exist specifying that wash powders must be to a great extent biologically degradable. It is, however, a fact that various compounds of tenside are poisonous in concentrations over 0.1 mg/l. This is, among other effects, because the change in the surface tension of water also influences the transfer of substances in and out of organisms.

But marine life proves to be very resistant in classical toxicity tests. It is only at concentrations above 5–100 mg/l that half of the mussels *(Mytilus edulis)* studied over periods of 4 days died; at 1.5 ml/l of the tenside TAE 10 EO even over a period of 5 months mussels are not hindered in their biological functions and produce offspring (Granmo and Jørgensen 1975). But even in concentrations of only 0.5 mg/l, the percentage of fertilized eggs is definitely smaller and the development of the larvae proceedes slower. The transition from trochopora larva to the veliger stage is only successful in concentrations of under 0.5 mg/l. Such concentrations do not occur in the sea except in unusual circumstances and then in those immediate areas where domestic effluents are introduced into seawater.

The city of Marseilles pours around 4 t of detergents into the Mediterranean daily. In the vicinity of the point of introduction, the concentration of detergents is 0.5 mg/l; at a distance of 6 km the concentration drops to 0.01 mg/l.

In those areas where the concentrations are over 0.1 mg/l, the brown alga *Cystoseira* ceases to be present in the shore areas. Farther away, at concentrations of 0.02–0.05 mg/l, the first signs of a change in the flora become noticeable (Bellan 1976). In the areas of the mouths of the Elbe and Weser River in Germany, concentrations are approximately 0.3 mg/l, so that here too effects should be noticed. However, it is difficult to distinguish the effects of detergents in nature from the influences of other effluents (see Chap. 2.9).

2.8 Residual Heat

Marine life is found in temperatures that range from $-2°C$ in the Antarctic and in deep sea regions, to $40°–50°C$ in tropical shallow areas with strong exposure to the sun's rays. However, temperatures between $32°$ and $34°C$ have a fatal effect on most tropical marine life. Basically, it cannot be said that marine life flourishes better at low temperatures than say, at $28°C$; on the contrary, metabolism, growth, and reproduction are more intensive at higher temperatures. Why then is the warming of bodies of water by the residual heat from power plants and other industrial plants a problem that is so passionately argued?

As is the case with the introduction of domestic effluents, the overall order is essential. When large amounts of warm water heat up a small body of water, the maximum temperature which water organisms can stand is quickly surpassed. Naturally, this depends on the particular adaptability of the organisms. In deep ocean water, even at scarcely 100 m, many animals are subject to a narrow range of cold temperatures similar to those in a freshwater wellstream. Year in and year out, the temperature varies but little, and animals and plants are damaged and die if the surrounding temperature should vary even minimally. Fortunately, the organism community in river mouth areas and placid sea coasts is adaptable to a wider range of temperatures. There, the water temperature varies with the season even more than the $1°$ to $18°C$ range found for example in the open North Sea. In intertidal areas it falls to temperatures below the freezing point in icy winters and rises to temperatures of $30°C$ and more under the summer sun. Those organisms which live close to the shore are capable of tolerating the vascillations of temperature, as long as they do not occur suddenly, and if they permit a period of adaptation.

When a new power plant is to be built in a coastal Central Europe area and seawater is to be used for cooling, the stipulation is generally made that the temperature difference that results after mixing is not to exceed $2°C$ and that in the peak heat season the water temperature is not to rise above $26°C$. Both of these extremes are around $1°C$ lower than the recommended temperatures for freshwater. Based on experience to date, they seem to be reasonable rules of thumb for temperate regions; at least until research into the effects of existing power plants furnishes a better basis of measurement. In the tropics, however, an increase of $1.5°C$ may be lethal.

It is a goal for ecologists to establish temperature standards for different bodies of water in order to be able to determine how water temperatures differ now from their preindustrialized state and to determine into which areas additional warm water can be introduced without reaching critical levels. In view of the high investment cost for power plant construction and the additional high cost of cooling towers which disperse the heat into the atmosphere, calculations relevant to water temperatures are of great economic importance. But only in recent years have the required funds been designated and released for programs and measurement and the development of hydrographic models (Figs. 20 and 21).

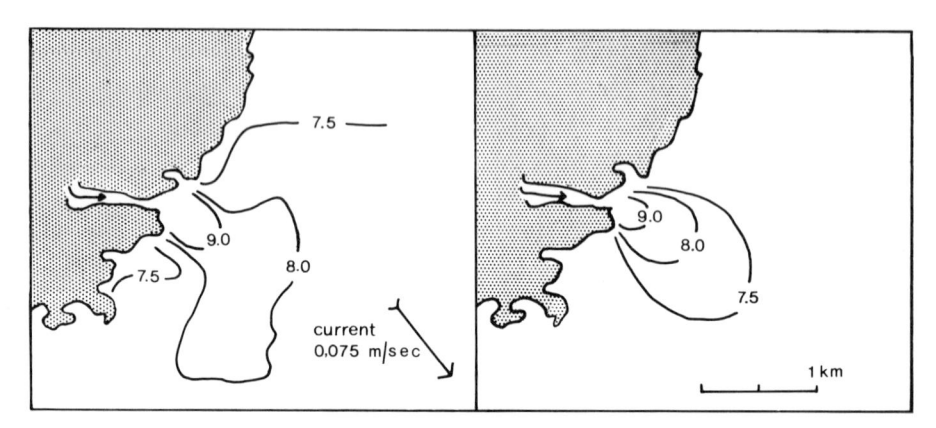

Fig. 20. The effects of warm cooling water can be calculated with numerical methods when simple hydrographic conditions prevail. The *right picture* represents the prognosis of the numerical model for a set of hydrographic and meteorological conditions, the *left picture* represents the seawater temperatures measured in situ. The example is the power plant Oskarshamn (460 MW) on the Swedish coast of the Baltic which heats up 22 m^3/s of cooling water by 10°C. (Data of Weil from Ehlin 1974)

In many cases in past years, physicists have not been able to predict accurately the processes of water mixture and temperature exchange. Even much more uncertain is the research on the biological processes, and the answer to the question of what will actually happen in the future in the possibly concrete event that the water temperature in an estuary near a power plant were to rise by 2°C. Organisms living there are as a rule not necessarily damaged or killed by such an increase in temperature. But that does not mean that the biological community goes on living unchanged, because it is to be expected that when the spectrum of temperature as an ecological factor alters it affects the composition of the community of organisms. Species which like high temperatures would benefit, and northern species might possibly be driven out if the period of growth is extended by higher temperatures. That need not be catastrophic, as the results of aquaculture in warm water have demonstrated. But a certain change in the composition of the community of organisms is to be expected, and species may arise which have been drawn in from warmer ocean regions. The tube-building polychaete *Mercierella enigmatica*, which is indigeneous to all warm water

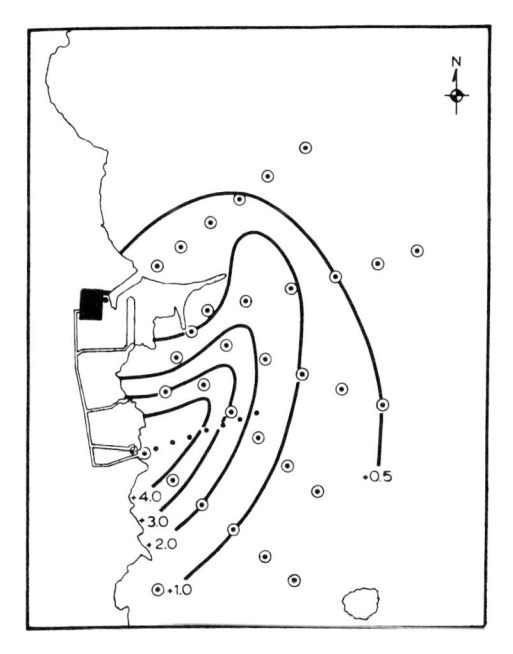

Fig. 21. In summer, seawater temperatures of 28°C can regularly be observed in many tropical and subtropical shallow water coastal areas. It is a fact that even with tropical organisms temperatures between 28° and 30°C have some adverse effects, and that temperatures between 32° and 34°C often are lethal. That means tropical marine animals live very close to the upper temperature limit of their existence, so that an unusually hot summer may have catastrophic consequences. The discharge of warm cooling water into tropical areas is therefore risky.

In 1967, the power plant Turkey Point was established in Biscaine Bay, Southern Florida, with a 432 MW-unit which heats up 6 m^3/s of cooling water by about 5°C. Close to the outlet algae disappeared, later calcareous algae and sea grass as well. Finally the area where the temperature was 4°–5°C higher than normal was covered by a carpet of bluegreen algae. In the areas where the temperature was less than 3°C higher than normal, sensitive algae disappeared, but other algae survived, and the fauna was mostly the same as before. The map presents the isolines of average temperature increase above normal (Zieman and Ferguson-Wood 1975)

areas, has been found in various harbor basins of Central Europe where cooling water is introduced: since 1975, for example, in the harbor of Emden on the North Sea (Kühl 1977).

Changes in the composition of the organism community in warmed bodies of water depend strongly on chance and can only be predicted to a limited degree. Those who would maintain nature in coastal areas exactly as it exists today cannot permit any increase in the temperature of water. Such persons must accept the fact, however, that the community of organisms as it presently exists has not been so since the beginning of time, but rather represents the result of an historical development in the course of which water temperatures went through significant changes. Just stop to think that the post-ice age history of the North Sea and Baltic transpired in a mere 7000 years. The coastal areas close to the equator are proof that ecosystems function in ocean water at relatively high water temperatures as well.

However, every increase in temperature causes an increase of the flow of energy in the ecosystem similar to the increases that occur in chemical processes at elevated temperatures. Primary production — that is, photosynthesis — is increased modestly. But the consumption of plant food by animals and the rate of the degradation of dead organic matter by bacteria is increased even more. The result of this activity is increased respiration, in other words, increased consumption of oxygen. But even at higher temperatures, the ecologically harmonious relationship in balanced communities can remain balanced (Fig. 1).

In principle, one could estimate how an increase in temperature affects the various components of an ecosystem, because one can apply the reaction rate rule (reaction time, approximately doubled at a 10°C increase in temperature) to life-activity-temperature correlations as one does in chemical processes. They do not need to be applicable in each and every case, because it has become increasingly clear in recent years through physiological experiments that many organisms have plateaus in which the activity does not change beyond a certain temperature range. Knowledge about this is not yet detailed enough; and information about the interrelationships in the ecosystem is also as yet insufficient, so that accurate simulations of even simple ecosystems are still in the infant stage. Nevertheless, available knowledge is ample enough to model, for example, the effect of an increase in temperature of 3°C, and compare with measurements (McKellar 1977):

The Crystal River power plant with a capacity of 900 MW, has been in operation on the Gulf of Mexico since 1966. It heats cooling water at 40 m^3/s by 5°–6°C and releases it into the lagoon system off the coast. The natural range of the water temperature in the lagoons during the year is between 14° and 30°C. The winter temperature of the special lagoon into which the cooling water is released is 15°C; in the summer it is 33°C. The phytoplankton biomass was somewhat less here than in the control area but the turnover rate was 5 days, 24 h faster than it was in the unwarmed area. The zooplankton was somewhat more numerous in the heated lagoon, the benthos somewhat less than in the comparison area.

The computer simulation for the summer gave a primary production that was 10% higher and a zooplankton biomass that was 10% less than observed; the computer simulation therefore does not agree completely with the data gathered by observation. Thus, the computer simulation for an additional 4°C increase in temperature (winter 19°C, summer 37°C) might also be speculative: an unaltered phytoplankton biomass with a distinctly higher respiration results, in the spring, in a higher primary production. The zooplankton should diminish by 40% in the summer; in the winter it should be similar in number to that of the unheated lagoons. In the summer, there should be 40% fewer fish in the heated water; in the winter, however, there should be 60% to 70% more than in the control areas because fish migrate and always seek optimal water temperatures.

Basic marine biologiy certainly has a long way to go before such simulations achieve sufficient reliability that they can be used as a predictive instrument. This is because the adaptation relationships of the organisms are reciprocally complex. Under the restricted conditions in estuaries, the oxygen level assumes a key position. Increased temperatures work in the same direction as an increased organic stress. At higher temperatures, the metabolism of bacteria accelerates and their oxygen consumption increases. At the same time, the physical dissolubility of oxygen in warm water is less than in cold water. For this reason, the probability of anaerobic conditions and the dying of fish in the summer becomes greater due to the introduction of warmth into an organically stressed estuary. If warm water is going to be introduced without additional damage to the ecosystem, care must be taken through treatment plants that the organic load is lessened in domestic effluents.

There is, by the way, a harmful side-effect of cooling water passing through a power plant. High flow-through speeds require effective pumps; plankton is partially destroyed by these pumps. Supposedly, it is best to use larger quantities of water for cooling so that they can flow more slowly through the cooling system, because many estuary organisms can tolerate a brief period in water up to 10°C warmer than ambient water.

2.9 Gradual Coastal Changes

An impoverishment of the animal and plant population in the vicinity of coastal cities is being reported all over the world, predominantly in shallow areas. The disappearance in the Adriatic of the brown algae of the genus *Cystoseira*, which forms a belt of up to 1 m, is especially noticeable. But in the Norwegian fjords as well, sea weed is disappearing from the areas near towns. Other algae are spreading in their place: *Cladophora, Enteromorpha*, and blue-green algae. In the Mediterranean, various somewhat deeper regions in which eel-grass *(Posidonia)* formed lush meadows, are also being affected. In the areas around Marseilles as well as in the Adriatic, eel-grass has disappeared from the vicinity of many towns. In its place the *Ulva* and *Halopteris* algae have increased. Changes have occurred in the benthos fauna as well (Avcin et al. 1974; Katzmann 1974; Munda 1974). In the Øresund between Malmö and Copenhagen, 20 species of large algae have disappeared in recent decades. Others have increased in their place; for instance, *Cladophora, Enteromorpha, Ectocarpus, Pylaella,* and *Polysiphonia* (Wachenfeldt 1971). The branching green plant *Chara aspera*, which formerly was common in the belt of skerries around Stockholm, is no longer present there. At most, pitiful specimens are occasionally observed. 25 to 30 km from Stockholm, these plants develop normally (Pekkari 1973).

Even very small communities of humans disrupt the coastal vegetation. The island of San Clemente, occupied by 300 marines, lies 76 km off the California coast. Only around 100 m^3 of domestic effluents reach the ocean daily from the island. Despite this, several species of algae have disappeared from the shore areas near the drain pipe: *Egregia, Hydrolithon, Halidrys, Phyllospadix,* and *Sargassum.* One hundred meters from the drain point, these algae are present, as they are everywhere else, and represent around a third of the plant cover. But around the drain area, they have been replaced by *Ulva, Gelidium,* and blue-green algae (Littler and Murray 1975).

Although a change of algal communities is being observed along those coasts all over the world that are affected by effluents, no single cause has been found to be responsible for it up till now. The mixture of domestic and industrial effluents, which are introduced into the sea by many coastal towns, can have various effects: the water is warmed; drain water contains suspensions of the most varied sorts which settle on the bottom; the oxygen balance is upset and trace elements can have a direct and poisonous effect; oil drifts on the water's surface and affects the tidal areas in particular; and plant nutrients are abundantly present in effluents. The effect does not need to be immediate either. It can happen circuitously: the flora of microscopically small algae which settle in on larger algae may be fostered; or parasite attacks may be facilitated.

However, up to now, the plant population along open coastal areas of England have not changed. This even applies to the 48-km-long North Sea coast of Durham where the Tyne, Wear, and Tees Rivers, which are heavily polluted by effluents, discharge into the sea. The first collecting of algae goes back to the years 1793 and 1864, before industrialization reached its high point. At that time, 48 species of algae were labeled as common; today, out of 128 species known in all, 53 are so labeled. *Furcellaria fastigiata, Lomentaria clavellosa, Plocamium coccineum, Polysiphonia elongata, Desmarestia aculeata,* and *Pelvetia canaliculata* were previously common; today they are absent; *Asperococcus fistulosus* and *Spongomorpha aeruginosa* are rare today. Thus, the spectrum of species has shifted somewhat, but it is not yet known for sure whether or not effluent flows played a role in this shift or whether it has to do with natural fluctuations that transpire in the course of relatively long periods (Edwards 1975). It is good to know that the decrease in species diversity, that the change from sensitive to pollution-resistant species is confined to polluted estuaries and to the vicinity of coastal cities, and still is not evident on the open North Sea shore.

After the dying off of sensitive organisms, an affected coast does not remain a desert, rather robust algae are fostered, and their population is, as a rule, richer in biomass and productivity than the indigenous vegetation. Phytoplankton is also thicker near the points of effluent entry and even marine aminals profit from the abundance of food (Fig. 4); the shell duck *(Bucephala clangula)* and the mute swan *(Cygnus olor)* from year to year winter over in greater numbers right at the Tay Estuary effluent pipes (Pounder 1974). There is a similar development on dry land where the natural vegetation cover is continually changing. Wherever man dominates, we encounter roadside flora. The ecosystem here has been developed in such a way that a few species dominate.

The coasts of the world's continents and islands together add up to approximately 1,5 to 2 million km, depending how precisely the curves in the coastline are followed. If we take the mean width of a coastal region to be around 70 m, it then turns out that coasts account for only 0.1% of the land, 0.05% of the oceans or 0.03% of the total globe. Another calculation (Table 6) results in the statement that all regions of benthic algae, coral reefs, and the estuaries together represent only about 0.4% of the globe. Exactly the coastal areas, on the other hand, are the areas where human settlements are the densest, and where urbanization is increasing (Gerlach 1980). Animal and plant life is endangered everywhere there are human settlements. In protected shallow zones, aquaculture is fostered, fish ponds are constructed, and oysters are raised in the shallows. Swimming and beach activity is expanding along sandy beaches. Divers frequent rocky coasts and coral reefs. A piece of coral which took many years to become large and conspicuous is easily broken. In the Mediterranean, the Carribbean, and other places where spear fishing has not been regulated, large fish have become rare sight for skindivers. The Philippine Islands alone export 3–30 million

Table 6. Main environments of the globe. Especially valuable are those environments which have only limited extension, or which have a high plant production. The total area of the globe is 510×10^6 km^2, the total primary production is assumed to be 77.6×10^9 t/organic carbon per year. (Data from Woodwell et al. 1978)

	% of globe	% of global primary production
Rare environments		
Benthic algae, corals	0.1	0.9
Estuaries	0.3	1.3
Freshwater	0.4	0.5
Wetlands	0.4	3.5
Cultured land	2.7	5.3
Productive environments		
Prairies, savannah	6.4	14.4
Woods	9.6	42.7
Shelf sea	5.3	5.6
Abundant environments of low productivity		
Tundra, desert, rock	9.8	1.8
Ocean deeper than 200 m	65	24

coral fish annually to tropical fish dealers in Europe and the U.S.A. Low-lying stretches of land, like the northwest European marshes, are protected against storm flooding by dikes, in the process of which the original saltmarsh is destroyed. Water construction works are being built to protect the coast, to regulate navigable rivers, and as harbors. These are all influences that do not fall under the heading of marine pollution, but obviously contribute to the deterioration of our natural environment.

It is therefore high time to create nature reserves in coastal areas, too, so that at least in certain regions we can pass this natural landscape on to future generations.

Presently, the Netherlands, Germany, and Denmark are trying to maintain the Wadden-Sea as a natural reserve and feeding ground for nordic migratory birds. This will only be possible through a regional organization: by totally protecting certain areas, by trying in other areas to harmonize tourism and fisheries, and by reserving additional areas for maritime commerce or industrialization.

In the tropics, plans have been made to designate certain coral reefs as protected areas, especially certain ones in the vicinity of cities and areas of tourism. Economic reasoning is very much involved; an underwater landscape that is intact and in which fish are plentiful is an appealing tourist attraction, so one might as well spend money to build sewage treatment plants in the areas of tourism. Otherwise robust algae could greatly increase in number and overgrow the coral, as is happening in Hawaii with the *Dictyosphaeria* algae (Johannes 1975).

On the Canadian coast of Nova Scotia, sea urchins *(Strongylocentrotus droebachiensis)* increased dramatically and in spots destroyed 70% of the *Laminaria* kelp growth. The diminution of the lobster stock *(Homarus americanus)* is presumed to be the reason for this increase because lobsters feed on sea urchins. Because of fishing, the lobster count has fallen off by half in the last 14 years (Breen and Mann 1976). On the

Adriatic coast, sea urchins *(Paracentrotus lividus)* are presently on the increase, but whether a decreased consumption of them, altered algae vegetation, or a direct effluent inflow is responsible is not yet known (Ghirardelli 1973). Till now, no one has yet been successful in explaining the mass increase of starfish (*Acanthaster planci)* in the Pacific where they have destroyed considerable portions of coral reefs (Johannes 1975). Nor has anyone yet determined that it was marine pollution that caused the eel-grass *(Zostera marina)* in the area of the North Atlantic to fall victim to an attack of the parasitic fungus *Labyrinthula macroystis* (Rasmussen 1973).

Our thoughts are influenced by wishful thinking, even in this age that is enlightened by the natural sciences. Just as in the middle ages a rain of blood, fish, and toad, as well as other strange events in nature, were interpreted as harbingers of impending disaster, many people at the present are prone to connect the strange emergence of marine organisms with the effects of ocean pollution: The occurrence of especially large jelly-fish off Dutch beaches, and of blooms which occasionally develop from the diatom *Coscinodiscus,* or from blue-green algae, and the appearance of poisonous phytoplankton which regularly occur in many warm seas as "red tides" and which are also occasionally sighted in the North Sea. Serious symptoms of poisoning then occur in humans which can be traced back to eating shellfish in which the poison accumulates.

In August 1978 reddish-brown seawater occurred at the southern coast of Ireland. Trout in aquaculture farms died, as well as about half of the shore snails *(Litorina),* limpets *(Patella),* barnacles *(Balanus)* and actinia on rocky shores, and lug-worms *(Arenicola)* on sandy beaches. This was understood as a warning, and the consumption of seafood was forbidden. The event followed a period of very quiet summer weather without waves, even without swell from the Atlantic, and these conditions were optimal for the growth of the plankton algae *Gymnodinium* and *Gonyaulax* which, in their metabolism, produce the toxin. On this part of the shore there are no waste water effluents (O'Sullivan 1978).

Natural communities of organisms are not static, but are subject to vascillations in quantity. Man as a triggering agent is only partially responsible. Often, it seems that a coincidence of a number of factors in constellation produces a result. To take note of these is an important task, because the study of many different constellation factors will alone furnish the key to an understanding of the causes.

3 Industrial Effluents

3.1 General

Industrial effluents are as diverse as industrial products. The many different kinds and the various effects upon life and water quality are dealt with in textbooks on water pollution. Much experience from fresh water management can be transformed to marine conditions.

In this book, only two aspects of industrial effluents are dealt with: mercury and pesticides. This is because their importance for the development of marine pollution science is predominant. The more global aspects of marine pollution with mercury and with chlorinated hydrocarbons are, however, dealt with in Chapters 8 and 9.

3.2 Effluents Containing Mercury

Minamata is a small town of 50,000 inhabitants on the shore of Shiranui Sea, a part of the inland sea of Japan (Fig. 22). Many of its inhabitants make their living from fishing. The rest of the city is economically dependent on the Shin Nihon Chisso Hiryo Company which, along with other products, had been producing vinylchloride and acetaldehyde since 1952. In 1953, puzzling symptoms of illness began to be observed in Minamata's human and cat population. Then in 1956, the illness assumed epidemic proportions. It was assumed to be contagious meningitis, and the patients were isolated. An investigation committee was established. It found that all patients had similar symptoms. They experienced numbness in their lips and limbs. After two weeks, disturbances in their sense of touch, in speech, and hearing became noticeable. Their gait became irregular; they walked as if drunk, unable to stop suddenly or to turn around. The most noticeable thing in all patients was, however, a concentric narrowing of vision. 46 of the 116 officially registered patients died. Only a few were cured. Most of them experienced permanent disorders.

Judging by the symptoms, Professor Shimanosuke Katsuku of the Kumamoto University assumed it to be a heavy metal poisoning. It was ascertained that all the patients had eaten a lot of fish. It was also observed that cats showed signs of Minamata Disease. When it was found that the symptoms could be artificially produced in healthy cats by feeding them fish from the Bay of Minamata for 5–8 weeks, it was clear that one was not dealing with a contagious disease, but with food contamination. Fishing was outlawed in that area of the Bay of Minamata at the beginning of 1957. That year, no additional persons became ill; in 1958, only 3 died.

Fig. 22. Forty six persons died from mercury poisoning between 1956 and 1958, because they had eaten fish from Minamata Bay. Since 1952 the Chisso Work had introduced large quantities of mercury-containing wastes into Minamata Bay. *Points* indicate the homes of victims 1953–1958. In 1959 there were again 10 cases of mercury poisoning north of Minamata (location *open circles*), after the factory had introduced wastes into the Minamata River and some fishermen had caught fish there in spite of an interdiction (Tokuomi 1969)

The search for the toxic element lasted 3 years. The effluents were found to contain 60 different substances with toxic properties, among them numerous heavy metals. But at the outset, none could be linked to the Minamata Disease.

The plant itself hindered the investigation and successfully attempted to invalidate the resulting findings. Not until 1959 was it successfully proven that mercury in an organic compound was responsible: In 1960 it was proven that the effluents contained methylmercury. But the Japanese chemical lobby was able to prevent the findings from becoming official. It is assumed that a total of between 200 and 600 t of mercury were introduced into the Bay of Minamata by the Chisso Works. The discharging of effluents was not halted until 1968 (Harada and Smith 1975).

In 1965, the same sickness occurred in the area around Niigata at the mouth of the Agano River, Japan. Thirty persons became ill; 5 of them died with the symptoms of Minamata Disease. Effluents containing mercury from the Kanose Works at Showa Denko, 60 km upstream, had allowed the mercury content to rise in fish used for human consumption. When by 1968, the Japanese government still had not taken a decisive stand on the matter, the first pollution suits in Japan's history were filed against the chemical works. Concerned citizen and protest group activities played a large role in the following years (W.E. and A.M. Smith 1975), until damages were awarded on March 20, 1973, to the victims of Minamata.

Up to the year 1975, 798 patients were diagnosed as victims of the Minamata Disease. The Chisso Works paid damages in the amount of U.S. $80 million. An additional

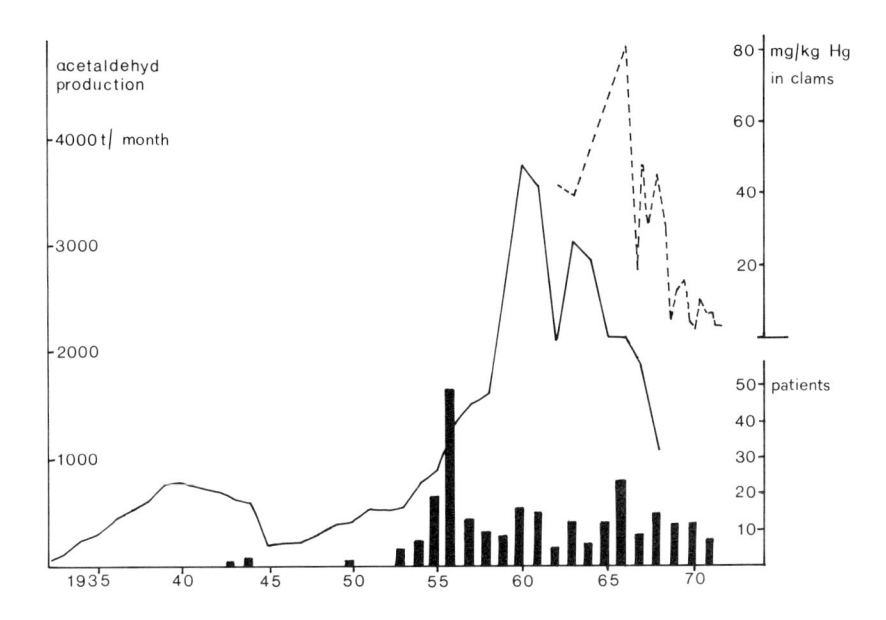

Fig. 23. History of the Minamata Disease. Between 1952 and 1960 the manufacture of acetaldehyde was greatly increased (*continuous line;* 1962 reduction of the production by labor struggle, 1968 stop of the production). *Black columns* indicate the numbers of patients suffering from Minamata Disease (first year of registration). Only after in 1966 was a waste water treatment effective, mercury concentrations in *Venus* clams collected a Koiji Island gradually decreased (*broken line,* in mg/kg dry weight), only after the stop of production were low values reached (Harada and Smith 1975)

2800 persons were still waiting in 1977 to be officially designated as victims of the Minamata Disease (Fig. 23).

Mercurychloride was used in the Minamata factory as a catalyst in the production of vinylchloride. In the washing process, a small portion of mercury goes out into the waste water. When producing acetaldehyde from acetylene, the losses were greater; mercuric sulfate in an organic compound with acetylene serves as a catalyst. 0.3–1 kg of mercury per t of acetaldehyde resulting from this process were released by the effluent, of this 15–50 g as the especially poisonous organic compound methylmercury (Tokuomi 1969; Ui 1969).

In the mud near the spot where the effluent containing the mercury was introduced into the Bay of Minamata approximately 2000 mg/kg of mercury were found in the sediment; further away, 12 mg/kg; in uncontaminated areas, only 0.4–3.1 mg/kg (Tokuomi 1969). Table 7 furnishes information about the mercury concentration in organisms from the Bay of Minamata and in affected cats and human beings. It must be taken into account when studying this table that the figures relate to the dry weight concentration; in the U.S.A and in Europe data are as a rule presented relevant to wet weight. If we assume a water content of 80%, then the dry weight concentration must be divided by 5 in order to be comparable with figures that are relevant to the wet weight ones.

Table 7. Mercury concentrations in samples from Minamata and from nearby areas without mercury pollution (Tokuomi 1969)

Material	mg/kg dry weight	
	Minamata Bay	Unpolluted area
Clams from the beach	11–39	1.70–6.00
Fish	10–55	0.01–1.70
Cats (from Minamata with mercury poisoning)		
Liver	40–145	0.64–6.60
Kidney	12–36	0.05–0.82
Brain	8–18	0.05–0.13
Man (from Minamata with mercury poisoning)		
Liver	22–70	0.07–0.84
Kidney	22–144	0.25–10.70
Brain	2–25	0.05–1.50
Hair	281–705	0.14–7.50

In other countries, too, industrial effluents have lead to considerable concentrations of mercury in fish for human consumption, but without fatal consequences for human beings. There is a chemical plant on the Adriatic near Ravenna, Italy, which produced acetaldehyde by the same process as in Minamata. Fish in the vicinity of the effluent drain pipe showed a mercury concentration of 1–2 mg/kg. But they were not eaten because oil they absorbed from other effluents ruined their flavor.

Paper and cellulose plants used to protect their products against fungi with mercuryphenylacetate or other organic mercury compounds. Liquid waste from such plants contained mercury. But in the meantime the application of fungicides containing mercury has been widely prohibited. On account of the effluents of a paper mill, the cod *(Gadus morhua)* in the Kragerö Fjord in Norway contained on the average 0.9 mg/kg of mercury relevant to wet weight (Underdal 1971). In Gunnekleivfjord in southern Norway close to a chlorine plant mercury concentrations in the sediment were 90–350 mg/kg, that is more than double the figure from Minamata Bay (Skei 1978).

In alkaline electrolysis plants in which chlorine gas and soda lye are produced, mercury elelctrodes are used in large quantities. Sweden has an annual production of 220,000 t of chlorine; 150–200 g of mercury per t were previously lost with the waste water. This amounted annually to approximately 25–35 t of mercury in Sweden. Canada's chlorine industry had to replace approximately 1000 t annually (Trites 1972). Twenty six percent of the mercury consumption in the U.S.A. was due to the chlorine-alkaline industry.

Forty areas in Sweden had to be closed to fishing, among them some coastal areas. In 17 states of the U.S.A., in Canada, Finland, and Norway, fishing was prohibited in certain bodies of water.

When over a period of many years mercury has been released into the marine environment, it accumulates not only in organisms but also in sediments. It will obviously take many years before, by dilution or by sedimentation, the concentrations go back to natural levels. There is a special affinity of mercury to the organic component of marine sediments (Fig. 24). One has to keep this correlation in mind when correlating mercury concentrations in the sediment with mercury sources.

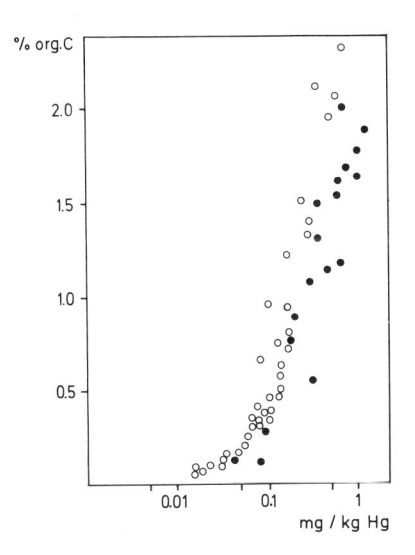

Fig. 24. Mercury that reaches the sea with waste water is accumulated in the marine sediment. The concentrations are higher with high contents of organic matter in the sediment. The diagram is from Swansea Bay, Great Britain, where mercury is released with effluents from a chlorine plant. Concentrations are higher in stations closer than 2 km away from the inlet *(filled circles)*, than in areas further away *(open circles)*. *Figures* refer to dry sediment (Clifton and Vivian 1975)

Since 1977, the European Community has been directing its efforts to preventing the release of more than 0.5 g of mercury per t of chlorine from new plants, 1 g from plants already in production, into the environment via waste water. Mercury emissions to the atmosphere, which now are mostly in the order of 10 g per t of chlorine, should be reduced to less than 4 g per t of chlorine. In European Community, the aim is to reduce mercury effluents into coastal waters so that fish do not have more than 0.3 mg/kg of mercury in their muscle tissue (see Chap. 8.1).

This seems to be technically possible and economically feasible and will result in minimal mercury contamination from industrial plants in coastal areas. The mercury content in the Thames was reduced by 75% in 6 years. However, in Great Britain many old factories are still producing and have difficulties in meeting European Community standards; they are struggling for a permit to discharge up to 8 g mercury per t of chlorine produced. Even in Denmark it is evident that the fertilizer plant Superfos in Fredericia has a permit to introduce waste waters containing mercury and cadmium into Little Belt, because the waste water treatment plant will not be ready before 1983.

To free coastal areas completely from contamination by mercury, additional means would have to be employed because mercury is even introduced into estuaries by domestic effluents. In Los Angeles, in 1971, it was thought that all sources contributing mercury to waste water effluents had been eliminated. Nevertheless, 3–5 kg of mercury were found to be present in the 1.29 million m^3 of effluents that daily pass through the Hyperion treatment plant (Fig. 35). Only through a complex and costly process would it be possible to eliminate from the waste water all the mercury that comes from dentistry, chemical laboratories, instrument manufacturing, instrument and fluorescent light bulb breakage, and from other sources, and which then presumably finds its way via atmospheric dust into effluents (Bargmann 1975; see Chap. 8.4). It is more realistic to resign oneself to the fact that effluents, even after treatment in a biological treatment plant, contain double (2 μg/l) the amount of mercury California

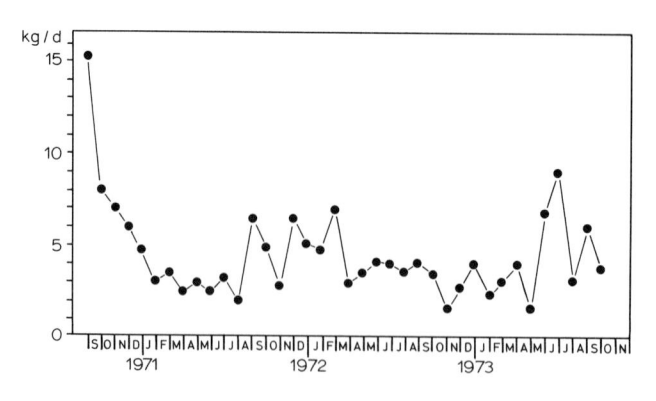

Fig. 25. Mercury in the effluent of water of the Hyperion sewage treatment plant at Los Angeles, U.S.A., after a stop was put to all known introductions of mercury in 1970. Figures in kg/day for an effluent volume of 1.29 million m^3/day (Bargmann 1975)

officials have actually determined to be the maximum permissable amount. If future research comes to the conclusion that even these small amounts are not tolerable, than the use of mercury in general will have to be prohibited or radically limited.

In the period of 1970, the River Rhine annually transported more than 90 t of mercury to the North Sea, partly dissolved in the river water, partly bound to suspended particles (Table 8). This amount is probably more than 20 times higher than the natural one which is due solely to the weathering of rocks. In the meantime, the load has been reduced as a consequence of pollution control. Sediments of the Rhine collected between Basel and the German-Dutch frontier in 1979 had only about one third of the mercury concentrations found in 1971 (Müller 1979). But this still is much more than the natural concentrations, and thus it happens that at the mouth of the Rhine 5–10 times more mercury is present in the seawater than in the open sea (Baker 1977), namely 30–60 ng/l versus 7 ng/l. It must be taken into account that these figures represent more recent analyses which arrive at lower concentrations than previous ones reporting mercury concentrations in the open, unpolluted Atlantic at 40 ng/l (see Table 2).

Table 8. Estimates of the amounts of heavy metals which, in 1970, the river Rhine transported downstream. Their annual values has been estimated to be around US $ 50 million. In the meantime, due to anti-pollution measures, amounts are lower, and mercury concentrations in the sediment of the river bed are now only about one third of 1970 values. In the river, approximately equal amounts are transported in dissolved form and with particles suspended in river water (Förstner and Müller 1974)

	Dissolved t/year	Suspended t/year
Zinc	11,000–25,000	7,000–20,000
Copper	800–3,000	1,400–3,000
Chromium	1,000–1,300	2,800–4,000
Lead	600–700	1,800–3,000
Nickel	800–2,500	200–800
Cadmium	100–500	100–200
Mercury	40–50	50–80

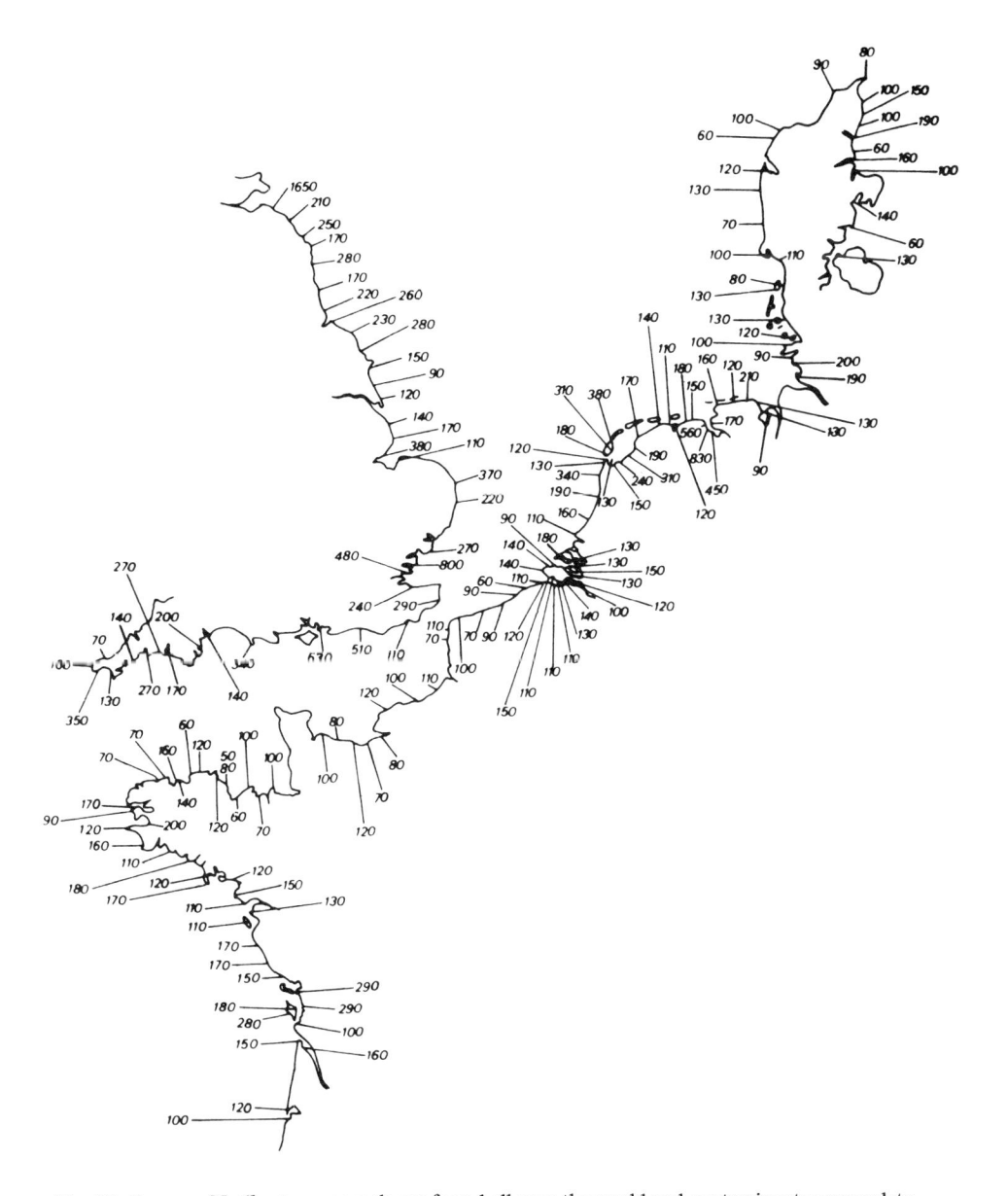

Fig. 26. Because *Mytilus*-type mussels are found all over the world and contaminants accumulate in the soft parts of their body, it has been suggested that a "mussel watch" be established, that mussels be analyzed in order to check on pollution of the environment. The mercury concentration in 30-mm-long northwest European mussels, *Mytilus edulis,* reflects the mercury contamination of the coastal waters. High amounts are to be found off the Rhine and the Thames. The figures refer to mg/kg of soft-body wet weight (Wolf 1975)

The higher the seawater concentration of mercury, the more mercury accumulated in the bodies of organisms. Mercury contamination of the northwest European sea waters is reflected in the mercury content of mussels. Large amounts, considerably more than 0.2 mg/kg, are present in mussels from the mouths of rivers and in the areas around industrial effluent pipes (Fig. 26). Likewise, differing levels of mercury contamination can be ascertained along the Atlantic coast of the U.S.A from the mercury content of the plankton (Fig. 27).

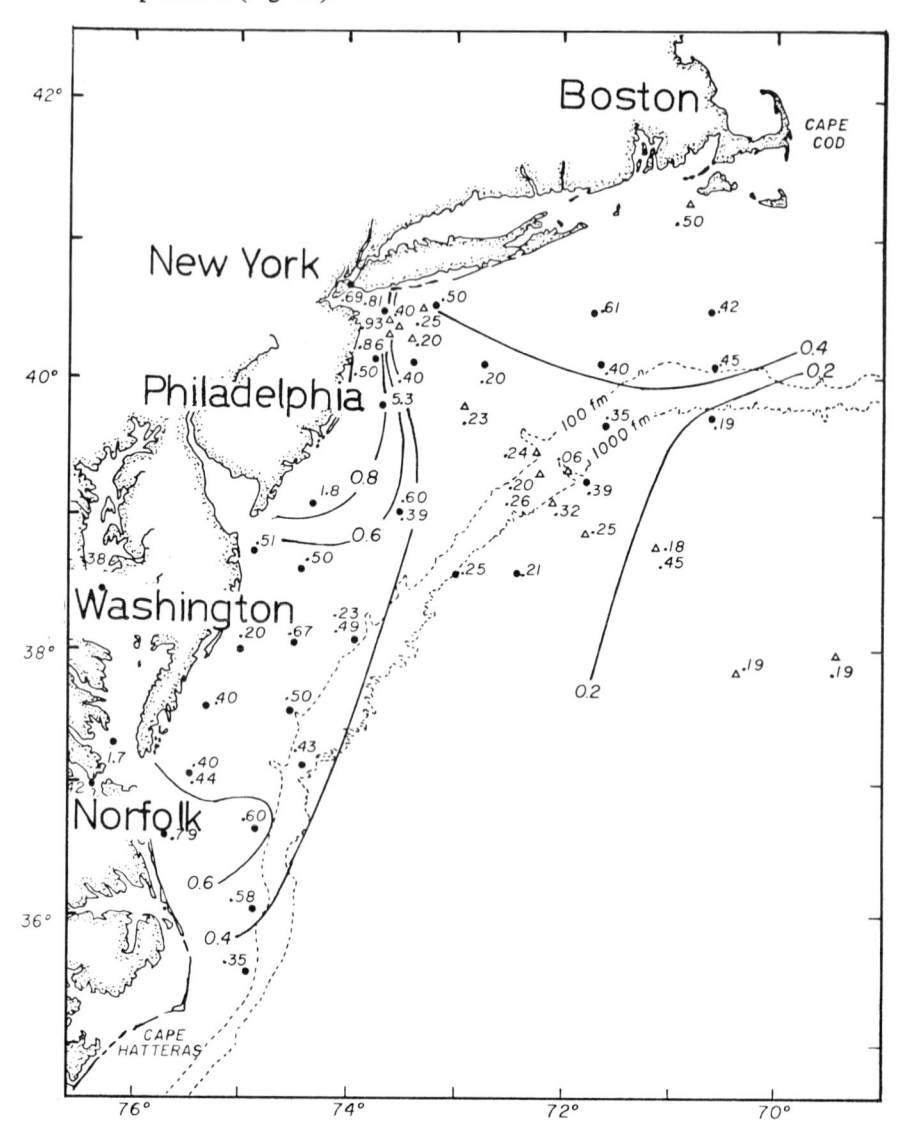

Fig. 27. Specimens of zooplankton taken from the surface water of the Atlantic close to the east coast of the U.S.A. show higher concentration of mercury than samples taken from areas distant from the coast. Specimens collected in 1969 are indicated by a *triangle*, those taken in 1975 by a *dot*. The figures refer to mg/kg of dry weight (Windom et al. 1973)

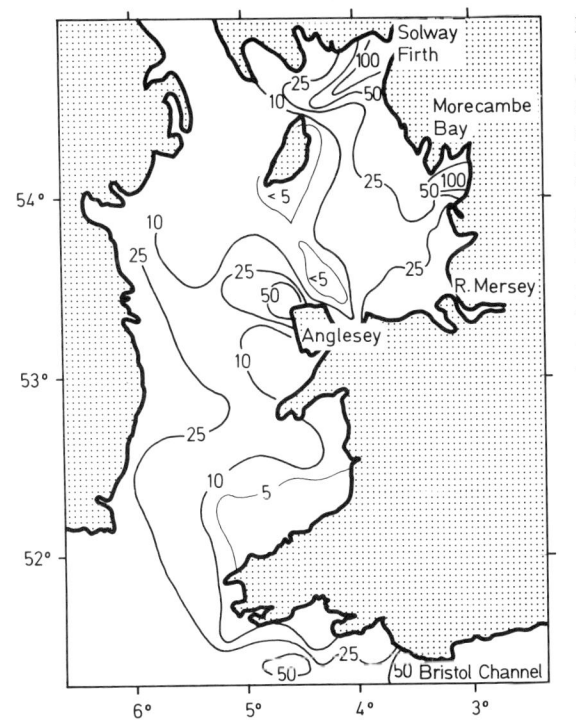

Fig. 28. High concentrations of dissolved mercury in the seawater of the Irish Sea occur in areas where sludge from Manchester sewage treatment plants is dumped (which contains up to 150 mg/kg mercury), where the River Mersey brings waste water from Liverpool into Liverpool Bay, in Morecambe Bay, in Solway Firth, northwest of Anglesey and in the Bristol Channel. Water containing mercury is transported anti-clockwise through the Irish Sea. Low mercury concentrations are found in those areas where during windy weather the sediment is brought into suspension. It is probable that in such regions mercury adsorbs to particles and then is sedimented. Normal mercury concentrations in Irish Sea sediments are 0.01–0.03 mg/kg. For example, in the Mersey Estuary and at the Fylde Coast, mercury concentrations in the sediment are 0.13–0.38 mg/kg dry sediment. These high concentrations persist in spite of the fact that dumping had been stopped several years ago. Figures in the map show mercury concentrations in seawater, as ng/l (Garder 1978)

In the Thames Estuary and at places in the Irish Sea where sewage sludge is dumped, edible fish contain 0.5 mg/kg of mercury. In these areas the mercury concentration in seawater is above 50 ng/l (Fig. 28), so that the concentration factor for fish musculature is in the order of 10,000 (Garder 1978). In the estuaries of the Weser and Elbe Rivers fish have similarly high mercury concentrations (Meyer 1972; Priebe 1976; Jacobs 1977); flounders *(Platichthys flesus)* have been found with above 1 mg/kg, eels *(Anguilla anguilla)* with 2 mg/kg and ruff *(Acerina cernua)* with sometimes close to 3 mg/kg. Such concentrations are above the limit set for edible fish in the Federal Republic of Germany (See Chap. 8.1). Hopefully it should be possible to bring down the concentrations in fish to below 0.3 mg/kg by stricter pollution control measures.

In areas, where fish is contaminated, particularly high concentrations of mercury are found in seals and marine bird life that feed exclusively on fish and are the final link in the marine food chain. In the liver of *Phoca vitulina* harbor seals caught off the coast of the Netherlands, 225–765 mg/kg of mercury, relevant to wet weight, was found; in seals from the North Sea coast off Germany up to 100 mg/kg of mercury is present (Fig. 29). In the meantime, the available data has become so massive that correlations can be made between mercury concentration and age on the one hand and contamination of the environment on the other hand (Fig. 30). It seems that a few months of eating invertebrates on English tidal flats results in elevated mercury concentrations in sea birds (Table 9).

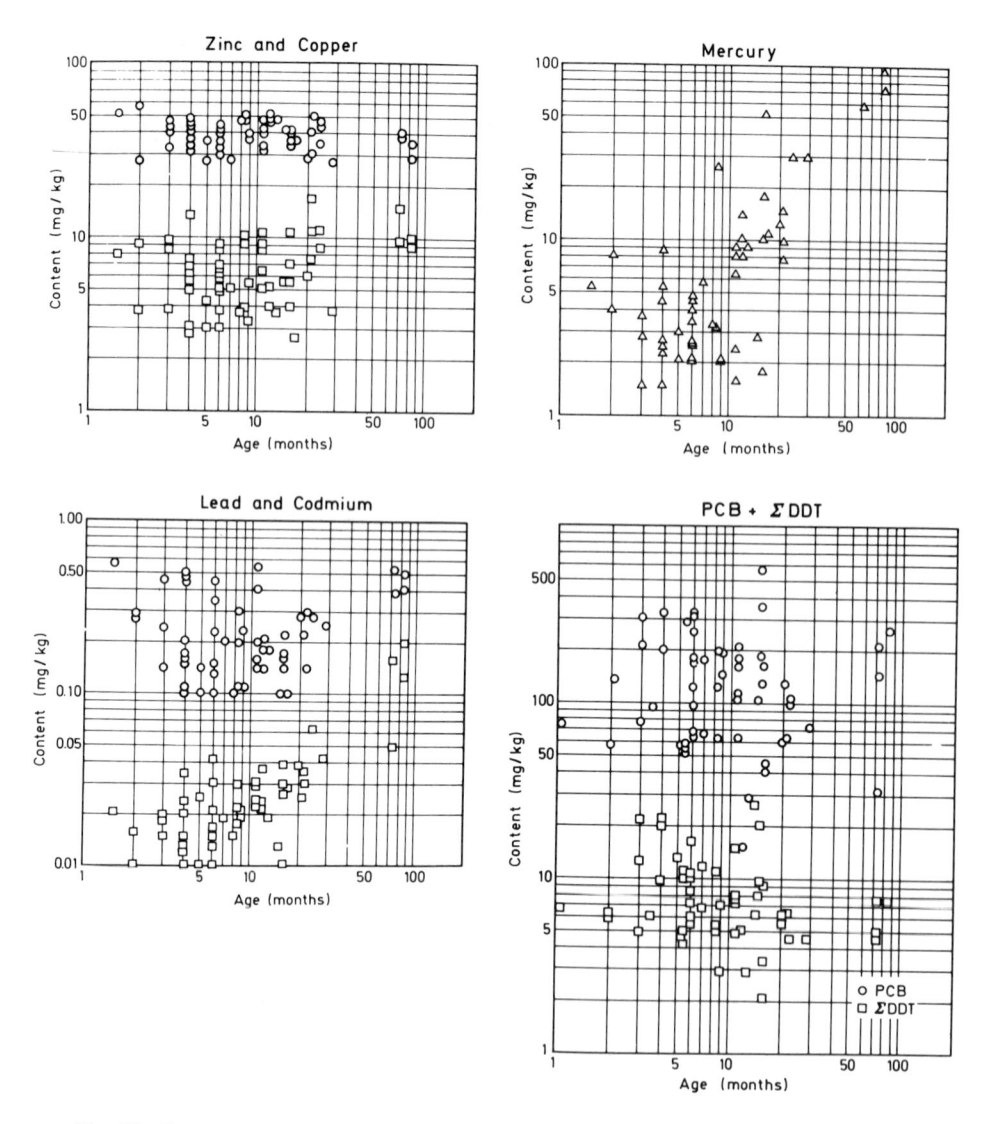

Fig. 29. Concentration of trace contaminants in 63 harbor seals *(Phoca vitulina)* either shot or found dead in the W. German Wadden-Sea 1974–1976. Data for DDT and PCB are from blubber, data for trace elements are from liver, concentrations in mg/kg wet weight. Concentrations of mercury and cadmium increase with age. There is not much difference in lead concentration between seals and their food, fish. Concentrations of cadmium, zinc, and copper, however, are significantly higher in seal than in fish, concentrations of DDT, dieldrin and lindan are about 100 times, concentrations of mercury and PCB's are about 1000 times higher in seal blubber or liver than in fish (Drescher et al. 1977)

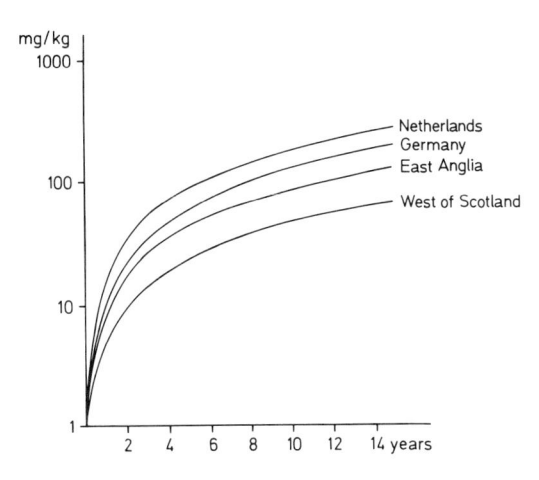

Fig. 30. With age, the mercury concentration in the liver of *Phoca vitulina* harbor seals increases. It is also clear that seals off the coast of the Netherlands contain more mercury than those off the German coast or off the southeastern coast of England and those off the west coast of Scotland. Data in mg/kg wet weight (Harms et al. 1978)

Table 9. Sandpipers hatch their eggs in the tundra areas of the Arctic far away from the sea. Their food there contains little mercury. For this reason, the mercury content in the body tissue of sandpipers is low when they migrate to the shallows of the southern part of the North Sea in the fall. The longer they feed there on *Corophium* crustaceans, *Macoma* clams, *Hydrobia* snails, and *Nereis* bristle worms, the higher the mercury concentration rises. Figures are for the knot, *Calidris canutus,* from the Wash area of southeast England. The range is given in parentheses (Parslow 1973)

Date	Mercury concentration in liver mg/kg dry weight
August	1.0 (0.4–1.5)
October	1.3 (0.8–1.7)
January	7.3 (6.6–8.0)
February	9.8 (5.1–12.9)
March	14.4 (5.6–24.9)

3.3 Effluents from Pesticide Plants

In 1961 on the coast of Jutland, north of Limfjord (Denmark), a chemical plant increased its production of the pesticide parathion (E 605). The effluents from this plant reached the North Sea through a pipeline. Dead fish began to be noted in the area off the mouth of the pipeline in 1964. Lobsters, too, were dying, up to a distance of 60 km along the coast. The results of tests showed that the effluents of the plant, even at a dilution of 1 to 50,000, were fatal for the *Homarus vulgaris* lobster. The effluents were even more toxic than the pesticide parathion itself. The plant was made to purify its effluents to the degree that *Lebistes reticulatus* guppies survived in them at a dilution of 1 to 50 (Boetius 1968).

In 1965, the number of *Sterna sandvicensis* terns on the island of Griend off the coast of the Netherlands decreased dramatically. Only 650 out of 20,000 breeder pairs survived; the dying birds showed definite signs of having been poisoned. The fish that the terns ate were poisoned by the effluents of a chemical plant near Rotterdam that manufactures the chlorinated hydrocarbon pesticies telodrin, dieldrin, and endrin. Between 1964 und 1967, the females of the *Somateria mollissima* eider duck were also found to be dying in large numbers. Obviously the poisonous chlorinated hydrocarbon substances were mobilized during breeding. In 1967, measures were taken so that effluents of that type no longer leave the plant. By 1974, the number of terns had risen back up to 5000 breeder pairs (Koeman et al. 1968; Koeman 1975).

Even in 1967, it was known that some organisms in the area around Los Angeles, U.S.A., contained a high concentration of DDT. Then in 1970, a thorough investigation showed that the livers of fish from Baja California contained an average of less than 1.2 mg/kg of DDT while 8 samples from the Santa Monica Bay were found to contain 370 mg/kg[1]. Farther to the north the amounts were back to below 1 mg/kg. In June, 1970, the canning of tuna fish had to be stopped because 13 mg/kg of DDT were found to be present in the meat. The fish were taken from the area around Los Angeles (Macgregor 1974).

In 1969, suspicions began to arise that something was the matter with the *Pelecanus occidentalis* pelican which breeds on Anacapa Island, 35 nautical miles west of the Santa Monica Bay (Table 10). Around 550 adult birds were found, but there were hardly any young. In many nests lay broken eggs, and their shells were found to be

Table 10. The story of the California pelicans. In 1969 and 1970 it became evident that brown pelicans *(Pelecanus occidentalis)* could not propagate on Anacapa and Coronado Norte Islands: their eggs contained high amounts of DDT, and the egg shells broke. In addition, sardines were scarce, but there is no proof that this is a consequence of DDT poisoning. In 1970 the industrial DDT pollution outgoing from Los Angeles was drastically reduced. Since 1971 pelicans again reproduced. In 1974 0.9 young birds per nest were reared; this figure, however, is still lower than the figure 1.2 which seems to be necessary to maintain the colony. In 1974, egg shells were still 16% thinner than in eggs collected before 1943, and egg shells broke more frequently than normal. In 1975 figures were similar to 1974: 0.88 young birds per nest on Anacapa Island, 113 mg/kg DDT in the fat of eggs, plus 120 mg/kg PCB (Anderson and Jurek 1977). It will be of great interest to follow up the next years, because figures will help to answer the question whether there is a further decrease of concentrations, or whether the 1974–1975 figures represent the pollution load typical for Pacific coastal regions (Anderson et al. 1975)

Year	Number of nests	Number of young birds	DDT in intact eggs (mg/kg fat)	DDT in sardines (mg/kg wet weight)	Abundance of sardines (1000 shoals per sea area)
1969	1125	4	907	4.27	140
1970	727	5		1.40	70
1971	650	42		1.34	80
1972	511	207	221	1.12	195
1973	597	134	183	0.29	275
1974	1286	1185	97	0.15	355

1 Figures refer to concentrations per wet weight and include metabolic DDT products, for example DDE and DDD (see Chap. 9)

decidedly thinner than usual. A DDT concentration of 79 mg/kg (or 907 mg/kg, based on the fat content) was found to be present in the eggs. In the pelican farther to the south, the amount of DDT decreased in proportion to the distance they were from the Santa Monica Bay.

The plant of the only company that manufactures DDT in the USA, the Montrose Chemical Corporation, is located in the Los Angeles area. In 1970, it furnished approximately two-third of the world's DDT need. From 1953 on, this plant had been releasing its effluents into the public sewage canal that flowed into Santa Monica Bay near White Point. Water samples were analyzed from the canal above the DDT plant's point of effluent introduction: DDT-concentration was 34 μg/l, corresponding to a daily amount of 3.3 kg of DDT. Below the point of introduction, the amount was found to be 297 kg per day. The cause of the DDT contamination of the fish, pelicans, and sea lions in the Santa Monica Bay was determined through these samples. In April, 1970, changes in the effluents system were begun, resulting in a reduction of the DDT-amount to 13 kg per day by October, 1971 (Macgregor 1974).

A good study case for the pollution by DDT effluents of an ocean area of over 100 km is the lantern fish of the family Myctophidae. They have been collected at many research vessel stations since 1950 and were later analyzed for their DDT content. In 1950, the DDT concentrations were around 0.5–1 mg/kg. Up to the period of 1961–1966, they rose to around 4 mg/kg (Fig. 31). After the effluents of DDT were drastically reduced in 1970, the DDT and DDD components in the meat of lantern fish have decreased while no distinct tendency had been established by 1972 for DDE (Fig. 32).

Reproduction by the California sea lion, *Zalophus californianus*, on San Miguel Island off the coast of California, U.S.A., had also been disturbed. More and more stillbirths had been noted since 1968; in 1971, 348 cases in a colony of 10,000–15,000 animals. The amount of DDT was much higher in stillborn than in live pups: 824 mg/kg as against 103 mg/kg in the fat; 25 mg/kg as against 6.7 mg/kg in the liver; 2.4 mg/kg as against 1.2 mg/kg in the brain (Delong et al. 1973). Stillbirths have been observed even after the ban on DDT. This can be due to high concentrations of mercury and PCB in the same animals which have high DDT concentrations (see Chap. 7.2 and 9.4). Or it could be that sea lions feed as well on demersal fish. The surface sediment of a 50 km^2 area of the ocean around the Los Angeles point of effluent introduction still contained 200 t of DDT in 1971 (Fig. 33).

Although the amounts of DDT which reach the ocean via the Los Angeles sewage treatment plant were reduced by a sixth, and the PCB (polychlorinated biphenyl) unit count by a twentieth between 1972 and 1975, 11 mg/kg of DDT and 1.1 mg/kg of PCB's were still found to be present in *Microstomus pacificus* sole in 1975 as compared with 17 mg/kg of DDT and 2.3 mg/kg of PCB's in 1972 (Young et al. 1977). Obviously, DDT remains present a long time in sediment and contaminates those organisms which live in contact with the bottom of the ocean.

These findings could explain why, in 1976, all sea gulls and cormorants of Los Angeles zoo died from DDT poisoning; sea gulls had 430 mg/kg DDT in their brain, cormorants 220 mg/kg. Such concentrations in the brain are comparable to concentrations which caused death in falcon fed experimentally with DDT-contaminated meat. In the zoo, the birds had, for years, been fed with queensfish *(Seriphys)* caught close to

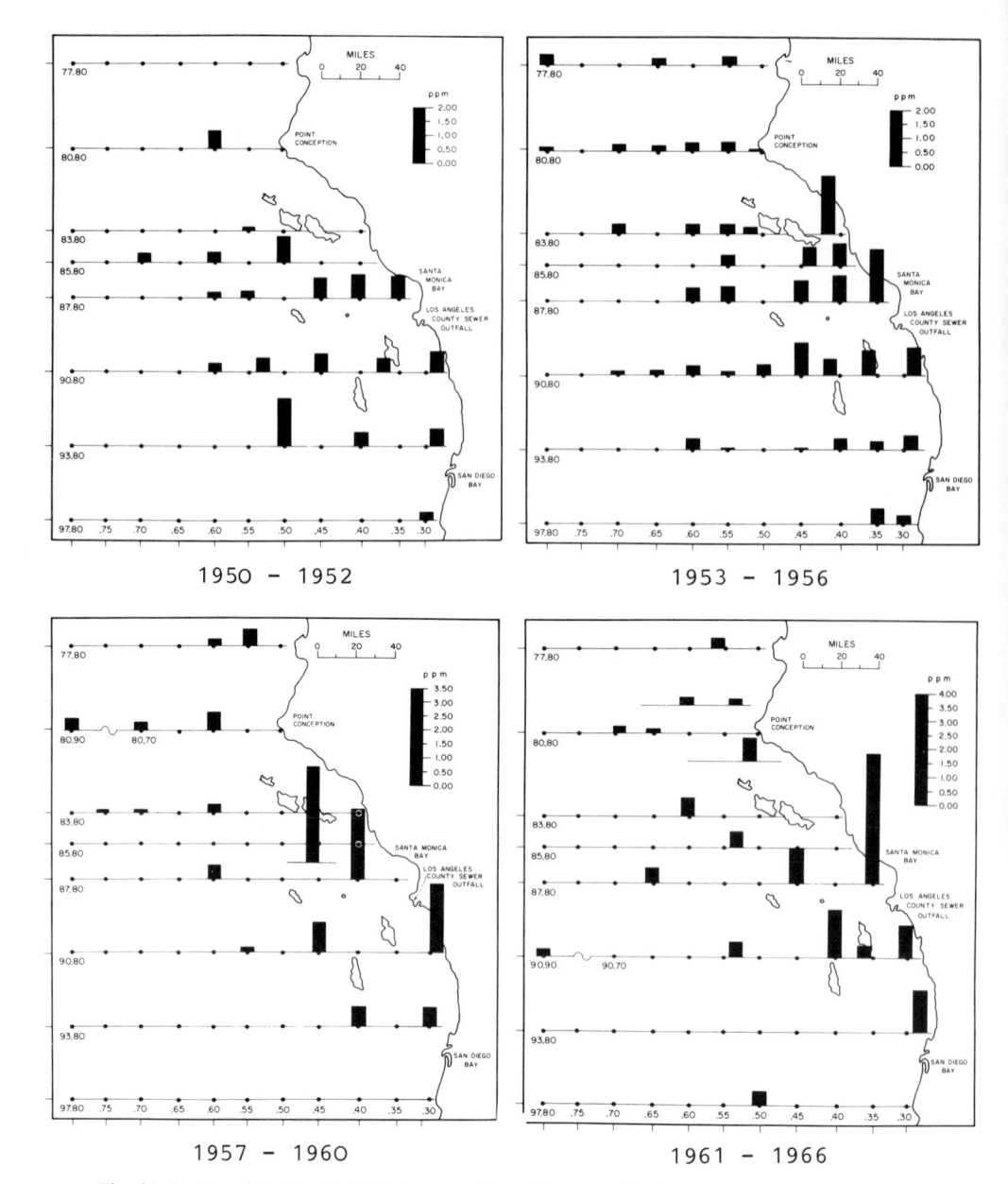

Fig. 31. Lantern fish (Myctophidae) were collected between 1950 and 1966 in a regular sampling program west of California. The DDT concentration in lantern fish (in ppm = mg/kg wet weight) reflects the pollution by the introduction of DDT-wastes from the Montrose Chemical Company at Los Angeles, U.S.A., 1953–1970 (Macgregor 1974)

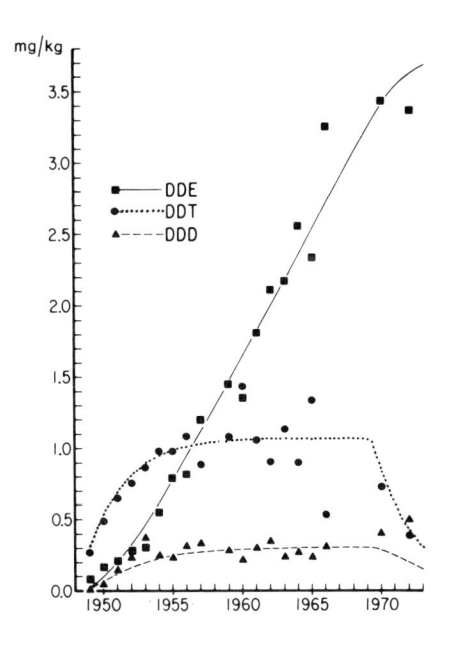

Fig. 32. DDT concentrations in the meat of a lantern fish 20 nautical miles west of Los Angeles (in mg/kg wet weight) calculated from data of Fig. 31. Between 1949 and 1955 there was a rapid increase of all DDT-related compounds, but later the true DDT, as well as the metabolite DDD, are equilibrated. The explanation could be that the degradation of DDT within organisms plus elimination and dispersal, had about the same order of magnitude as the continuing input of DDT. However DDE continued to increase. DDE is a persistent metabolite of DDT, and concentrations did not stop increasing before the stop of DDT pollution in 1970 (Macgregor 1974)

the outfall which brought DDT-contaminated waste water to the sea. This fish had a DDT concentration of 3.1–4.2 mg/kg (Clark 1978c).

Insecticides, naturally, do not reach the sea solely along with the waste water of chemical plants, but also when they are used in large quantities in the vicinity of the coast. When sand flies became a disturbance and were combated with DDT for the comfort of bathers on the coast of Southern California, U.S.A., mackerel were later found containing 2–17 mg/kg of DDT and had to be banned for human consumption

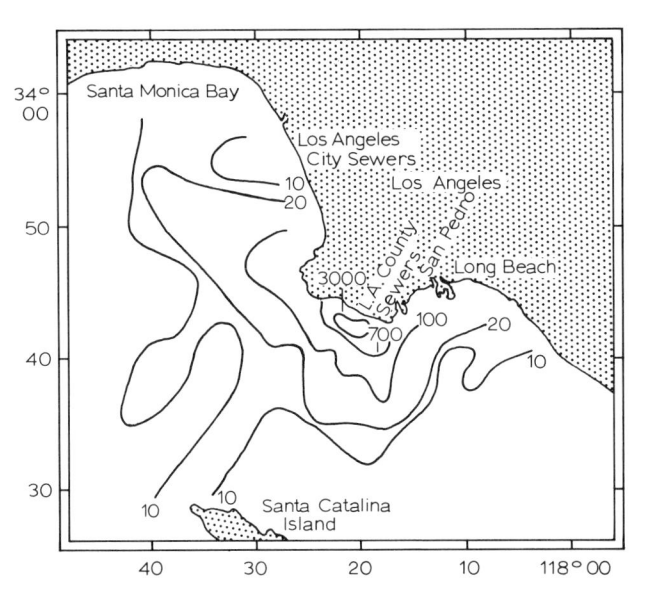

Fig. 33. 103 sediment samples from the area off Los Angeles, U.S.A, were analyzed for DDT in 1971. Amounts of up to 6600 mg/m^2 were found, with DDT specially concentrated in the upper 2–3 cm of the sediment. Altogether there are about 200 t DDT (including DDD, DDE) in a 50 km^2 area in front of the coast where up to 1970 large amounts of DDT were introduced. About 100 t are present in the 3000 km^2 area further away. Figures in mg/m^2 (Macgregor 1976)

(Stout et al. 1972). Mullets *(Mugil)* off the Pacific coast of Guatemala had amounts of DDT up to 36 mg/kg. After shrimp had first died off, dead mullets were observed. The cotton fields which were located directly on the coast had been intensively treated with DDT (Keiser et al. 1974). DDT was used in large quantities as a pesticide in the orchards all around the Sogne Fjord in Norway up until 1970. As a consequence, the liver of cod *(Gadus morhua)* from the Sogne Fjord contains DDT up to 10 mg/kg and may not be used for human consumption in the Federal Republic of Germany (Stenersen and Kvalvag 1972). Fish caught at the mouth of the Viskan River on the west coast of Sweden have such high amounts of dieldrin that they are not allowed to be sold. Pesticides are also present in the waste water of wool factories because in them textiles are treated against moths. PCB's can reach coastal waters when factories in the vicinity of coastal areas have leakages in their cooling systems in which PCB's are used as heat exchanger. One example is the Gulf of Mexico where *Penaeus* shrimp had 17–132 mg/kg of PCB's in their intestine gland (Nimmo et al. 1971). Mangrove trees are very sensitive to herbicides. During the Vietnam War, approximately 1000 km² of mangrove fields were defoliated which consequently caused the death of the plants (Odum and Johannes 1975). PCP, pentachlorophenol, seems to be a recent addition to the catalog of marine pollutants (Fig. 34).

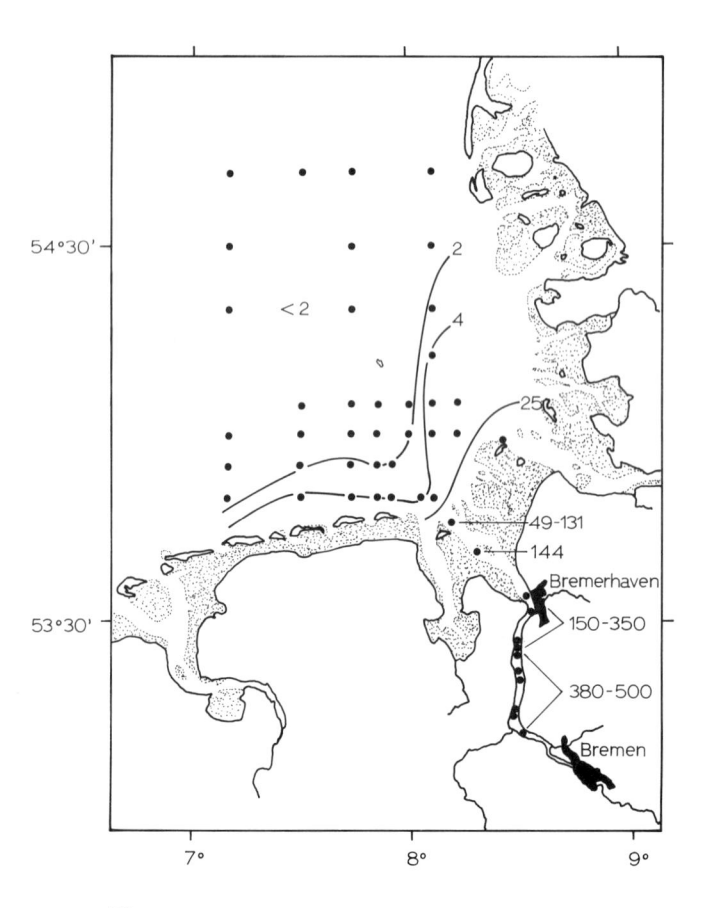

Fig. 34. After fungicides containing mercury compounds have been widely banned, pentachlorophenol (PCP) has been extensively used in wood protection and in paper fabrication, preventing fungi from degrading such materials. PCP is accumulated with varying concentration factors, for example 2600–8500 in the bristle worm *Lanice conchilega*. PCP is toxic for marine organisms: application of 76 µg/l in seawater had a damaging effect upon benthic macrofauna, while 7 µg/l had no effect (Cantelmo and Rao 1978). PCP concentrations in brackish water of the Weser Estuary are 100–500 ng/l, while concentrations in the German Bight are much less. At present one cannot tell whether sublethal effects occur under such small concentrations. Figures in ng/l (Ernst and Weber 1978)

4 Pollution of the Sea by Ships

4.1 Discharging of Wastes on the High Seas

Oil spills irritate the tourist at the shore and send the bird ecologist to the barricades. Unnoticed by the public, however, thousands of barrels with residue of chemical production in various industrial nations have been dumped in the ocean. In 1968, 14,000 t of waste matter from pesticide plants on the Mississippi were dumped into the Gulf of Mexico every month. This figure is just a fraction of the estimated total of 330,000 t of pesticide waste matter that was dumped by the U.S.A. in various ocean areas, together with 560,000 t of waste matter from oil refineries, 140,000 t from the paper industry, 940,000 t of various waste matter and 2,7 million t of waste acids (Fig. 35). Approximately 40 t of chlorinated hydrocarbons in this period were dumped in the Atlantic every month by the Federal Republic of Germany. In the period from 1963 to 1969, 38,000 barrels containing chlorinated hydrocarbons and 40,000 barrels containing cyanide compounds, arsenic, and other poisons from Great Britain were dumped. Containers repeatedly turned up in fishermen's nets, not only on the high seas but also in Dutch and German coastal water areas. This gives rise to the speculation that some boat captains made quick money by throwing the drums containing poison overboard shortly after leaving the harbor and saved themselves the trip out to sea.

These barrels contained a variety of matter (Greve 1971; Berge et al. 1972), primarily waste products from the production of PVC, polyvinylchloride. Often it is a mixture of short-chain aliphatic hydrocarbons known as EDC tar. We are dealing with more than 80 different substances, 15 of which have known cancer-causing effects. As far as amounts are concerned, dichlorethane (EDC) and trichlorethane predominate.

For 1974, a 10 million t world production of vinylchloride was reckoned with. As part of this production, around 400,000 t of EDC tar resulted as a by-product. The north and central European industrial areas contribute approximately 75,000 t of EDC tar to this figure (Jernelöv et al. 1972).

In 1970, Scandinavian scientists analyzed the aliphatic chlorinated hydrocarbons that occur in EDC tar in the seawater of the North Atlantic and the Norwegian Sea. The findings were alarming, because if analyzable concentrations are present in seawater, then the total amount of waste that has been put into the ocean must be considerable.

EDC tar consists partly of the same vaporous chlorinated hydrocarbons with 1–2 carbon atoms which reach the ocean via the atmosphere (see Chap. 9). The concentrations measured in seawater in 1970, therefore, do not necessarily come solely from chemical waste dumped in the ocean. At that time, however, the news was received with justifiable alarm.

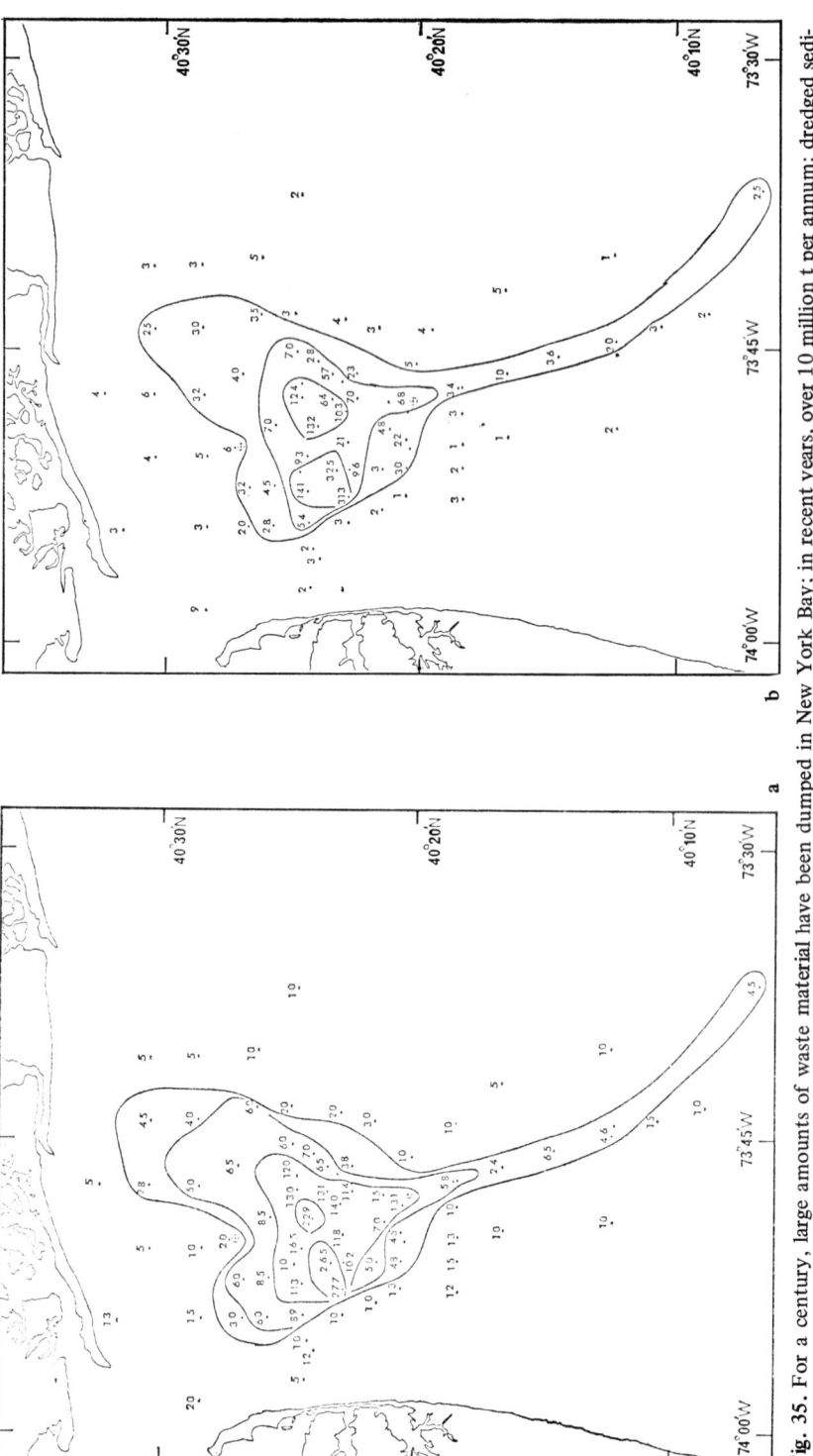

Fig. 35. For a century, large amounts of waste material have been dumped in New York Bay; in recent years, over 10 million t per annum: dredged sediment, construction refuse, sewage sludge, and waste matter from chemical plants. The heavy metal concentrations in the sediment in the dumping area are correspondingly high and enable us to determine the direction of the predominating water current. a lead; b copper; c zinc; d chromium. Figures in mg/kg of dry sediment (Carmody et al. 1973)

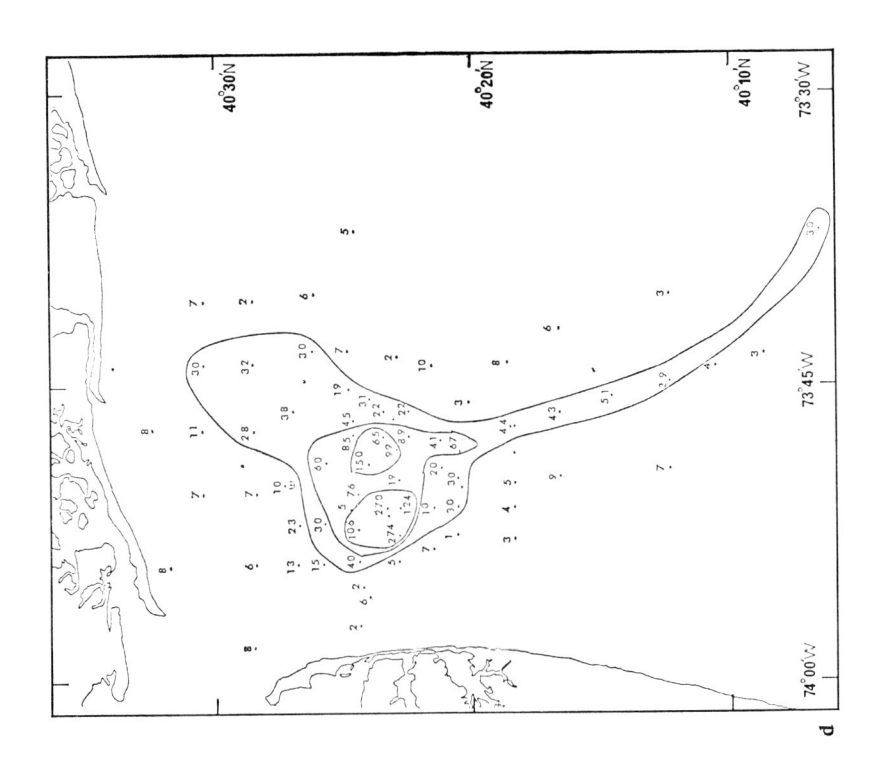

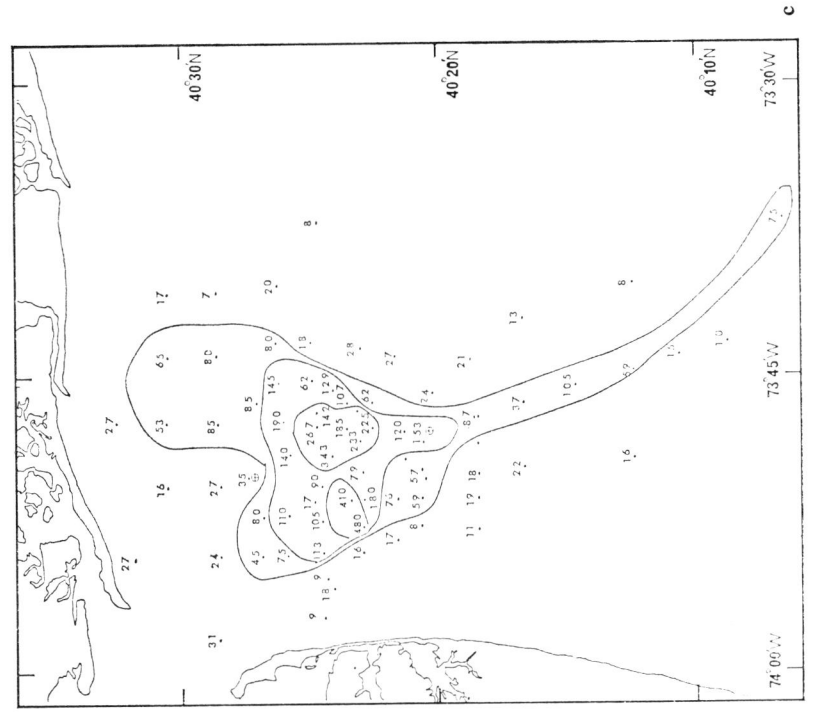

Fig. 35 c,d

In July of 1971, the coastal freighter "Stella Maris" left the harbor of Rotterdam loaded with 600 t of waste material from the Dutch vinylchloride plant AKZO with orders to dump its load near the Halten Bank, 60 nautical miles west of Norway. When this became known, the Scandinavian governments intervened, objecting to an intended dumping so close to an important fishing area. Because of this, the "Stella Maris" was rerouted to a position south of Iceland. It intended to bunker at Thorshavn in the Faeroe Islands along its new route, which Faeroe fishermen vigorously prevented. The British government objected to bunkering in Stornoway on the Hebrides. Ireland threatened to take action with a warship, should the freighter intend to enter Irish Waters. On July 26, 1971, the ship returned to Rotterdam without having discharged its cargo. The barrels were stored there until a waste incineration plant for special waste materials was completed (Anon 1971).

Since then ships have been converted and outfitted with special burning facilities on board and such waste matter is burned at sea. This has the advantage that the gasses given off do not affect humans. The hydrochloric acid that results from this burning probably has no adverse effect on marine life in the North Sea because the seawater is well buffered with bicarbonates.

It is not well known what dangerous wastes other than hydrochloric acid may be released with the gasses; their total amount, however, should be lower than 0.1% of the original amount to be burnt. About 50,000 t of waste chlorinated hydrocarbons from the German industry have been burnt on the sea, mostly from the 999 gross tons, 72 m long incineration vessel "Matthias II" which has been in operation since 1972 with a capacity to burn 10 t of wastes per hour. In 1979, a new incineration vessel "Vesta" of 999 gross tons has been launched, which can carry about 1300 t of liquid waste and burn it at temperatures of $1200°-1300°C$. A sea area 40 nautical miles east of the Doggerbank in the North Sea has been approved for the incineration at sea, and the performance of burning is in general better than the figure of 99.9% which is set by the *Regulations for the Control of Incineration of Wastes and other Matter at Sea,* a recent annex to the London Convention (see Chap. 10).

It is to be feared that we will never know how much waste matter has been dumped on the high seas in past decades and is still resting on ocean bottoms, war materials being among them. For decades, obsolete munitions and other dangerous materials were commonly dumped in the ocean. In 1958, the U.S.A. loaded the mothballed war ship "William Ralston" with 8,000 t of mustard gas and sank it. In the period up to 1968, an additional 12 ships loaded with munitions were sunk (Reed 1975).

When Baltic Sea fishermen occasionally find a poison gas grenade in their nets, dumped there after the Second World War, one gets goose pimples. More than 80 Danish fishermen had accidents in the decades after the war, and there are many casualties among German fishermen, too, who had contact with war gas bombs and grenades, specially in the area east of Bornholm and southeast of Gotland. Eleven thousand four hundred t of gas ammunition, however, which had been dumped after the war south of Little Belt in only 30 m of water depth, apparently have disappeared. Bombs and grenades recovered in 1971 were empty, and obviously the gas had leaked out after corrosion of the metal, and had mixed with seawater. Similarly, there is no indication of damage from the about 70,000 to of ammunition sunk inside 30 vessels at 700 m depth in the Norwegian Trench south of Norway in 1945.

Authorities sometimes react nervously when events take place that do not actually justify such a reaction (Anon 1975). For example, the 115,000 t tanker "Enskeris" left Finland in March, 1975, with a load of around 100 to of arsenic compounds packed into 500 barrels and destined to be thrown overboard on the high sea of the South Atlantic. This aroused public anger, and the governments of Brasil and South Africa protested. As a result, the Finnish government ordered the ship back. In the meantime, international agreements have been worked out which either prohibit the dumping of poisonous compounds or make dumping dependent on a thorough analysis of the situation before a permit to do so is issued (see Chap. 10).

4.2 Waste Acids and Ferrous Sulfate of the Titanium Pigment Industry and the Problem of the Indicator Communities

The impetus for a thorough-going scientific study of the problems of marine pollution in the coastal area of the Federal Republic of Germany was the 1966 plan by Kronos Titan Ltd. to establish a plant on the Blexer Groden, near Nordenham in the Weser Estuary, for the manufacture of titanium dioxide as an important paint pigment. Titanium ore (ilmenite) is to be cleaned by means of sulfuric acid. As the waste products of this process, ferrous sulfate and 18% sulfuric acid results (Table 11). Since further use of these waste products at that time was not profitable, the company intended to ship them out into the North Sea by special tanker and dump them into the propeller wash in a way similar to that used for several years in the Bay of New York and off the coast of the Netherlands.

Table 11. Chemical reactions in titanium pigment production

I. Separation of titanium pigment (TiO_2) from ilmenite ($FeTiO_3$) by treatment with sulfuric acid. In this process 1.85 g $FeSO_4$ are produced per g of TiO_2

$$FeTiO_3 + 2\,H_2SO_4 \longrightarrow TiOSO_4 + \boxed{FeSO_4} + 2\,H_2O$$

$$TiOSO_4 + 2\,H_2O \longrightarrow TiO(OH)_2 + \boxed{H_2SO_4}$$

$$TiO(OH)_2 \longrightarrow TiO_2 + H_2O$$

II. Introduction of ferrous sulfate into sea water

$$FeSO_4 + 2\,H_2O \longrightarrow Fe(OH)_2 + \boxed{H_2SO_4}$$

$$4\,Fe(OH)_2 + O_2 \longrightarrow \boxed{2\,Fe_2O_3\text{ aq.}} + 4\,H_2O$$

III. Introduction of sulfuric acid into seawater. $Ca(HCO_3)_2$ is used to express the alkalinity of the seawater

$$H_2SO_4 + Ca(HCO_3)_2 \longrightarrow \boxed{CaSO_4} + 2\,H_2CO_3$$

$$H_2CO_3 \longrightarrow H_2O + \boxed{CO_2}$$

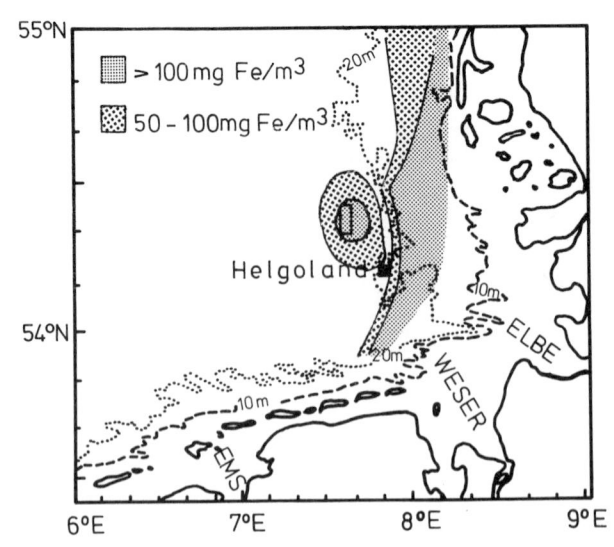

Fig. 36. Since 1969 about 1800 t per day of sulfuric acid and ferrous sulfate have been introduced into the seawater 11 nautical miles northwest of Helgoland. The graph represents iron concentrations in seawater analyzed in September 1970, 16 months after the beginning of the waste disposal activity. There is a discrete increase of iron concentration in the area, however the order of magnitude is similar to the concentration in nearshore regions and in the areas off the rivers Elbe and Weser (Weichart 1972)

Since the jurisdiction of the Federal Republic of Germany — and of the State of Lower Saxony — then ended at the 3 nautical mile limit, no legal steps could have been undertaken to prevent the practice of a company's getting rid of its effluents on the high seas in international waters. But since it was already forseen in 1966 that the freedom of the seas would in the future have to be limited in favor of the public welfare (see Chap. 10), executives and research institutes were called in and were asked whether or not that sort of waste disposal would cause ecological damage. The question was a new one for German marine science. The answer turned out to be somewhat vague. Damage could not be prophecied nor could safety be guaranteed. An area of 12.5 square nautical miles, approximately 11 nautical miles northwest of Helgoland was assigned for dumping (Fig. 36).

Since May of 1969, an average of 1800 t, after 1976, about 2000 t per day of wastes from the titanium pigment industry has been transported to the dumping area. This waste consists of 12% sulfuric acid, 14% ferrous sulfate, and approximately 3% mineral components, including 9 t per year of nickel, 66 t of vanadium, and 120 t of chromium (but practically no mercury). Even before effluents began to be introduced, the bottom fauna of the area was studied (Stripp and Gerlach 1969). It is a sandy bottom, approximately 28 m deep. In 1967–68, on the average 375 individuals of so-called macrofauna per m^2 lived there, that is, those bottom animals which can be filtered out of the bottom material by washing it with a sieve whose meshes are 1 mm wide. The wet weight of the animals was 16 g/m^2, which is relatively little. The area was selected as a dumping ground precisely because of its relatively low population. In 8 years of effluent introduction, there have been no changes in the bottom fauna which can be attributed to the damaging effect of effluents (Fig. 37, see p. 61). Also, the fish fauna of the affected area, in 1972 seemed unchanged; 18 species of fish were present (Dethlefsen 1973). This result was, at that time, not surprising since no essential changes of plankton and bottom fauna had been observed in the Bay of

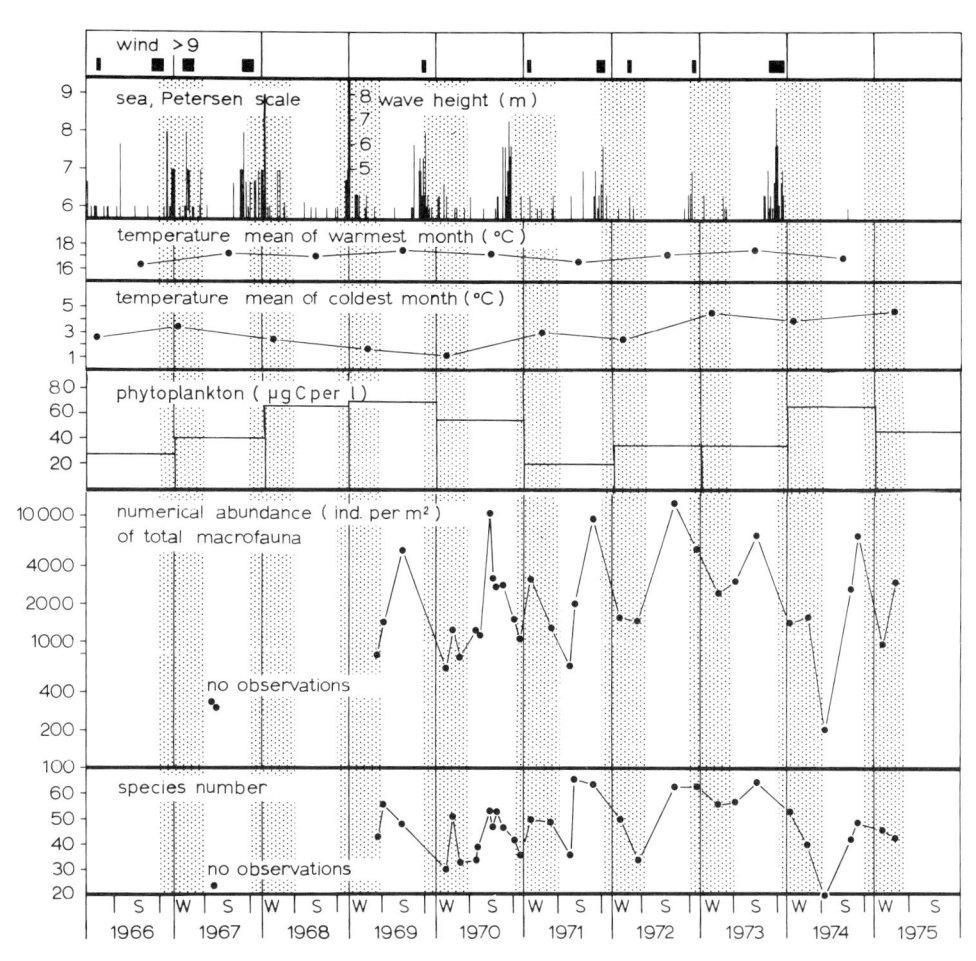

Fig. 37. The graph shows that the daily introduction of 1800 t of titanium pigment industry wastes did no damage to the macrofauna living in the area northwest of Helgoland. Figures are from the station in the center of the dumping area, 1967–1975. The number of species found and the number of individuals per m^2 fluctuates very much between summer and winter samples. Apparently there was a general increase of figures from 1969 to 1973, that is, after the waste disposal started in 1969. However, most probably this increase is caused by relatively quiet winters without severe storms, but with sedimentation of suspended material and with settlements of small animals. There were heavy storms in the winter of 1973/74 which eroded the sea bottom at 25–30 m depth in the German Bight and probably caused a reduction of macrofauna figures. However, there might be other factors to be of importance: the increase of winter sea water temperatures, fluctuations in phytoplankton stocks (Fig. 12) and fluctuations of demersal fish which predate on macrofauna (Rachor and Gerlach 1978)

New York, even though over 50 million t of waste acid had been dumped there since 1948 (Vaccaro et al. 1972).

Recently it has been tried to correlate a fish disease with the dumping of wastes from the titanium industry northwest of Helgoland. In the surrounding of the dumping area some samples of dab *(Limanda limanda)* had up to 6% fish with epidermal papillomes, a tumor of the skin, while in other areas of the German Bight and the North Sea the percentage of this disease in dab was about 0.5% (Dethlefsen and Watermann 1980). Unfortunately, up to the day in June 1981, while I am finally correcting this book, basic data in this interesting correlation have not been published, and experts are discussing matters with different points of view. It could well be that a closer look into the basic material allows explanations other than the supposed correlation with dumping, for example with the effluents of the polluted Elbe river (Gerlach 1981). In principle, however, it could well be that dab with its very delicate skin is a more sensitive indicator than invertebrate macrobenthos living in the sediment. Future scientific investigations hopefully will make the case clear. The public discussion of the possible correlation between dumping of wastes in the North Sea and fish disease, including activities of the Greenpeace organization, has forced the German and Dutch Governments to some activity; probably the licence for dumping will expire soon, in the Netherlands in March 1982.

Disease is a concomitant of life for any species of organism; symptoms of a disease as such can be quite normal under undisturbed, natural conditions. However, any departure from normal environmental conditions, natural accidents, or man-made impact produces a degree of stress and may contribute to an increase of diseases. It is well-documented that human activity has increased stress for fish in estuarine and nearshore waters because concentrations of toxic heavy metals (see Chap. 3.1) or toxic organics (see Chap. 3.2) have increased, or because the oxygen regime is burdened with degradable organic substances (see Chap. 4.3). Examples of good correlations between the frequency of fin rot and ulcer in fishes and shell disease in crustacea, and with pollution have been given (Sindermann 1978). However, there are many diseases where up to now such correlations could not be demonstrated beyond doubt.

Back to the period of 1900 there are reports about the "cauliflower disease" of eel *(Anguilla anguilla)* in the lower reach of the Elbe River. But such tumors in the head region of eel seem to occur nowadays more frequently, in 12–28% of all eel caught in summer, but only in Central Europe, not in West Europe and the Mediterranean region. It is not known what causes the skin defects; stress from reduced oxygen concentrations seems to trigger its outbreak (Peters 1981). On the Pacific coasts of Canada and the northern U.S.A. flatfish tumors are more frequent close to larger cities and close to sewage inlets. But again there is no direct proof that pollution causes these diseases. Apparently geographical areas exist in which flatfish or eels have a certain chance of developing skin tumors, and pollution stress may increase the chance. But outside those areas populations of such fish are not affected by tumors even if they inhabit polluted estuaries (Stich et al. 1977).

To make any firm association of the disease with environmental pollution requires knowledge of the history of the disease and the pollution situation, a following-up of future trends, knowledge of the life history of the disease, a baseline study to compare polluted with unpolluted situations, and, if possible, experimental reproduction of the disease by exposure to contaminants (Sindermann 1978). In the case of the epidermal papillomes of dab in the region of Helgoland Bight where wastes of the titanium pigment industry are dumped, up to now only the first attempts have been made to make a case out of a baseline study. Surprises may occur with future research. At this moment one should not misinterpret the evidence from the invertebrate bottomfauna (Fig. 38).

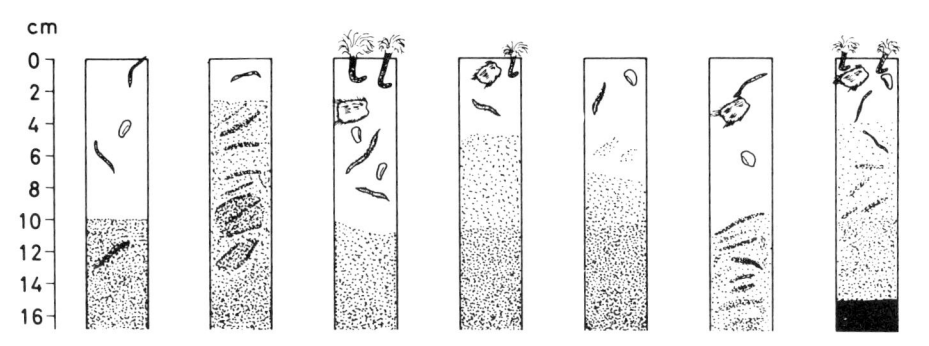

Fig. 38. Cores from the 28-m-deep sandy area in the German Bight which later became the waste disposal area for the wastes from the titanium pigment industry. Cores were sampled after a heavy gale struck the area in February 1967 and eroded the sediment even at that water depth. There is a 3–10 cm thick layer of freshly deposited sand *(white)* above the older, grayish sediment *(dotted)*. In a schematic way the position of some macrofauna representatives is indicated (Hickel 1969)

It cannot be expected that the bottom fauna of a 28-m-deep area in the North Sea will remain unchanged over a period of decades, and reflect in an ideal way unchanged pollution stress. According to the season, the spectrum of species changes in correspondence with the reproductive rhythm of the various species. Over a relatively long period of time, increases and decreases result from stocks that do not successfully reproduce in the same numbers every year. In the winter of 1962–63, 5 years before the beginning of the bottom research in the dumping area, very low water temperatures had a devastating effect on the bottom animals. In particular, certain species of molluscs had practically disappeared in wide areas of the Bay of Helgoland. Thus, ecological niches were freed which were at first partly filled by various worms until species of molluscs were gradually able to repossess the area. In February of 1967, that is, just before the beginning of the bottom fauna research, a strong gale stirred the bay up so strongly that the sea bottom was eroded. After the weather had settled down, a 10-cm layer of sediment was deposited (Fig. 38). It can be calculated that the strength of waves 100 m in length and 6 m in height is enough to produce an orbital current of 2 m/s at a depth of 20 m which is enough to stir up the sediment (Gienapp 1973). Since 1967, the number of small, easily injured worms has increased in the dumping area. To associate this increase with the effect of the acids which have been dumped would, however, be wrong. Presumably, the next strong gale will destroy this fauna again and leave behind just the robust, heavy animals and those which are able to retreat into the farily deep layers of sand.

The erosion of the bottom during storms probably also causes especially large amounts of nutrient elements to enter the sea water and cause a more luxuriant growth of algae than usual. This could have a favorable effect on the bottom animals that secure their food from seawater as filter feeders. Chance, too, can also play a large role in bottom animal communities. Most bristle worms (Polychaeta), echinoderms, and bivalves produce pelagic larva for propagation. These larvae are not able to settle directly on the sea floor but must drift around in the water for a few days or weeks before they are ready to metamorphize. Tidal currents in the German Bight transport larvae of benthic animals continuously, so that it is certain that they are not able to settle down where their parents lived. A community of benthic animals is therefore dependent on larvae from elsewhere ready to metamorphize when they pass over the area. For many species that are distributed over an expansive area, one can assume that their larvae are present everywhere in the plankton in the corresponding season and that they are able to settle anywhere that offers them the possibility to live. However, with species that are distributed in patches, the larvae are transported in clouds by the water, and it is a matter of chance whether such a cloud will settle in a certain area or not. Sometimes the corresponding species is missing right next to it, even though the conditions for life might be favorable. Marine biologists then try in vain to deduce what reasons there may be for the various animal settlements he encounters in his research.

One must be familiar with these complicated alternating relationships if one has to diagnose possible pollution effects as a cause for changes in the stocks of organisms. This is easy in the case of very strong effects (Fig. 4), but in the case of subtle effects, only research conducted over a long period of time can lead to significant results. This limitation of field results holds true for macrobenthos (now 11 years of evidence) and for fish diseases (to be confirmed by future research) as methods to monitor pollution effects northwest of Helgoland.

If it should be true that daily amounts of 1800 t of waste acid have no effect on the organisms of a marine area, then this finding is of great importance, because it contradicts laboratory results. Experiments in which waste acid was added to various marine organisms showed an adverse effect of industrial effluents from the titanium plant, even at dilutions of 1 part to 50,000. Obviously however, such concentrations occur only in the immediate area of the propeller wash. A dilution of the waste fluid of 1 to 1000 is immediately accomplished by the ship's propeller. In the mixing process, this figure is reduced to 1 to 20,000 in 2 h. The acid effect of the waste does not penetrate to the sea bottom at 28 m depth where a pH indicator was installed. In the surface water of the area the reduction of the pH is less than 0.1 (Weichart 1975b).

If waste acid gets into seawater, a portion of the natural alkalinity of the seawater is required for neutralization, and the amount of sulfate in the seawater is increased. We do not know much at present about the influence of reduced alkalinity or increased sulfate content on marine aminals. Ferreous sulfate is hydrolyzed in seawater. While absorbing oxygen, iron $^{2+}$ oxidizes to iron $^{3+}$ and forms flakes of ferreous hydroxide (Table 1). There is ample oxygen available north of Helgoland in surface water so that there is no danger of a shortage. However, the iron concentration in the seawater has measurably risen. Values were analyzed that correspond roughly to those found to be produced by nature close to the coast and in estuarine areas (Fig. 36). Apparently a balanced situation is thereby attained, and just as much iron is transported to the north by currents as is introduced by effluents (Weichart 1975a). The concentration of chromium in the fine fraction of the sediment is about twice as high in the surrounding of the disposal area in Helgoland Bight, compared with the regions farther away.

The news about possible correlations between skin tumors of a flat-fish, the dab, and the wastes of the titanium dioxide plant have again intensified the discussion whether these wastes are dangerous to marine life or not. Some years ago, in the Mediterranean, it was decided that they are dangerous, and Italian factories are supposed to keep to stricter rules than in other areas. This is the consequence of Corsican and Italian fishermen being engaged in massive protest against wastes from the titanium pigment plant in Scarlino, Italy. It has 3000 t of wastes per day: 1000 t of sulfuric acid, 300 t of iron, 23 t of titanium, 0.5 t of vanadium and 0.1 t each of cadmium and lead. Dumping takes place 30 nautical miles northwest of Corsica on water 1250 m deep and in an area without fishing activity of any consequence (Perez 1976). These wastes differ from the wastes dumped into Helgoland Bight by relatively high amounts of cadmium, which could have some adverse effect. Otherwise it is not easy to accept that wastes from the same type of frabrication should have different effects in North Sea and Mediterranean. In many other countries plants exist for the fabrication of titanium pigment, and most of them release their wastes into the sea. However, there are no published reports available on the effects of these wastes upon marine life.

One thing may be taken for granted: if a factory obtains a permit to dispose of the wastes via dumping or a pipeline, and if another factory does not have this permit and is forced to develop alternative means of waste disposal or waste recycling, both factories have different economic chances on the market due to different expenses for waste treatment. Since 1977, therefore, guidelines exist set by the European Community concerning the wastes from the production of titanium dioxide pigment. These guidelines are to insure that possible adverse effects should constantly be monitored, and that the amounts of wastes shall be reduced if necessary. By July 1980 there should have been some progress to reduce the quantity of wastes from titanium pigment fabrication. However, this is only a guideline, not a strict rule, and it seems that for obvoius economic reasons some nations and some factories do not keep to it.

Altogether, in 1977 about 900,000 t of titanium pigment were produced within countries of the European Community. Only 15% were produced by chloride treatment which releases less wastes into the environment, but needs high-quality titanium ores which are not generally available. But even with the sulfuric acid treatment, there are ways to recycle or use the wastes. About one third of the ferreous sulfate produced from the Nordenham-Blexen plant by now is sold to waste water treatment and drinking water purification plants, or as a material to increase the fertility of agricultural soils. In principle, it would even be possible to produce high-quality sulfuric acid from waste acid and ferreous sulfate by introducing additional sulfur. The result, however, is twice as much sulfuric acid as is used in the pigment industry, and at a price which is not economic. But the example points a way as to how recycling could eventually reduce the wastes to be dumped into the sea.

The pigment titanium dioxide replaced the toxic white zinc and lead pigments formerly used. The introduction of the nontoxic titanium pigment certainly improved the quality of man's environment.

4.3 Discharging of Dredge Spoil and Sludge

Conventionally, dredge spoil from harbors and from shipping channels is dumped into the sea. The same is done with a variety of earth material, partly used as land fill when reclaiming shallow coastal areas for urban or industrial purposes. Waste water treatment plants release rather clean treated effluents, but have to dispose of the sludge. Waste water treatment plants close to the sea conventionally dump the sludge into the sea. In all these cases, the material dumped at sea cannot be classified as highly toxic, because toxic substances are only in trace concentrations. But because large quantities of material are concerned, the effect of such disposal upon the coastal ecosystem has to be carefully evaluated.

Great quantities of sewage sludge from sewage treatment plants in New York and New Jersey have been dumped onto the Continental Shelf off these two states, anually up to 4 million m^3 containing 200,000 t of organic solids. This import of organic matter into the sea is about ten times the amount of organic substance produced by phytoplankton photosynthesis on the shelf off New York. An area of 36 km^2 is affected, where, if distribution is even, 4 mm of sludge settle annually. Anaerobic conditions in the sediment result; poisonous hydrogen sulfide (H$_2$S) form, and the fauna dies out or is restricted to the few species that are able to resist (compare Fig. 4). In some years anoxic conditions spread from the waste disposal area southward. Oxygen concentration in the seawater of the disposal area is in general 2–3 mg/l lower than in the surroundings; in summer, however, sometimes it is only about 2 mg/l in a large area of New York Bight, so that a further reduction in oxygen concentration means anoxic conditions (National Academy of Science 1975a).

In July 1976 dead fish, crustaceans, and molluscs were reported from the sea area south of the waste disposal locality in New York Bight. Subsequent hydrographical investigations (Fig. 39) revealed an area of 165 km^2 with less than 2 mg/l of oxygen in the seawater close to the sea bottom, and some patches with anoxic conditions and even with hydrogen sulfide in the lower half of the 35 cm deep water. A total of 12,000 km^2 was concerned. It may well be that the even worse conditions in the dumping area triggered this large-scale catastrophe, where 143,000 t of *Spisula* clam died in an area of 6,700 km^2. However, this assumption is not certain.

The year 1976 was an extreme situation. Spring and summer were very warm, without storm action, therefore a thermocline could develop better than in other years, and light warm water at the sea surface was separated by a stable transition zone from the heavy, cold water at the sea bottom. The bottom water, therefore, was cut off from atmospheric oxygen. Second, there was an extremely luxuriant plankton bloom which occurred just below the thermocline. When the plankton algae died off and sank into deeper water, they had, for degradation, an enormous oxygen demand (see Chap. 2.1). The deep water therefore was quickly depleted of oxygen, the anoxic situation was established. Certainly natural processes and the addition of surplus organic substance to the oxygen-consuming degradation process acted in the same direction. However, with the present knowledge at hand, it is impossible to decide what was the man-made effect, and whether, under similar meteorological conditions, the anoxic situation would not also have been established without pollution.

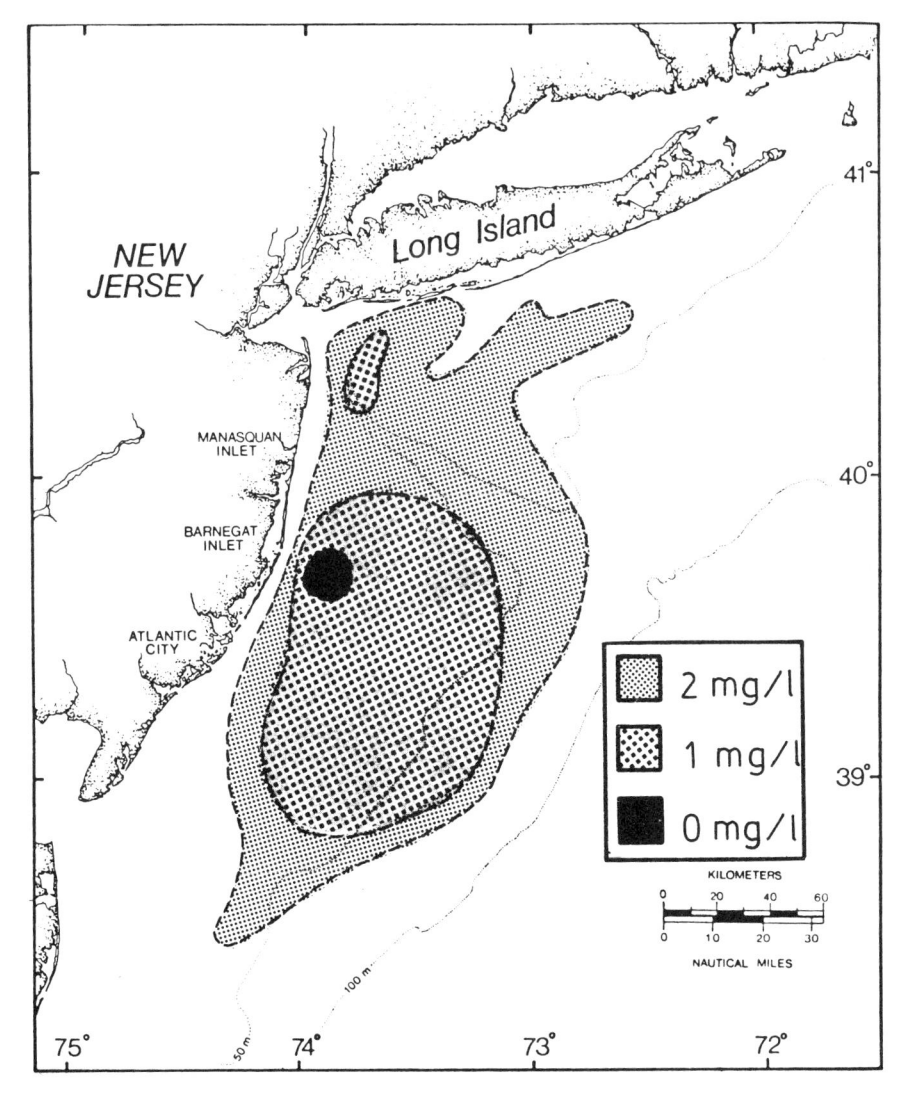

Fig. 39. Maximum extension of the area with poor oxygen conditions (below 2 mg/l) or without oxygen in the sea area south of the dumping ground for sewage sludge in New York Bight, summer 1976 (Steimle and Sindermann 1978)

Annuallly, approximately 1 million m³ of sludge are dumped into the Clyde River from Glasgow, Scotland, at a depth of 80 m. Affected by this are approximately 10 km² of Irish Sea bottom which have been made uninhabitable for the Norway lobster *(Nephrops norvegicus)*. It is estimated that the fishing industry is losing an annual income of £ 75,000 because of this. On the other hand, additional effects on fish counts have been insignificant. The effects on phytoplankton are approximately like those of other areas close to the coast; the production of plankton in affected sea areas is larger than usual, and phosphate is present in such large quantities that it is not depleted by the plankton algae in the summer (McIntyre and Johnston 1975).

Where there are strong tides, such far-reaching changes do not take place on the ocean floor. That is the case with London's sludge, 5 million m^3 of which are annually dumped in the mouth of the Thames, and with the sludge from the Manchester-Liverpool area, about 0.6 million m^3 in 1971, with about 40,000 t of dry material per year, which is dumped in Liverpool Bay, Irish Sea. A working group concluded that even an increase to 250,000 t of solids is unlikely to produce changes beyond a tolerable limit (Department of the Environment 1972). Great Britain dumps 25 times as much sludge as the Federal Republic of Germany. Since 1962, approximately 0,3 million m^3 per year of treated sludge have been dumped by the city of Hamburg in an area northeast of "Elbe 1" light ship in the Helgoland Bight. Nearly every day a barge was loaded with 1200 m^3 of sludge containing about 8% solids, towed for about 11 h to the site, and emptied. Unloading took about 2 h so that the sludge is spread over a larger area. The costs of this procedure are said to have been DM 7 per m^3. An investigation of the area, in 1971, showed an increase of organic matter in the sediment, and high concentrations of the clam *Abra alba,* and of the polychaete *Pectinaria* (Caspers 1975). But in the summer of 1975, a large part of the bottom fauna died out. It is still not certain whether sludge is mainly responsible for this, or whether a natural phenomenon is involved and a lack of oxygen extended in wide eddies in Helgoland Bay during the especially placid summer weather (Fig. 40). The basic situation seems to be similar to New York Bight (Fig. 39). From 1980 on, the sludge disposal so close to the mouth of the Elbe river has no been longer tolerated, sludge disposal in 1981 took place west of Helgoland, in deeper water. The policy in the Federal Republic of Germany is to prohibit dumping of sludge in coastal waters. This means that the city of Hamburg either has to bring the sludge to the edge of the continental shelf, or find means to dispose of it on land (Dethlefsen 1981).

Sewage sludge is a mixture of substances — including clay particles, organic remnants, and trace elements of the most varied sorts, that have very different effects. With anaerobically treated sludge from Hamburg sewage treatment plants, about 50 t of solid material was introduced into the region of the light ship "Elbe 1" every day; about half of this amount is organic matter, and it contains, per year, about 500 t of zinc, 18 t of copper, 13 t of chromium, 11 t of lead, 2.4 t of nickel, 0.5 t of cadmium, 0.5 t of silver, and between 36 kg and 270 kg of mercury. These amounts will double when a new sewage treatment plant comes into operation, in 1981, and if dumping should continue. Sewage sludges have different concentrations of heavy metals; in general, calculated by dry weight, they contain 10 times more copper, 30–50 times more lead, zinc, and cadmium, and 200 times more silver than average Earth crust material or unpolluted marine sediments. Where industrial effluents are included in sewage, concentrations may be 10–100 mg/kg of mercury and cadmium, 20–200 mg/kg of nickel, 1–4 g/kg of lead and chromium, and 1–15 g/kg of copper and zinc (National Academy of Sciences 1975a). Sludge dredged from the bottom of harbors and from heavily polluted estuaries may have even higher heavy metal concentrations. This is the reason for rather high concentrations in the sediment of dumping areas (Fig. 35 and 74).

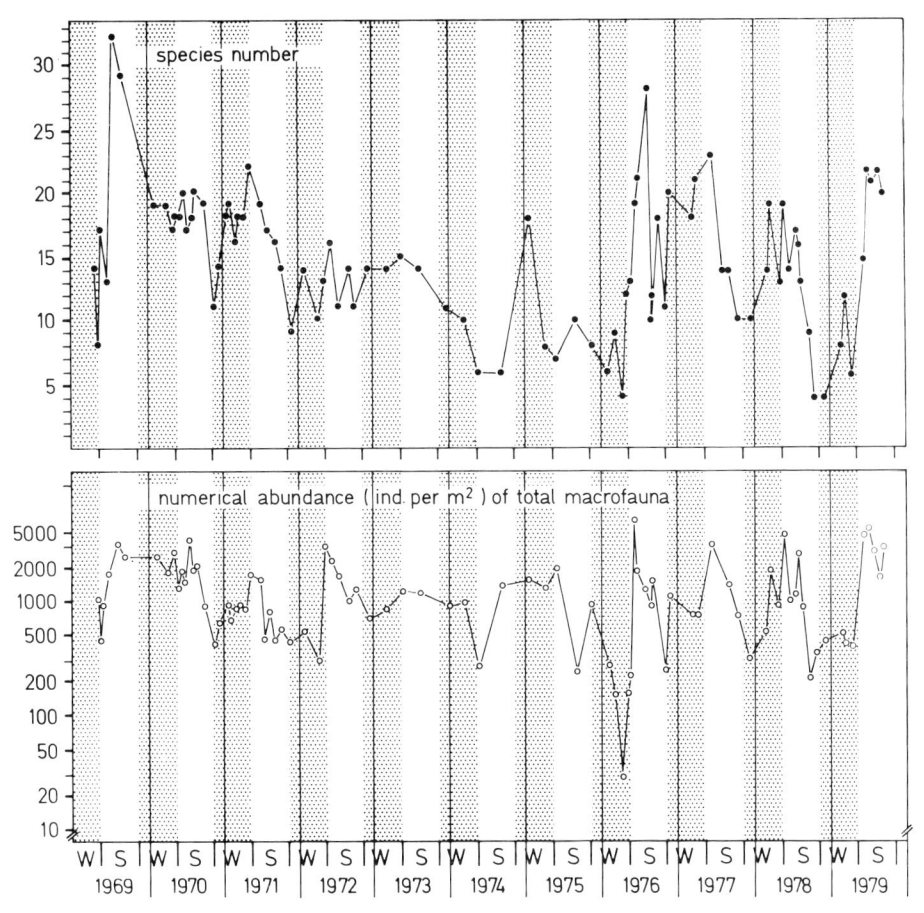

Fig. 40. Since 1962, the city of Hamburg has dumped about 0.3 million m³ of sewage sludge every year into Helgoland Bight. Since 1969, a macrofauna station has been monitored which is situated a few km west of the dumping site and which may be influenced by fine material from the sewage sludge. Up to 1973 a rather diverse fauna was observed. Between 1974 and 1979 there were periods of drastic reduction with only four species *(Nucula, Diatylis, Nephthys, Ophiura)* surviving, sometimes without any animals at all, and periods of new settlements. These occurred possibly as a consequence of special events, a heavy storm in January 1976, and a breakdown of the fauna due to the very cold winter 1978/79, and there were only temporarily successful recolonization efforts in the early summer months which declined in late summer of the same year. The time series is not long enough to decide whether natural, climatically induced stress factors, or man-made pollution stress is more responsible for the fluctuations. In calm summer periods light brackish water from the rivers forms a surface layer in Helgoland Bight which prevents the deeper water from contact with the atmosphere and with the oxygen-producing phytoplankton. Then the bottom layer may suffer from oxygen deficiency. This stress could be more severe when luxurious phytoplankton growth due to eutrophication delivers surplus dead organic matter to the bottom layer and consumes oxygen for degradation, and when sewage sludge containing high concentrations of organic matter is dumped into the region (Rachor 1978; supplemented with data from Rachor 1980)

As long as polychlorinated biphenyls (PCB's, see Chap. 9) were widely used in industry, paints, and lubricants, waste water from cities contained traces of PCB's, and because PCB's are rather persistent, they still contaminate the environment and will be found in waste water for a reasonable period of time. In the treatment of waste water, PCB's in the main get into sludge; the treated effluents therefore are not at fault. The sludge from New York waste water treatment plants, in 1975, contained an average of 3.4 mg/kg PCB's calculated for dry matter in the sludge. As a consequence, PCB concentrations in the sediment of the dumping area for sludge in New York Bight are above 1 mg/kg (dry weight) in the upper 5 cm sediment layer, compared with 1–100 µg/kg in the area farther away (West and Hatcher 1980). One has to assume that such PCB concentrations originating from the dumping of sewage sludge accumulate in marine organisms and fish (Lawrence and Tosine 1977), and PCB's will enter the land vegetation if sludge is disposed on land and used for agricultural purposes.

PCB's range under "organic substances difficult to degrade", like many others, phenolic substances for example. However, they make up only a small fraction of the total organic matter in sewage sludge. When sewage sludge is anaerobically digested, for example in Hamburg treatment plant it contains organic substances equivalent to a Chemical Oxygen Demand (COD) of 37–51 g/l. The COD is determined by wet oxidation of the sludge with dichromate, therefore it includes the large quantities of organic substances which are not readily degradable and cannot be measured as oxygen demand during microbial degradation (Biochemical Oxygen Demand, BOD, during a 5-day period, see Chap. 2.1). BOD of Hamburg waste water treatment plant digested sludge is only 3.5–10 g/l, or roughly 10% of COD. There is not enough knowledge available regarding the nature and fate of the substances in nature which contribute to the bulk of COD. Certainly they are nontoxic, have properties like cellulosis, and resemble the material which composes the organic fraction of marine sediments. The degradation of this material within the sediment, by action of the sediment bacteria and fungi, is slow, and to a large extent anaerobic in the deeper sediment layers. Effluents from paper mills, which contain wood fibers and other cellulosis and lignine material, are in many respects similar to sludge.

4.4 Toxic Substances in Antifouling Paints

Shipping adds to the heavy metal pollution of some coastal areas which are used as harbors and for yacht clubs, because the ship's bottom is painted with toxic paint to prevent algae and sessile marine animals like barnacles and mussels from fouling the hull. Some years ago, mercury was used extensively in antifouling paint, and, for example in 1969, 12% of the mercury used in the U.S.A. went into such paints (see Chap. 8.4). Toxic organic substances like DDT and PCB's have been used also. In the meantime, even if there is no regulation for the do-it-yourself yachtsman, shipyards in the Federal Republic of Germany have to obey regulations regarding dangerous materials, and no antifouling paint is allowed which contains compounds of mercury, arsenic, DDT, HCH, PCB's and PCT (for explanation, see Chap. 9.1), except when a special permit is granted by the authorities.

Antifouling paints which use tin or copper as toxicants, the latter mostly as copper-oxyde, or more modern types of organometallic compounds are legal. Chromium, lead, and zinc are important constituents of bottom primers, cadmium is used in some paints, and zinc in sacrificial anodes to prevent steel hulls from corrosion. At the coast of southern California, U.S.A., between Santa Barbara and San Diego, 37,500 recreational boats are moored, and their bottom is painted annually with a paint that contains, on the average, 600 g of copper per liter. That amounts to 180 t of copper which are used annully and which in some way enter the environment, mostly the marine environment. The figure is significant compared with copper input into the area with municipal outfall of the Los Angeles region (510 t), with surface runoff (40 t) and with dry aerial outfall (30 t). Copper from boats and ships, therefore, is the reason for elevated copper concentrations for example in the mussel, *Mytilus edulis.* In Newport Harbor, with a high concentration of recreational boats and repair yards, copper in mussel muscle is about nine times higher than in mussels from the open shore, and values for cadmium, chromium, lead, tin, and zinc are also elevated (Young et al. 1979).

It may be that the more modern types of hard antifouling are effective with less quantities of heavy metals and therefore do not contaminate the marine environment to such an extent; however, detailed investigations are lacking, and at least in the Federal Republic of Germany one obtains no information about the heavy metal concentrations of the content, if one buys a tin of antifouling.

4.5 Garbage from Ships

The wealthy countries of the world produce, exclusive of kitchen leftovers, 1.6 kg of garbage per capita per day. There are human beings on passenger ships on the world's oceans for 17 million per capita days, and they produce 28,000 t of garbage per annum. Merchant marine crews are said to produce 0.8 kg of garbage per day which consists of 63% paper, 15% metal, 10% textiles, 10% glass, and 1% plastic and rubber. On the average, there are 9000 large merchant marine vessels on the seas. With a 40-man crew, this means 140 million per capita days and 110,000 t of garbage per annum. For the 3 million pleasure craft, 103,000 t of garbage are estimated, 340,000 t for the 120,000 fishing vessels, and 74,000 t of garbage for navy vessels. If the waste residue thrown overboard from a ship's load, and sunken vessels from accidents on the high seas are added to these figures, we arrive at an estimated figure of approximately 6 million t of refuse annually stemming from navigation for all the oceans of the world (National Academy of Science 1975a).

Naturally, coastal areas are the ones that are primarily affected. Part of the garbage decomposes, part of it sinks to the bottom of the ocean. Some refuse, however, floats on the surface of the ocean for a long period of time until washed up on a beach. For this reason, ongoing national and international efforts are underway to prohibit ships from throwing garbage overboard (see Chap. 10). It is good to know that since 1980 the Helsinki Convention forbids even small vessels to throw garbage overboard in the Baltic Sea.

The observations made by scientists on a research voyage in the North Pacific show that the pollution by civilization has reached a phenomenal level. While their vessel traversed 150 km in 8 h, they surveyed 12.5 km^2 of ocean surface. They counted 6 plastic bottles and 22 pieces of plastic, 4 glass bottles, and 12 fisherman's bobs, 1 piece of rope, 1 weather balloon, 1 piece of finished wood, 1 shoe brush, 1 sandel, 3 pieces of paper, and 1 coffee pot (Venrick et al. 1973).

Tourists at the beach are not only annoyed by clumps of tar, but by all sorts of plastic material as well. Floating pieces of plastic foil are a danger for small craft, because they can wrap themselves around the boat's propeller. At a specific density of 0.92, pieces of plastic wrap float on the surface. Under the influence of the sun's rays, the material changes and becomes heavier, and when small calcareous encrusting Bryozoa settle on them they sink to the bottom where they are a nuisance. Fishermen in the Skagerak almost always find pieces of plastic foil in their nets when they pull them up from a depth of 180–400 m (Holmström 1975). Threads and ribbons can wrap themselves around fish and marine birds, and small pieces of plastic have been found in the intestines of many marine animals.

5 Oil Pollution

5.1 The Composition of Petroleum

Petroleum has been formed from the organic remains of dead organisms over different periods in the Earth's history. These organics have been preserved till the present in fossil form in places where the absence of oxygen has prevented their decomposition. Because different factors figured in the formation and settling of the different sediments, the quality of petroleum is different everywhere it is found. Thousands of different compounds may be present in petroleum. 200–300 compounds are present in any particular grade of crude oil. More than half, 50%–98%, of petroleum consists of hydrocarbons, i.e., compounds that contain carbon and hydrogen alone. The following components can be identified in hydrocarbons (Fig. 41):

a) Alkanes (paraffines), which are related to methane, ethane, etc. that are also present in oil deposits. Compounds with 5–7 carbon atoms are liquid; those with a higher number of atoms are in a solid state. There are paraffins with over 60 carbon atoms. Besides straight alkanes, there are branched isoalkanes. Alkanes are relatively nonpoisonous and are biodegraded by many types of microbes. However, the more

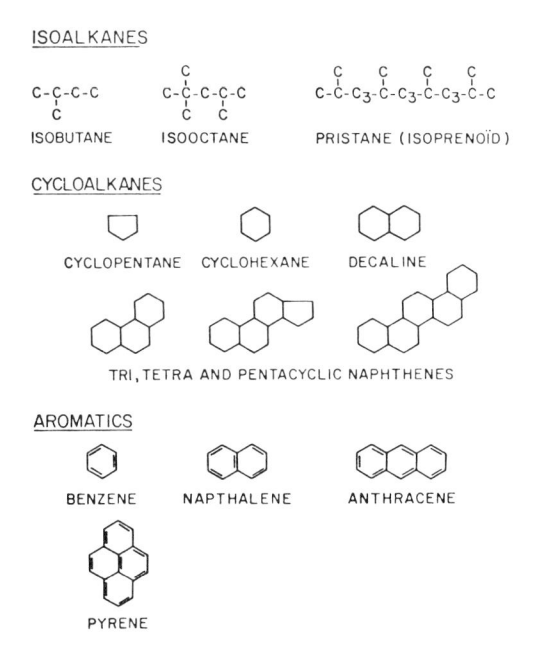

Fig. 41. Structure of some petroleum hydrocarbons

branches present, the more difficult is the biodegradation. The smaller the number of carbon atoms, the more volatile the compounds are, the better they dissolve in water.

b) Cycloalkanes (naphthenes) with 5–6 carbon atoms arranged in a ring comprise 30%–60% of petroleum. In addition to cyclo-pentane and cyclo-hexane, there are also bicyclic and polycyclic naphthenes. These compounds are very resistant to microbial degradation.

c) Aromatic compounds account for 2%–4% of petroleum. Volatile compounds in ring form (for example, benzene, toluene, and xylene), bicyclic aromatic hydrocarbons (principally naphthalene), tricyclic (like anthracene and phenanthrene), and polycyclic ones with more than three rings (pyrene, for example) are present (polynuclear aromatic hydrocarbons, PNAH). There are some microorganisms which are specialized in the biodegradation of these compounds.

Besides hydrocarbons, various other compounds occur in petroleum. Of these, those containing sulfur are the most important. The sulfur content of petroleum can amount to up to 10%. Compounds containing sulfur are presumably responsible for fish and mussels tasting oily after having been in contact with petroleum. In addition, there are fatty acids (up to 5% oxygen in oil), nitrogen compounds (up to 1% nitrogen in oil), vanadium, and nickel.

It is known that a number of these compounds are biodegraded by microorganisms. As for the rest, the biochemical processes of microbiological degradation are complex, and a petroleum component can be approached in different ways by different taxa (Higgins and Burns 1975).

The different components of crude petroleum are isolated in a refinery above boiling point. Gasoline has a boiling point of under 200°C and contains compounds with 5–12 carbon atoms. Medium distillates like kerosene, diesel oil, and light heating oil boil at between 169°–375°C and contain compounds with 9–22 carbon atoms (as water-soluble, poisonous components, they contain principally naphthalene). Gas oil, heavy heating oil, lubricating oil and greases boil at higher temperatures and contain compounds with 29–36 carbon atoms. Residual oils with even higher boiling points have asphalt-like characteristics.

5.2 Fate of Oil on the Surface of the Ocean

Tanker or drilling platform accidents are spectacular. In such cases, great amounts of crude oil reach the surface of the ocean. Depending on the composition of the crude oil, their fate is different. The following factors have an effect on the quantity and the quality of oil on the surface of seawater.

a) **Spread**: Oil spreads out on the boundary layer between the seawater and the atmosphere. If the quantity is small, an oil film forms that reflects the color spectrum. 100–200 l of oil are sufficient to cover 1 km^2 of sea surface with a film which is about 0.1 μm thick. Light oils spread faster than heavy oils. Within 10 min, 1 t of Iranian crude oil spreads to a slick of 48 m diameter and 0.1 mm thickness (Warren Spring Laboratory 1972). Naturally, the layer on the surface is thicker in massive oil spills. As a rule, plans are only made to combat an oil spill if the thickness of the oil layer is not much below 0.1 mm, unless sensitive beaches, coastal stretches, or bird life have to be protected.

The oil film on the surface moves with water currents and wind drift. This makes it possible to predict its drift. It can roughly be estimated that the film will move at 60% of the water current rate and 2%–4% of the wind speed. This rate of drift is of ecological significance because the oil film is displaced vis-à-vis the body of water in such a way that fresh quantities of seawater come into contact with the oil and can release components from it, without accumulating high concentrations of these toxic substances.

b) Evaporation: As the number of carbon atoms increases, the vapor pressure of the various hydrocarbon compounds decreases. In this respect, paraffins do not differ significantly from aromatic compounds (Butler 1976). Evaporation is strongest during the first few hours. Spilled crude oil can only catch fire in the first half hour. After that, there is an insufficient amount of easily volatile compounds in the layer of oil on the surface of the water. At the end of one day, half of the compounds with 13–14 carbon atoms are vaporized, at the end of three weeks, half of the compounds with 17 carbon atoms. The vaporization process extends over a period of months, possibly years, and eventually leads to more viscous and, in the end, tar-like clumps.

The extent to which oil vaporizes off the surface of the water depends on many other factors. When cold, only a small quantity of oil vaporizes. Wave activity fosters vaporization, but at the same time leads to the more rapid formation of water-in-oil emulsion. Oil no longer floating on the surface of the water cannot vaporize either. Gasoline vaporizes quickly and disappears completely from the surface while heavy heating oil hardly vaporizes at all. Almost all North Sea oil consists of around 50% hydrocarbons with up to 12 carbon atoms. These oil components evaporate within 24 h (Table 12).

Table 12. Crude oil from the Ekofisk field in the North Sea is relatively light. 7% of it has a boiling point of under 100°C, 11% between 100°–160°C, 15% between 160°–250°C, 19% between 250°–350°C, and 47% above 350°C. How effectively an oil slick will be reduced by evaporation and dispersal under different weather conditions can be calculated. Correspondingly, the extent to which measures must be taken to combat an oil slick can also be calculated (Blaikley et al. 1977)

State of the Sea	Loss of oil through evaporation	Loss of oil through dispersal		
		Day 1–3	Day 4–5	Day 6 and later
Calm	25–35%	10–30%	5–15%	0–5%
Medium	30–40%	20–40%	10–20%	0–7%
Rough	35–45%	30–50%	20–30%	0–10%
Very rough	35–45%	40–60%	25–35%	0–10%

Oil that evaporates from the sea's surface continues to exist as an air-polluting element. It is assumed that a portion of it is returned to the ocean by rain. In the case of a serious oil spill, however, evaporation is the most important natural factor which makes large quantities of oil disappear from the surface of seawater.

c) Dissolution: Just as with evaporation, the dissolution of petroleum hydrocarbons depends on the number of carbon atoms. Roughly speaking, 10 mg/l from a compound with 6 carbon atoms, 1 mg/l from a compound with 8 carbon atoms, and 0.01 mg/l from a compound with 12 carbon atoms dissolve in distilled water; solubility in seawater is somewhat less. The situation becomes, however, more complicated when there is a complex mixture of different compounds which, as regards solubility, have reciprocal effects. From American crude oil an amount of 46 mg/l dissolves in seawater within 8 days, while the concentration derived from light fuel oil is only 7.5 mg/l after 5 days, and from heavy fuel oil, 2.3 mg/l (Benjamin and Polak 1973). It has not been sufficiently explained to what extent it is a case of genuine dissolution and whether colloid micelle or adsorption to small particles plays the major role.

When oil is decomposed under the action of ultraviolet sunrays by oxidation water-soluble fatty acids and fatty alcohols are produced which are easier for microorganisms to degrade than the original hydrocarbons.

d) Emulsification: When the seawater surface is agitated by wind, and especially when foam and spray result, water is absorbed by the masses of oil. Some oils very quickly absorbe more than 50% of their weight in water and form brown masses dubbed "chocolate mousse". Within 7 h, North Sea oil absorbs up to 80% water. Then the oil slick breaks up into little masses. Besides water-in-oil emulsions, oil-in-water emulsions (dispersions) form naturally or particularly under the influence of added dispersion chemicals (dispersants). There, tiny drops of oil will be found floating around in the water.

On the whole, emulsification and dispersion are very effective means for reducing large quantities of floating oil. Of course, oil is not eliminated from the marine environment by emulisfication and dispersion; it is merely removed from the ocean surface and distributed in the water mass where its poisonous effect persists.

e) Biodegradation: At least 90 strains of marine bacteria and fungi are capable of biodegrading some components of petroleum. Some algae also have this ability. In part, we are dealing with specialists that prefer to live off petroleum hydrocarbons. They are generally found in the seawater of areas that regularly suffer oil pollution. They are rare in other areas, fewer than 100 germs per liter of seawater, and they increase only when the amount of petroleum hydrocarbons increases because of an oil accident. In the North Sea, the number of oil-degrading bacteria is generally low, and fairly large concentrations have only been found in the vicinity of the Ekofisk and Forties oil-producing platforms (Oppenheimer et al. 1977). Naturally, it takes time after an oil spill until an optimal number of oil-decomposing bacteria is present.

The efficiency of these bacteria depends on a number of environmental factors. According to some experiments, at low temperature biodegradation is very slow. At freezing point temperature, they only degrade around 10% of the amount they would at 25°C. However, there are other experiments which indicate that sometimes degradation at low temperatures is not much less than at warm temperatures. As a rule petroleum alone does not contain enough nutrients like nitrogen and phosphorus to make a good growth of oil-degrading bacteria possible. For the decomposition of 1 mg of oil, the microorganisms require as much nitrogen as is normally contained in a liter of coastal seawater. In addition they require 3.3 mg of oxygen which corresponds to the content in 0.4 l of seawater. The amount of organic substances dissolved in seawater also plays a role, possibly also the amount of suspended particles in seawater to which bacteria can attach themselves. Because the amount of nutrients is low (Table 13), the biodegrading of oil proceeds much more slowly in the clear region of tropical seas than it does in estuaries and coastal water. By adding nutrients in an experimental environment, the ability to biodegrade can be increased considerably.

Table 13. Amount of biodegradation of the individual components of crude oil at water temperatures between 13° and 24°C which are prevalent during all seasons along the coast of South Caroline, U.S.A. Half-life time = 1/2 incubation time/amount which is mineralized to CO_2 during the period of incubation (Lee 1977)

Hydrocarbon	Zone	Concentration in μg/l	Amount of biodegradation in μg/l per day	Half-life time in days
Benzene	Estuary	24	0.330	37
Toluene	Estuary	6	0.041–0.058	45–65
Hexadecane	Estuary	25	0.100–0.130	85–105
Heptadecane	Estuary	20	0.140	70
	Coast	20	0.034	295
	Open sea	20	0.003	3350
Naphathalene	Estuary	30	0.870	17
	Coast	30	0.330	115
	Open sea	30	0.012	1250
Methylnaphthalene	Estuary	30	0.250	60
	Coast	30	0.000	interminable
	Open sea	30	0.000	interminable
Anthracene	Estuary	15	0–0.070	above 145
Benzo(a)pyrene	Estuary	5	0–0.002	above 1750

In order to create a better growth of oil-decomposing bacteria as a means of combating oil, the possibility of spreading synthetic fertilizers on the surface of the ocean has been considered. It has also been suggested that the germs of oil-degrading bacteria, together with the corresponding nutrients, might be spread over oil slicks so that the decomposition can begin more quickly. However, this is probably not very realistic because freshly spilled oil is also poisonous to microorganisms (Zobell 1973). It can be assumed that microbial degradation only begins to take place after oil at the surface has aged and lost part of its highly volatile, poisonous components by vaporization.

There is increasing evidence that bacterial degradation of petroleum hydrocarbons is faster, when oil is not floating on the surface of the sea, but is dispersed in form of 1 μm droplets in the water column, by natural forces of by chemical dispersants

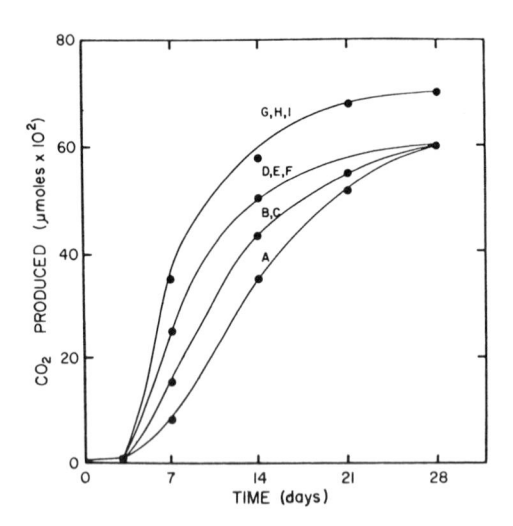

Fig. 42. There is evidence that dispersants of low toxicity have a positive effect upon the efficiency of oil-degrading bacteria in seawater. The graph represents experiments where 0.5 ml/l of commercially available dispersants have been added to a dispersion of 10 ml/l American crude oil in seawater from the U.S.A. Atlantic coast, containing optimum concentrations of phosphate and nitrate. The efficiency of bacterial degradation is demonstrated by the amounts of CO_2 produced (Atlas and Bartha 1973b). *A* Control without dispersant, *B* Corexit 7664, *C* Pyraxon, *D* Magnus, *E* Marine cleaner 8555, *F* Corexit 8666, *G* BP 1100, *H* Smith herder, *I* Shell oil herder

(Fig. 42). The modern dispersants have so little toxicity that, by providing huge oil-water interfaces, they enhance the degrading activity of microorganisms; for example the dispersant Corexit 9527 is readily biodegradable itself (Traxler and Bhattacharya 1978).

Many microorganisms that biodegrade petroleum release substances into the environment which act as dispersing agents so that dispersal of oil also plays a role in the natural process of biodegradion.

In experiments, scientists have attempted to determine how quickly microorganisms decompose different grades of oil: up to 2 g/m^2 and 1 g/m^3 per day. Often however, these tests have indicated far smaller amounts (Floodgate 1973). Presumably, a clear distinction has not been established between loss of oil through evaporation and through microbial decomposition.

Researchers have also added single petroleum components to seawater and studied the reduction down to carbon dioxide (Table 13). In the process, half-life periods of weeks and months result. In some cases only a very slight reduction is noted. However, the question here, too, is whether unrealistic values are being obtained when the individual components are analyzed one by one. It would be better to mix radioactively marked components in with the complex petroleum and to observe how the reduction takes place under these conditions, and to observe degradation in the surroundings of an oil clump. Because when an oil clump floats in the ocean, components are released and come into contact with the seawater. Microorganisms which grow on the surface of a clump of oil could have different growth conditions compared with petroleum components dissolved with seawater.

Aromatic compounds with many carbon atoms and cycloalkanes are difficult for microorganisms to attack. Presumably, there are yeasts and fungi which specialize in this, but they probably work so slowly that clumps of tar last for years on the surface of the ocean.

Asphaltenes are persistent, too, and of course the elements vanadium and nickel are left over when petroleum components are degraded. Sometimes it is possible to identify the oil quality from which a certain tar ball derives, by analyzing the vanadium-nickel percentage.

Little is known about the fate of petroleum hydrocarbons in the metabolism of animals, and whether animals play a role in petroleum degradation. Marine animals living in oil-polluted seawater exhibit a higher concentration of certain enzymes, the mixed function oxydases (MFO) and this seems to be an indication that the animal organism reacts on contact with petroleum hydrocarbons.

f) Mechanical Reduction: Clumps of tar as big as peas are floating on all the seas. They have different origins. Tarry residue is scraped off the walls of a tanker when it is cleaned. This residue is flushed out at sea. Clumps of tar are also left over when oil ages that has been floating on the surface of the sea. The volatile components disappear through evaporation, dissolution, emulsification, and microbiological degradation, and those components remain which are difficult to attack. If water samples are studied more carefully, minute particles, 20–80 μm in size, and which have the same composition as the clumps of tar on the surface can be found floating in the water, even at depths of 100 m. It seems that clumps of tar decompose on the surface, finally turning into fine particles (Morris et al. 1976).

g) Sinking: Floating clumps of tar often form the substratum for sedentary marine animal life, for goose barnacles *(Lepas)* and other barnacles, for example. The calcareous parts of these animals can reduce the buyocancy of tar to the extent that the clumps sink. When other organisms with calcareous or siliceous solid parts also settle on clumps of tar, the specific gravity can become heavier than the seawater. In the aging process, some kinds of heavy oils attain a corresponding density.

Apparently, it is more common for fresh oil to reach the bottom of the ocean than was previously assumed. When drops of oil are dispersed in seawater, they are absorbed just like other floating particles into the intestines of filter-feeding plankton animals. Together with other indigestible materials, they appear regularly in the fecal pellets of copepod plankton. Even if they consist of up to 7% oil, these balls of excrement have a heavier specific gravity than seawater and sink to the bottom (Conover 1971). They are absorbed by filtering bottom animals.

There is a remarkable affinity of oil to particles suspended in the seawater, and oil-seston-compounds form which may be heavy enough to sink to the sea bottom (Boesch et al. 1974). It is still an open question whether the final fate of oil released with an oil accident is microbiological degradation, or incorporation into the sediment. Deep sea sediments contain about 1 mg/kg or 50 mg/m^2 in hydrocarbons; this amount is comparable to the amounts of oil estimated in the water column or drifting at the sea surface as tar balls (Butler et al. 1977).

h) Beaching: Floating clumps of tar and fairly large amounts of oil collect on the beaches when an oil slick is driven up on the coast by the wind. Oil can then easily get worked into the sediment. Penetration into the interstitial system between grains of sediment is particularly intensive when dispersant chemicals are used. Oil is persistent when it is buried so deeply in the sediment that it no longer has any contact with the oxygen-containing surface stratum. It is persistent because bacterial biodegrading of oil rarely occurs under anaerobic conditions. After all, fossil deposits survived for millions of years under anaerobic conditions.

8000 t of heavy fuel oil was spilled, and covered 300 km of the coast of Chedabucto Bay in Nova Scotia, Canada, when the tanker "Arrow" went agorund in 1970. Five years later, oil was still traceable in the sediment. Since oil streams out of the sediment into the adjacent shallow areas at low tides, animal life in the shallows is still affected by it. The prediction is that it will be more than 150 years before this oil has disappeared (Fig. 51; Vandermeulen and Gordon 1976).

i) Persistence and Wide Ranging Ocean Distribution: Hydrocarbons with many carbon atoms, particularly cycloalkanes and aromatic compounds, hardly evaporate, are insoluble in water, and are not usually biodegraded by organisms. Because of this, the preconditions are met for these compounds to be persistent, for holding them in the environment of the ocean for a long time, and for their accumulation over the years when the amount of added oil components outweights the amount of natural sedimentation. It is speculated that part of the oil components that during World War II got into the Atlantic by torpedoed tankers is probably still present in the ocean, but quantitative models are still lacking that would furnish information on whether the concentration of persistent petroleum hydrocarbons is presently on the increase in those seas with tanker traffic or whether the new regulations are causing a decrease in oil pollution.

Aside from remote areas, for example the Beaufort Sea in the Canadian Arctic before the oil boom, tar can be found floating on all the oceans of the world. In 1975, there was 0.3 mg/m^2 in the North Sea (Oppenheimer et al. 1977), 10 mg/m^2 in the subtropical areas of the Atlantic, for example Sargasso Sea. The figure stands at up to 500 mg/m^2 for the Mediterranean. Perhaps 0.1 mg/m^2 is the figure for all the oceans of the world. This means 36,000 t. Other estimates go up as high as 150,000 t. The analyses of millimeter-sized to potato-sized tar balls drifting on the Atlantic Ocean in the region of the Canary Islands provide hints that they derive from the washing of tankers on their way back round the Cape of Good Hope (Ehrhardt and Derenbach 1977).

Hydrocarbon analyses in seawater are available for various ocean areas, but they are difficult to compare because the results differ, depending on how the water samples were analyzed. From ocean water that is not particularly polluted, 0.5–5 μg/l of hydrocarbons have actually been extracted. There is some probability that in the Atlantic Ocean and in the Pacific Ocean the average concentration is about 1.5 μg/l. But hydrocarbons analyzed are a mixture of petroleum and biogenic compounds, and the water samples have not been filtered, so that part of the hydrocarbons may occur in particulate form (Brown and Searl 1977).

It has not been explained up till now what percentage of these hydrocarbon amounts is traceable to marine pollution by petroleum. Petroleum is a product of nature which came into being during the geological past. Petroleum components, also, are not unknown to recent vegetation. It is certain that many organic compounds contained in petroleum are continuously being formed by organisms living today. This is the case with land vegetation, in particular where forests of conifers give off large quantities of volatile terpenes into the atmosphere. Likewise in forest fires, numerous hydrocarbon components reach the atmosphere. Through rainfall, they can enter the ocean. Marine vegetation also produces hydrocarbons, and hydrocarbons are found in all kinds of marine animals. The hydrocarbon concentration in marine organisms is above 50 mg/kg dry weight, probably with an average of 400 mg/kg. If marine plant production is 32% of the world plant production (Table 6) or about 24×10^9 t of organic carbon per year, equivalent to 60×10^9 t of organic matter, then a minimum of 3 million t of hydrocarbons are produced every year by the marine vegetation, about as much as man introduces in form of petroleum into the world oceans (Table 14). Such calculations, however, are rather speculative and there are many question marks. It seems that recent marine organisms can synthesize only hydrocarbon compounds other than straight alkanes with up to 22 carbon atoms, but seawater contains many more compounds with higher numbers of carbon atoms than could have been introduced with petroleum. Either they are rather persistent and have a long life time in the sea, or there are unknown sources of such compounds in the sea (Koons and Monaghan 1977).

Hydrocarbons formed by vegetation stand out by the predominance of even-numbered carbon atoms in a series of compounds with different carbon atom counts. In petroleum hydrocarbons, there is no difference in even and uneven numbers. However, some recent bacteria are also producing hydrocarbons without showing a preference for even numbers of carbon atoms. One compound produced by vegetation is pristane, while phytane is practically absent. In petroleum hydrocarbons 1.5 to 2.5 times more phytane is present than pristane. Finally, it is characteristic in hydrocarbons produced by recent organisms that few components predominate while petroleum hydrocarbons distinguish themselves by a complex mixture of aromatic compounds and cycloalkanes. It is clear that this mixture has not come into being recently (Morris et al. 1976).

Among the polycyclic aromatic hydrocarbons found in the marine environment are carcinogenic compounds like benzo(a)pyrene. It seems that even in the cleanest oceanic seawater concentrations of benzo(a)pyrene are about 0.01 μg/l, in nearshore coastal waters up to 0.05 μg/l. About 0.02 mg/kg are found in organisms, referring to wet weight, and figures for sediments are 0.05 mg/kg, referring to the organic substance in sediments. Fortunately, it seems that edible fish do not accumulate benzo(a)pyrene. Concentrations are very low. However, this can be misleading because there is evidence that benzo(a)pyrene in fish is metabolized to other compounds which are cancerogenic, too. Levels in most shellfish samples were less than 0.1 mg/kg wet weight, which is about the level in charcoal-grilled meat (Dunn and Fee 1979).

There are some signs to suggest that natural production by ocean bacteria and vegetation is more important in benzo(a)pyrene and other polycyclic aromatic compounds than petroleum which contains around 100 mg/kg of polycyclic aromatic hydrocarbons (Andelman and Snodgrass 1974). Locally in dock constructions on the other hand, wood treated with tar-derived impregnations and coatings is of significant relevance to benzo(s)pyrene concentrations. While freshly caught lobsters *(Homarus vulgaris)* in Canada had less than 1 μg/kg in their meat, after they were kept in tidal ponds constructed of creosoted timber, they contained up to 281 μg/kg of carcinogenic hydrocarbons, more than in any other foodstuff known. Elevated concentrations may also result from sewage effluents (Fig. 43), and in the polluted Severn Estuary

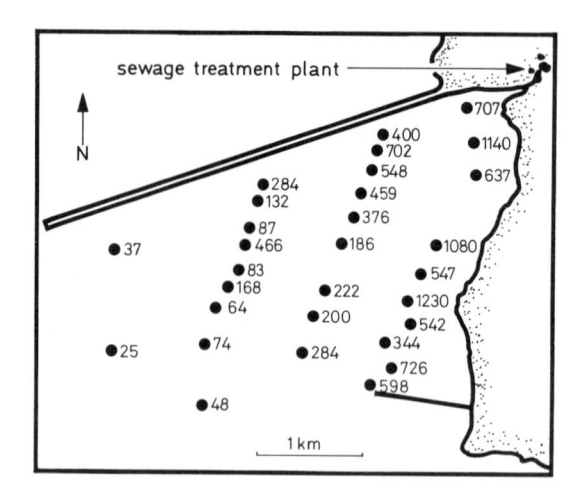

Fig. 43. The distribution of benzo(a)-pyrene in the sediment along the coast of Vancouver, British Columbia, Canada, indicates that the introduction of waste water plays a role. The Iona Island treatment plant processes 320,000 m^3 of effluents per day. In μg/kg, the data indicate how much benzo(a)pyrene the organic matter in the sediment contains (Dunn and Stich 1976)

(England) concentrations of up to 0.5 mg/kg have been found in the sediment (Thompson and Eglington 1978). In such situations, a monitoring of shellfish could be necessary, but the reason for the high contamination does not seem to be petroleum.

5.3 Sources of Marine Pollution by Petroleum

After public opinion was aroused in 1970 by the news that 5–10 million t of petroleum pollute the world's oceans per year, the estimated amounts were recalculated for recent years. But because it obviously is not easy to get hold of the exact amounts, there are still considerable gaps between the various estimates (Table 14).

In 1971, 2500 million t of petroleum were transported all over the world[2]. Of this amount, 400 million t came from off-shore wells; 1400 million t were shipped to their destination by 6000 tankers. In 30 years, the amount has increased more than tenfold. This explains why oil pollution has hardly decreased despite the many attempts to reduce it.

Because they are said to contribute only 0.3 million t per year to ocean pollution, accidents play only a minor role. However, when there is a tanker accident, the damage is particularly obvious, and the public is justifiable interested in it. However, other sources of oil pollution deliver much larger amounts of oil to the sea.

It must be emphasized that there are coastal areas where off-shore oil deposits are so close to the bottom of the ocean that crude oil escapes and fouls the water and the ocean surface. As far as the overall picture is concerned, just as much petroleum reaches ocean water from such natural undersea oil seeps as through accidents.

---

2 1 t of petroleum corresponds to about 1.16 m^3 or 7.3 barrels, because 1 m^3 contains 6.3 barrels with each 42 US gallons

Table 14. Estimates of the amount of crude oil and petroleum hydrocarbons that annually reached the ocean in the period 1969–1972. (National Academy of Science, 1975b, regarding 1969–1971, and Grossling 1977 regarding 1972)

	Million t 1970	1972
Natural escape from offshore seeps	0.6	0.6 ?
Via the atmosphere	0.6 ?	?
From rivers	1.6	
Rain water from city surfaces	0.3	2.8
Domestic effluents	0.3	
Industrial effluents	0.3	
Oil refineries	0.2	?
Offshore drilling and pumping	0.1	0.3
Shipping, excluding tankers	0.5	?
Servicing of tankers in docks	0.2	?
Small and outmoded tanker traffic	0.8	0.4
Modern, load-on-top tanker traffic	0.3	
Shipping accidents, excluding tankers	0.1	?
Tanker accidents	0.2	0.2
Total amount	6.1	about 4.3

It was known even before the Spanish conquistadors came to the coast of California, U.S.A., that oil eruptions constantly took place in the Santa Barbara Channel. They are still taking place today. However, no damage to the marine fauna has been observed (Straughan 1976). The variety of coastal fauna is not different from that in unaffected areas. A thorough regular survey over more than 2 years revealed even higher densities of macrofauna, especially deposit-feeders (Davies and Spies 1980).

The amount of hydrocarbons and other petroleum components which reach the ocean via the atmosphere cannot yet be accurately determined with reference to the whole picture. Some estimates have been set as high as 10 million t a year. Not all heating oil and motor fuels are completely burned. Part of the hydrocarbons reach the atmosphere in an unburned state. The chemical industry annually manufactures 12 million t of benzene and 8.4 million t of toluene. Part of these amounts also reach the atmosphere. Apparently, photochemical reduction does not play a significant role. Microbial biodegradation also proceeds at a slow rate. This may explain why 1 μg/l of benzene was found in the analysis of a number of coastal water samples and 20–40-fold higher concentrations in fish (National Academy of Science 1975a). Currently, it is estimated that only as much oil (0.6 million t) reaches the ocean via the atmosphere as through natural seeps.

Apparently, the largest amounts of oil from land reach the sea by rivers along with domestic effluents and with the dirt washed off the streets and into sewers. Estimates run up to a total of 2.5 million t and are well-grounded because every inhabitant in the developed world uses 6–27 g of oil and lubricants daily. Around half of this amount can be found in sewage water despite official efforts to keep sewer lines free of waste oil. Industrial effluents introduce 0.3 million t into the oceans of the world and refineries nearly 0.2 million t.

Even when there are no spills, some oil continually escapes from the many small leaks in off-shore wells and on the pumping platforms so that 0.1 million t of oil escapes from off-shore drilling and pollutes the sea. On a world-wide scale, drilling for oil has been done at about 20,000 offshore locations during the past 30 years. In the North Sea, since 1964 drilling has been done at about 1000 locations, and

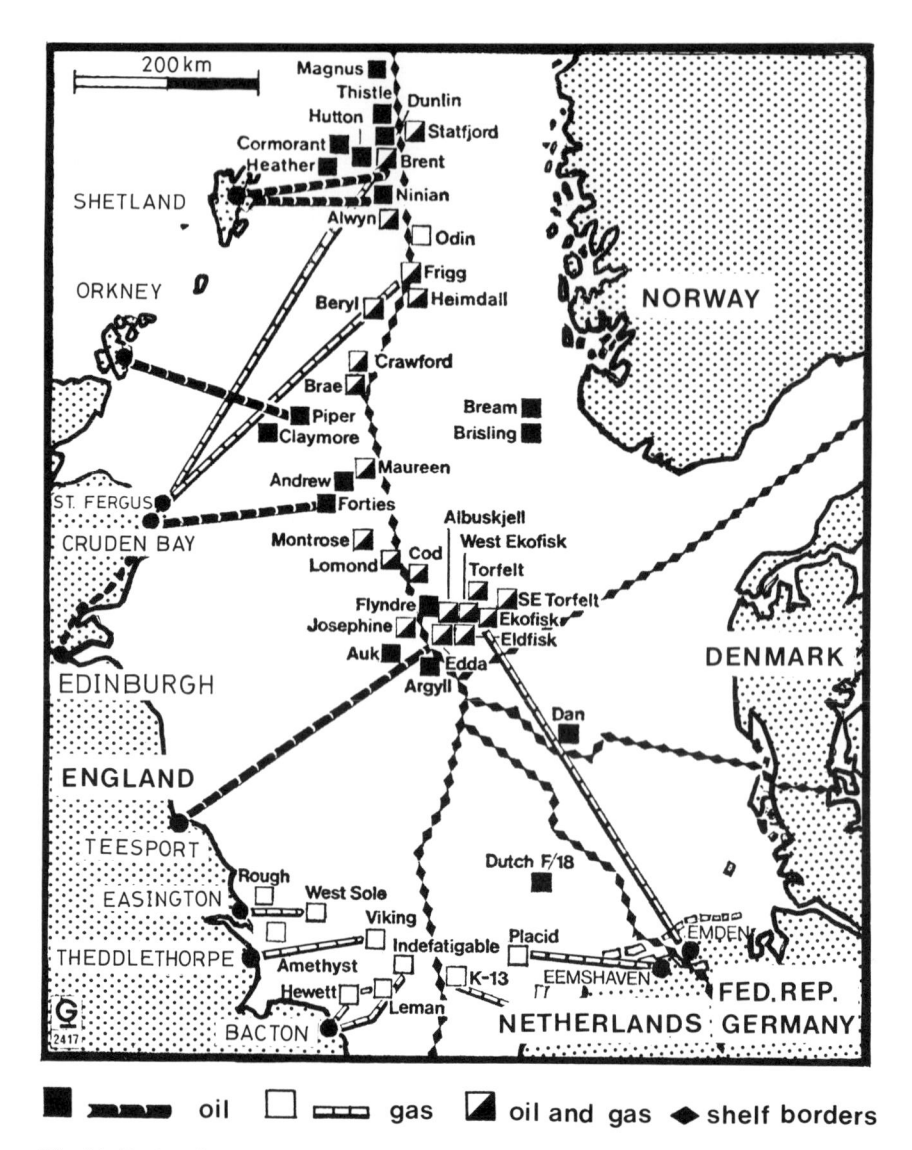

Fig. 44. Exploration and pipelines for oil and gas in the North Sea, (From information provided by H. Schoeneich in Jahrbuch für Bergbau, Energie, Mineralöl und Chemie, 1977, and from newspaper information)

there are 350 producing wells established (Fig. 44). Normal, accident-free navigation polluted the seas with 0.5 million t. This amount may now be less because international controls have become strict (see Chap. 10). Likewise pollution by oil tankers discharge into the sea is declining.

It is estimated that 0.77 million t of oil reached the oceans in 1971 from smaller and outdated tankers while large, modern, load-on-top ships only discharged 0.31 million t. If the amount of oil carried by these modern ships had been transported by the types of tanker that were still standard a few years ago, they would probably have lost 4.2 million t of oil at sea (Fig. 45).

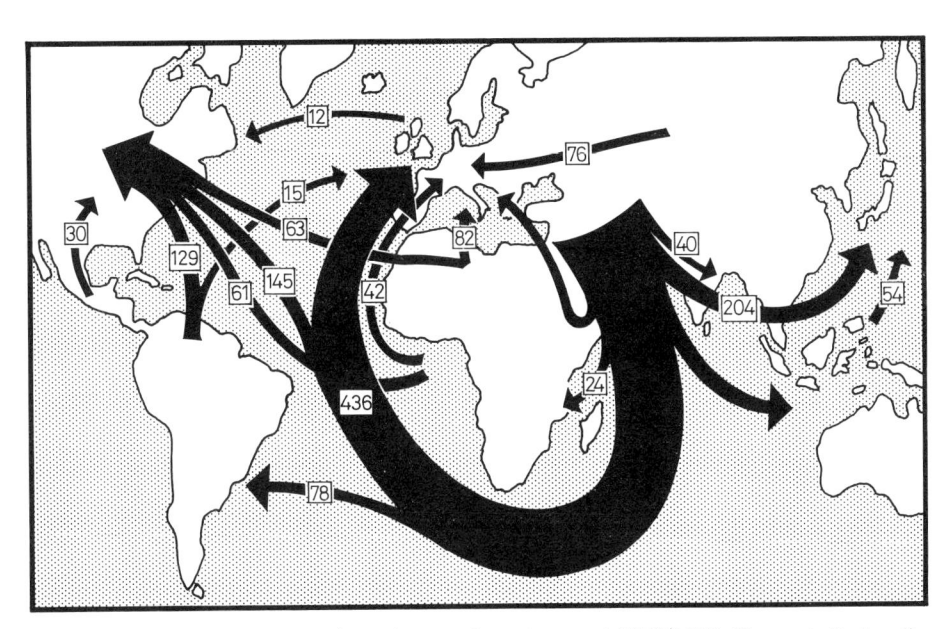

Fig. 45. The most important sealanes for petroleum transport 1978/1979. Figures indicate million t per year. (From BP statistical review of the world oil industry and newspaper in formation)

When a new tanker is filled, 0.3% of the oil sticks to the walls after the first trip. After unloading, the tanks are washed by hot water sprayed at high pressure. The empty tanks are also partially flooded with seawater for ballast on the return trip. Formerly, cleaning water with oil residue was pumped into the sea during the return voyage, and ballast water was pumped out shortly before docking in an oil habor. This explains the large amounts of oil which reached the sea from old-fashioned tankers. Modern load-on-top tankers have special tanks in which water is separated from oil. This oil should be unloaded in a harbor for reprocessing. Till now, there were not many such facilities. Also, ship owners had to pay harbor, canal, and customs costs for the uncleaned oil. So naturally the temptation was great to discharge the oil into the sea despite laws against it (Moss 1971).

Accidents cannot completely avoided even when every precaution is taken to avoid them. At present, the oil industry calculates that an accident will happen on a statistical average every thousandth harbor approach or after 50 years of regular tanker service. This comes out to an average loss of 87 g per t of oil shipped. Accidents on oil platforms account for a loss of 72 g per t of oil (Beyer and Painter 1977). Other calculations indicate that a loss figure of 2.2 g per t oil is due when loading or unloading oil. However, this figure can be reduced. In the British oil harbor of Milford, only 0.6 g per t of transferred oil are lost. Large mishaps are omitted from this figure, and speculation is still imprecise whether or not a 100,000 t oil accident every 25 years in the area of the North Sea and a 400,000 t oil accident every 50 years must from a statistical standpoint be reckoned with (Johnston 1976b), or that one tanker accident of above 135 t is due for every 437 million t of oil shipped through the North Sea (Department of the Environment 1976). Probably the many tanker accidents of the past years make new calculations necessary.

About 75% of all tanker accidents are caused by human error. Accidents occur more frequently on tankers which sail under cheap flags and are registered in states where safety regulations and crew requirements are not so rigid as in states with high standard (Clark 1978a). Three thousand one hundred and eighty three accidents with tankers above 2000 gross tons have been registered between 1969 and 1973, which resulted in 452 cases of oil pollution. Causes of the accidents were 123 times running aground, 126 times collisions, 94 times structural weaknesses of ship or equipment, 46 times ramming the pier, 11 times a breakdown of the engine, 17 times fire and 31 times explosions. As a consequence of these accidents about 1 million t of oil was released into the sea (Grossling 1977). Oil accidents registered in 1974 off the coasts of the U.S.A. refer to 53% crude oil, 11% fuel oil, 11% heavy fuel oil, 7% diesel oil, 6% gasoline, 5% chemicals and 4% other qualities (Boyd et al. 1977).

The Federal Republic of Germany imports 41% of its oil needs through North Sea harbors. In 1977, 26 million t were imported through Wilhelmshaven, Emden, and the harbors on the Weser River, 14 million t through Brunsbüttel and the Elbe River harbors.

Worldwide, a total of 216 "large" oil spills were noted between 1960 and 1975. Only one accident of this size took place in the coastal area of the Federal Republic of Germany (Van Gelder-Ottway and Knight 1976): On February 20, 1966, the Norwegian tanker "Anne Mildred Brövig", loaded with 39,000 t of Iranian crude oil, collided with a British ship in thick fog between light buoys P6 and P7 on the Elbe-Humber route in the North Sea west of Helgoland and caught fire. Stern down, the tanker later sank to the 37-m-deep North Sea bottom. Approximately 16,000 t of oil escaped. Of this, 200–3000 t were combated with 70 t of dispersants, costing approximately DM 900,000. Because the direction of the wind was favorable, the rest of the oil quickly disappeared without causing damage to the German coastal areas (Drost 1966).

Since that time, the coast of Germany has also been fortunate. True, the Kuwaiti tanker "Al Founta" did ram the oil pier in Wilhelmshaven in 1974 when a reverse maneuver error was made. 40 t of oil did spill out, but the tanker was able to make its way back out into the open North Sea, despite the fact that the water was running away with the tide. The tanker's draft was so deep that it just barely kept off the bottom at high tide in the 300 m large Jade channel. In the fall of 1976, the Liberian "Energy Vitality" with 200,000 t of crude oil got stuck in the Jade, as did the

Norwegian "Grimlevand" with 60,000 t of oil in the mouth of the Elbe, but tugs were able to pull them back into the navigation channel. In May 1977 it was the tanker "Classic", in December 1977 the Greek "Nicos J. Vardinoyannis" that run aground in the Jade, but only about 100 t of oil were lost. In April 1978 the British tanker "Camden" ran aground, but there was no oil leaking out, and there was again no oil accident, when in February 1979 the "Esso Hawaii" stuck to the ground when leaving Wilhelmshaven. In summary, the German coast had a really high number of tanker groundings, but a surprising low oil pollution record. By the law of probability, a fairly large oil spill is long overdue in the North Sea area.

How accidents happen and what consequences they can have can be learned from the case of the "Torrey Canyon" which went aground on the Seven Stones Reef northeast of the Scilly Islands on March 18, 1967, in daylight and in good weather at a speed of 17 knots with a load of 117,000 t of crude oil from the Arabian Persian Gulf (Fig. 46). Six of its 18 tanks were split open; 30,000 t of oil escaped and floated toward the English Channel as an oil slick 20 nautical miles wide. Part of this amount of oil reached Brittany between Les Heaut and the Bay of Lannion on April 11, 1967. The stormy seas following the accident caused about 18,000 t of additional oil spill into the sea. A strong west wind blew this oil toward the coast of Cornwall between March 24 and 26. On March 26, the wreck broke up in heavy seas and an additional 40,000–50,000 t escaped.

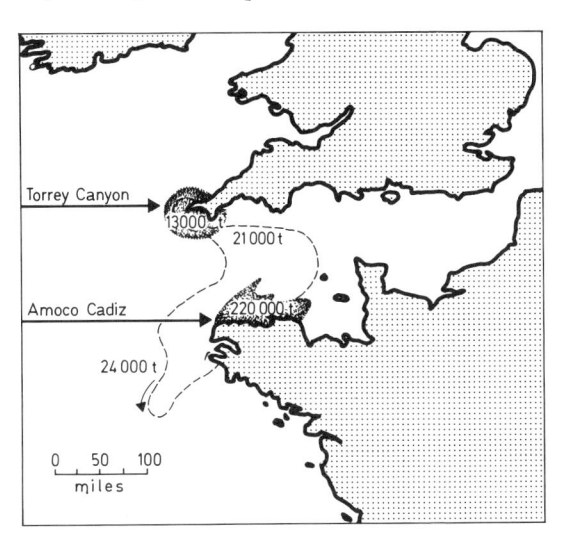

Fig. 46. On March 18, 1967, en route to Milford, the tanker "Torrey Canyon" went aground on a rift northeast of the Scilly Islands. The course of the slick which escaped from the wreck on different days is indicated on the map. Also shown is the area influenced by the accident of the tanker "Amoco Cadiz" which went aground at Portsall, Brittany, on March 17, 1978

At first, this huge quantity moved in the direction of the coast of England. Literally at the last moment, however, the wind shifted and blew from northerly and northeasterly direction for the next 30 days. The oil was driven toward the Bay of Biscay and only a small amount, less than 300 t, reached the coast of Brittany between Pointe du Raz and Crozon 5 weeks later (Smith 1968).

On the other hand, the blowout on April 22, 1977, on the oil rig "Bravo" in the Ekofisk field in the central North Sea (Fig. 44) was comparatively harmless. A measuring device was stuck in the well and impaired the flow of oil. The crew tried to alleviate the damage, and to safeguard the well they pumped barytic sludge into the well, then set a blowout preventer on the hole. Unfortunately, it was put on incorrectly and not tightened securely. A mixture of sludge, oil, and gas shot out of the well 20 m above sea level at a temperature of 75°–90°C and formed a 30-m-high geyser. A specialist from Texas was successful in capping it on the fifth attempt and on the morning of April 30, the danger was past. If this operation had not succeeded, an off-center hole would have had to be drilled in order to alleviate the pressure.

During those 8 days, about 13,000 t of oil and 19,000 t of gas escaped. Only 750 t of oil could be removed from the sea surface with mechanical tools. Chemicals to disperse the oil were quickly available, enough for 4000 t of oil within 36 h, however dispersants were not used. There was no continuous oil slick on the sea surface, but smaller patches which developed quickly into water-in-oil emulsions with 71% water, The emulsion disintegrated one day later, 32 km away from the site of the accident. Oil was visible, at least as a film, over an area of 3000 km^2. It started to drift in easterly direction, then went 120 km to the north, until a northerly wind drove the oil nearly back to the "Bravo"-platform. Estimates are 100 t of oil left at the sea surface after 45 days, drifting about 120 km south of "Bravo". The oil disappeared in mid June in southwesterly direction and has not been observed since then. Drift charts thrown into the sea at the site of the accident were washed up after 100 days on Dutch and German shores, but no oil.

No damage to marine life has been reported. Fortunately, there were not many sea birds in the area and the mackerel, whose eggs float right under the surface, had not yet spawned (Bourne 1977; Auduson 1978; Mackie et al. 1978).

On March 16, 1978, a hydraulic pipe in the steering engine of the 233,000 t tanker "Amoco Cadiz" from Liberia broke, and the ship was out of steering control 25 nautical miles off Britanny, France. If the captain had asked for assistance, nothing serious would have happened, because the tugboat "Simson" was only a few hours away. But much time was wasted with phone calls between the captain and his company in Chicago, before the German salvage tug boat "Pacific" was contracted. She was not strong enough to tow the supertanker in a Beaufort 9 gale. The vessel ran aground 1.5 nautical miles away from the small fishing village Portsall (Fig. 46). The "Amoco Cadiz" contained 223,000 t of light crude oil (30%–35% aromatic compounds), and 4000 t of bunker fuel. Within 11 days after the wreckage about 90% of the oil was lost in the sea, because all attempts to unload the cargo failed. The vessel broke and was bombed at the beginning of April in order to release the rest of the oil.

Estimates are that 40% of the oil evaporated (90,000 t). Onshore winds brought the smell of oil to large areas of Britanny, and not only the men fighting the oil, but people further inland complained of dizziness, headaches, and nausea. Birds flew away. During the days of the oil accident, westerly winds prevailed and transported the brownish water-in-oil emulsion, the "chocolate mousse", toward the shores. Estimates are that 80,000 t of oil landed at the shore. Later, southerly winds drove the oil toward the Channel Islands. Chemical dispersants, mostly BP 1100, were sprayed onto the oil from 6 British vessels (750 t of dispersants) and from 30 French vessels (1300 t of dispersants). Mechanical devices to collect the oil from the sea surface had only limited success (Spooner 1978).

During the first 17 days of the oil accident, 72 km of shoreline were contaminated. Estimates are that at the end of April 10,000 t of oil were left on the shore, but the oiled shores were distributed over 320 km of coastline. 6000 men, mostly military personnel and up to 1500 volunteers worked with the oil and removed about 100,000 t of material, mostly sand and seaweed, including about 25% oil. By June, 1978, before the tourist season started, most beaches appeared to be clean. However, in many cases, the view was deceptive, because a considerable amount of oil had been buried under the sand. There were 1.5 million tourists less than normal, the oyster beds were ruined. But at the end of May seaweeds could be harvested, and fisheries were never seriously impaired. The biological damage to the shore biota was considerable; what long-term damage was caused, the future will tell (Hann et al. 1978).

Only 5 weeks after the "Amoco Cadiz" accident, on April 28, 1978, a near-accident happened in the English Channel. The Kuwait tanker "Al Fahia" with 267,000 t of crude oil bound for Wilhelmshaven, happened to sail on the wrong side of the seaway. The vessel could be warned by British sea police when it was only 2 nautical miles away from the dangerous Fall Bank. It is said that a failure in the giro compass was reason for the navigational error (Anon 1978).

Since then many tanker accidents have occurred, and some accidents with oil platforms. The largest spill ever recorded was the blowout of the oil rig "Ixtoc I" in the Gulf of Mexico, off the state Campeche, Mexico, on June 3, 1979. During the first period about 4000 t of oil per day went into the sea. Efforts to stop the oil resulted only in a reduction of the flow, and when finally after 9 months, on March 24, 1980, the hole was closed with a 30 t concrete top, approximately 0.5 million t of oil had polluted the area. The damage was considerable, and also the money spent in clean-up measures. However, in October 1980 newspapers recorded that the recovery of the Gulf of Mexico from the oil pest was quicker than expected, and fisheries and tourist industry are more or less normal again.

5.4 Effect of Oil on Marine Life

To date, very little is known about the effect of petroleum components that reach the sea via the atmosphere or via rivers. The possibility cannot be excluded that the fairly substantial amounts of oil that reach the sea through effluents are responsible for the expansion of an impoverished, altered coastal environment and the disappearance of sensitive types of organisms close to the outfalls. These are also the observations that are being made in the immediate area around oil refineries in coastal regions (see Chap. 2.8). These are, however, local manifestations, and it is difficult to demonstrate that oil alone is the main reason for such environmental deterioration, because normally other stress factors act simultaneously. One should be specially careful with generalizations, because it has not been possible, up to now, to demonstrate adverse effects upon marine life in the surrounding of natural submarine petroleum seeps, as they occur in the Santa Barbara Channel (see Chap. 5.3), or in areas which are chronically polluted from oil exploration (Lake Maracaibo, Venezuela, coastal areas of Louisiana, U.S.A.). It is possible that organisms living under chronic petroleum stress adapt to these conditions and are more tolerant of toxic effects of oil than organisms from unpolluted areas (Mertens 1977).

Numerous studies have been conducted to learn about the toxic effect of oil on marine organisms. There are, however, major experimental problems when complex oil samples are investigated, as they are represented by all kinds of crude and refined oil. Stirring oil with seawater results in a labile system with particulate, colloidal and dissolved hydrocarbons. The different toxic components of oil are volatile in different degrees. Therefore toxicity experiments yield quite different results, when undertaken with freshly prepared oil-seawater systems, or with an aged medium where some time was available for volatile compounds to evaporate from the system, or to transform in another manner (Figs. 47 and 48). Toxic components with few carbon

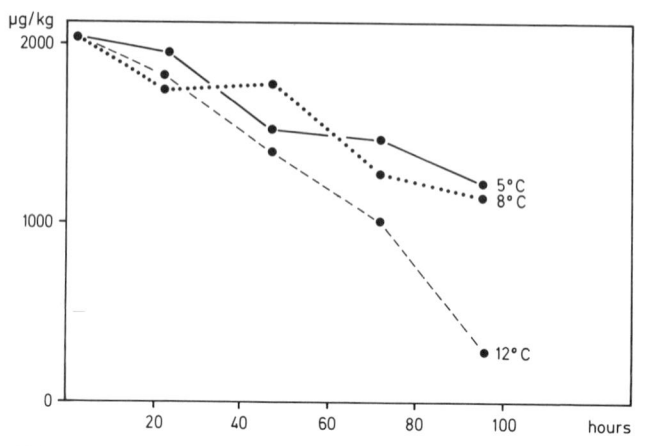

Fig. 47. Concentration of total aromatic hydrocarbons, derived from the water-soluble fraction of a crude oil, in seawater after different periods of aging. No aeration was applied. The disappearance of aromatic compounds from the system is faster with higher seawater temperatures (Cheatham in Rice et al. 1977)

Table 15. Toxicity of different aromatic hydrocarbons to marine animals. Results are from 96-h experiments (96 h LC_{50}) and indicate, at which hydrocarbon concentration, in mg/l, 50% of the experimental animals are killed (Neff in Rice et al. 1977)

Compound	Polychaete: Neanthes arenaceodentata	Crustacea: Palaemonetes pugio
Benzene		27.0
Toluene		9.5
Trimethylbenzene		5.4
Xylene		7.4
Naphthalene	3.8	2.4
Methylnaphthalene		1.1
Dimethylnaphthalene	2.6	0.7
Trimethylnaphthalene	2.0	
Phenanthrene	0.6	
Methylphenantrene	0.3	

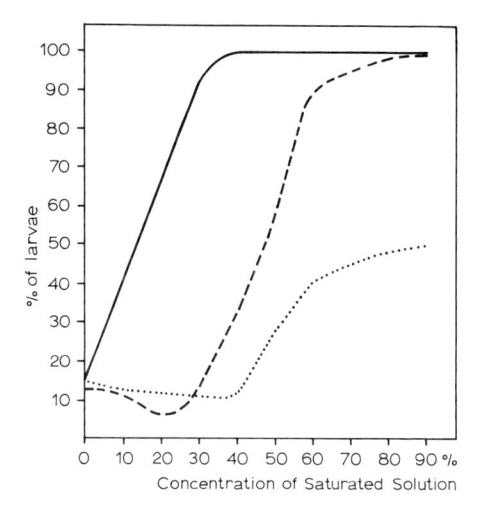

Fig. 48. In toxicity experiments with the water-soluble fraction of fuel oil (type: Exxon Bayton no. 2) nauplius larvae of the barnacle *Balanus eburneus* are impeded so that instead of swimming in the water they sink to the bottom of the experimental vessel. Toxicity is strongest with a fresh extract produced by stirring 1 part of fuel oil in 8 parts of seawater. This results in 21 mg/l of soluble compounds in the seawater (saturated solution, see *unbroken line* in the graph). Toxicity is much lower when after producing the solution this is kept in a sealed vessel in a refrigerator for 24 h *(broken line)* and even lower, when the solution is stored for 3 days in an unsealed vessel, so that evaporation can act *(dotted line)*. The graph shows how many larvae sedimentate at the bottom of the experimental vessel after 1 h of exposition. Results are similar when instead of fuel oil water-soluble extracts of naphthalene, toluene, or benzene are investigated (Blundo 1978)

atoms are very volatile and disappear quickly from the system, as do the bicyclic and tricyclic aromatic compounds. Particularly toxic are naphthalene, methylnaphthalene, phenanthrene, and trimethylbenzene (Table 15). Especially high amounts of naphthalenes reach the seawater when diesel oil or similar refinery products are introduced into the sea. Only 170 mg/l diesel oil mixed with seawater leads to an increased activity of the enzyme benzo(a)pyrene-mono-oxigenase in the fish *Blennius*, a sublethal reaction which, by the way, demonstrates that the cancerogenic component benzo(a)pyrene could have been formed within the fish (Kurelec et al. 1977).

It is well-documented that even 1 mg/l of oil dispersed in seawater or 1 µg/l of water-soluble oil components can harm sensitive organisms. For example, they may impede that the larvae hatching from fish eggs are healthy (Fig. 53; Hyland and Schneider 1977; Kühnhold 1977). When a fairly large oil slick covers an ocean area, and when fish eggs drift near the surface, it can be expected that fish larvae hatched will die. This was observed when on December 15, 1976, the tanker "Argo Merchant" with 29,000 t of fuel oil went aground on Fishing Rip, 29 nautical miles from Cape Cod, U.S.A. Up to 250 µg/l dispersed oil components were analyzed in the seawater, and drops of oil in zooplankton organisms. Mortality of the eggs of cod *(Gadus morhua)* and pollack *(Gadus pollachius)* was ascertained (Grose and Mattson 1977).

Oil components in seawater interfere with sex behavior of marine animals (Table 16), and they may have an effect on chemical orientation of marine organisms.

Table 16. Even in concentrations in which crude oil shows no damaging effects in experimental situations, subtle effects can influence the reproduction of organisms. In precopulation, the male and female amphipodes clasp each other for a fairly long period of time and only complete their copulation during the next moulting procedure. If the partners are separated during precopulation, they usually find their way back to each other quickly. But in seawater into which very small amounts of Venezuelan crude oil have been introduced, this behavior patter is disturbed in *Gammarus oceanicus* (Lindén 1976)

Concentration of oil	Number of pairs out of 10 that precouple after separation Period of oil effect in days			
	1	2	4	8
0 (Control)	10	10	10	10
1	10	10	10	10
10	0	0	2	5
20	0	0	0	0
40	0	0	0	0

If *Pachygrapsus crassipes* crabs are exposed to a water-soluble extract of crude oil for 24 h, they do not react as usual, with quick movements of their mouth appendices to the amino acid taurine which usually signals food to them. On the other hand, it is said that oil-soaked rags and bricks attract lobsters into traps as well as any other bait, and there are experiments which show that the perception of chemical stimuli is interfered with by petroleum components, not only in predator crabs and snails, but also in mobile bacteria.

Signal substances which are synthesized in an organism and released into the water in order to inform another organisms of the same species are called pheromones. Watersoluble extracts from crude oil cause the pheromones produced by the female *Pachygrapsus* crab to be ineffectual and they do not generate seeking movements and the mating position as is usually the case when they reach the male (Takahashi and Kittredge 1973). In some cases these animals must be kept up to 6 days in clean seawater before the pheromone effect is normal again. Instead of using crude oil, the same inhibiting effect can be temporarily achieved by using various monocyclic aromatic compounds; it can be longer-lasting using naphthalene, anthracene, and other polycyclic aromatic hydrocarbons at concentrations of $1-10\,\mu g/l$. It is not yet known whether or not the hampering effect of oil in chemical signaling plays an essential role between marine organisms when petroleum enters the seawater chronically or through marine accidents (Fig. 49).

It is of economic importance when small traces of petroleum components are responsible, fish and mussels acquiring an oily taste and being rejected by the consumer. The taste can only be eliminated after the fish and mussels have been kept in a tank of clean seawater for a long time, but the process is not dependable. Fortunately, com-

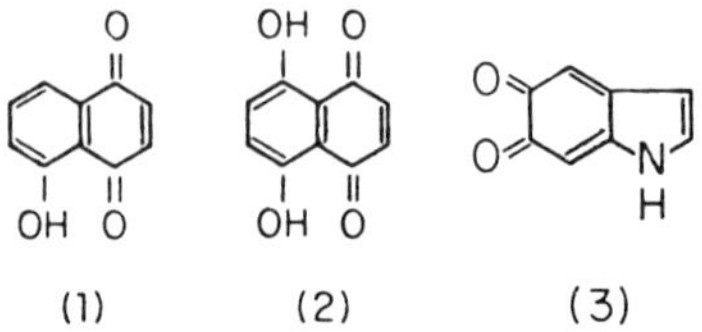

Fig. 49. Various naphthochinones are known to act in chemical signal transfer: (1) juglone, (2) naphtharazine, (3) indole-5,6-quinone. In theory they could be disguised or inhibited in their performance by similar compounds released from petroleum into seawater. Especially the refined petroleum products are rich in olefinic compounds (Kittredge et al. 1973)

pared to the contributions by man from other sources, the amounts of cancer-causing polycyclic aromatic hydrocarbons that can be transfered to fish and other marine edibles via petroleum seem to be insignificant (see Chap. 5.2).

Of greater significance are the adverse effects that marine bird life suffers from oil pollution. Right after World War I, the Royal Society for the Protection of Birds in Great Britain began a study of oiled birds on the coasts. As a result, the International Conference on Oil Pollution of Navigable Waters, at which 14 nations took part, came into being in 1926. This was the first attempt on an international level to achive a reduction of oil pollution by ocean traffic. Finally, the 1954 IMCO convention on oil pollution (see Chap. 10) also goes back to the activity of British ornithologists in the British Advisory Committee on Oil Pollution of the Sea.

There have been numerous surveys about the number of birds that die annually as a result of oil pollution (Bourne 1976). In the period of 1950, there were said to have been 50,000–250,000 sea bird casualties annually in the area of the British Isles. The figures for the North Sea and the North Atlantic are now estimated at 150,000–450,000 annually. An observation network has been organized along many European

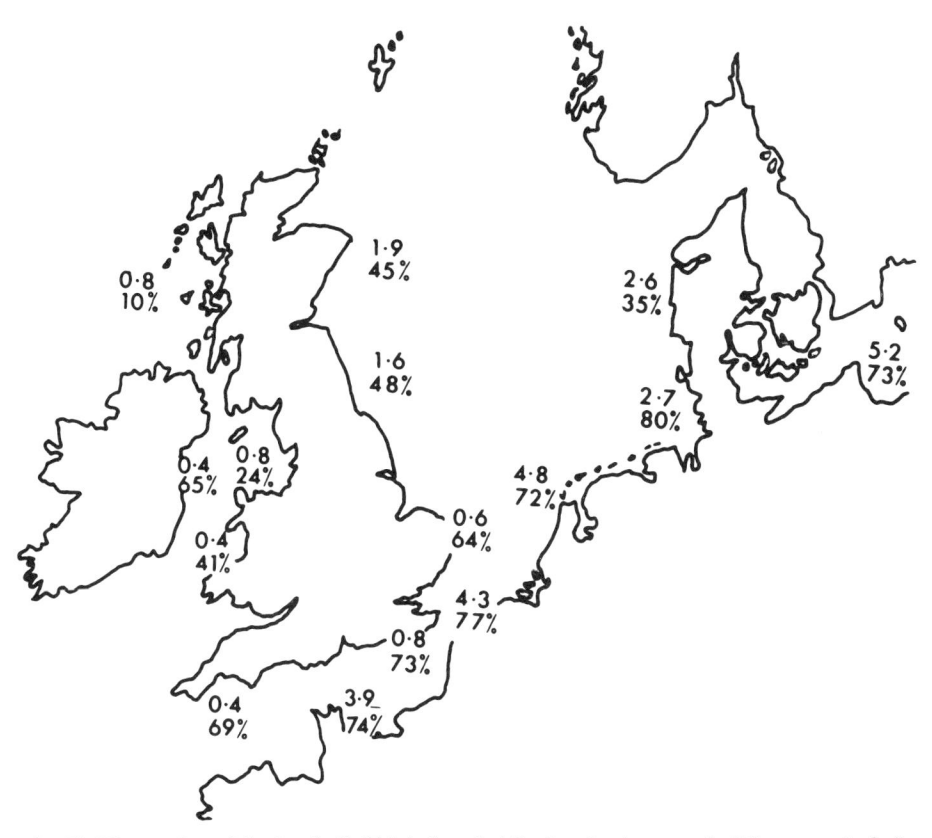

Fig. 50. The number of dead and oiled birds found at the beaches is a record of the amount of oil pollution of the sea. The map shows the situation around the British Islands and in the North Sea, 1967–1973. Figures are the average number of birds found dead per km of shoreline, and the percentage of dead birds which show distinct traces of oil (Bourne and Bibby 1975)

Table 17. It seems that from 1970 onward the North Sea has become cleaner in regard to oil pollution; the number of birds killed by oil accidents and other release of oil on the Danish west coast and in the area of the Danish Islands declined in this period (Joensen and Hansen 1977)

	Cases of oil pollution causing the death of		
	more than 100 birds	more than 1000 birds	more than 10,000 birds
1968/69	6	2	1
1969/70	17	2	1
1970/71	7	2	1
1971/72	9	2	1
1972/73	4	2	1
1973/74	2	1	–
1974/75	6	1	–
1975/76	4	–	–

coastal areas and once a winter a count is taken to ascertain how many dead birds have been washed up on shore and how many of them show distinct signs of oil (Fig. 50). But only 10%–25% of all oiled birds wash up on shore; many sink to the sea bottom. Between 1958–1962, 12 oiled birds per km of beach were counted annually along the coast of the Netherlands, 47 between 1963 and 1968 (Parslow 1971). Since then, at least in Denmark, the count seems to have gone down. It is hoped that this is already the result of increasingly strict regulations on ocean discharging of oil (Table 17), but figures reported in winter 1980/81 were high again; according to newspaper information more than 60,000 birds were killed by oil in the Skagerrak.

Sea birds landing on an oil slick absorb oil in their feathers. Water then penetrates their insulation, which is normally insured by their feathers, and the birds freeze or drown because they lack buyocancy. When they try to clean their feathers with their beak, they ingest toxic oil components into their digestive tract. When breeding birds get oiled feathers, toxic components from the oil may penetrate the egg shell. Not more than 0.02 ml of fuel oil on the surface of the egg of the eider duck, *Somateria mollissima,* are necessary to prevent breeding success (Albers and Szaro 1978).

How great a toll sea birds actually pay for oil pollution in relation to their breeding potential cannot yet be determined because accurate data for the total number of North Atlantic birds is not available. It is also not yet known to what extent damage to the reproduction of bird populations is done as a result of pollution by chlorinated hydrocarbons (see Chap. 9.4). It is probable that at the very least the reduced numbers of puffins *(Fratercula arctica)* and guillemots *(Uria aalge)* in Brittany and in southwest England stems from oil, likewise the reduction in the number of old-squaw ducks *(Clangula hiemalis)* that spend the winter on the Baltic Sea.

Wherever an oil accident occurs, bird-lovers make efforts to clean oiled birds in order to rescue them. In 1967 as a consequence of the "Torrey Canyon" accident, about 8000 birds were cleaned, mostly guillemots and puffins. However, most of the treated birds died, either from the oil or from the cleaning procedure. Therefore, in November 1967 the University of Newcastle-upon-Tyne (England) established a Research Unit on the Rehabilitation of Oiled Sea Birds (Clark 1978b). In the U.S.A. the International Bird Rescue Research Center of Berkeley was established (Int. Bird Rescue Research Center 1978). Both research units advise the use of detergents, not solvents;

both think that the most important item is shelter, warm water, and appropriate food, conditions which can only be provided by professional units which have made preparations to deal with oiled birds before an oil accident occurs. But even with such preparations the percentage of birds rescued will be very low; amateurs should not try to do better.

The most important damage from oil accidents occurs if oil or "chocolate mousse" are driven by wind and tides onto the shore. The effects are different according to the type of shore. In principal, oil will be physically broken into minute particles on wave-exposed beaches, and the particles will accumulate in protected areas, like other detritus and clay material. In protected areas without heavy wave activity, the largest oil amounts are deposited right at the high water line. The oil remains stranded if the wind is onshore, and the next tide brings more oil in. The coastal areas below the high tide level do not suffer so much from the oil because the water enters between sediment and oil slick and tends to float the oil free, so that relatively little oil becomes incorporated into the sediment. According to the different type of coast and the different wave exposition, different prognosis for the effect of an oil accident have been worked out (Fig. 51, Table 18).

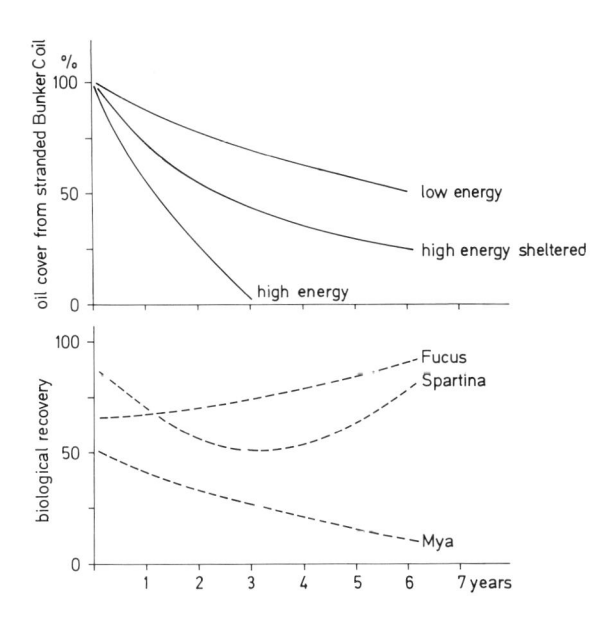

Fig. 51. On February 4, 1970, the Liberian tanker "Arrow" sank in Chedabucto Bay on the Atlantic coast of Canada; about 10,000 t of fuel oil (Bunker C) were released into the sea, about 1700 t polluted the beaches of the region. Cleaning operations were intensive, nevertheless even after 7 years considerable remnants of the oil are to be found. From such observations, an "oil spill vulnerability index" has been worked out (see Table 18). Microbial breakdown after the "Amoco Cadiz" accident was much faster (Atlas et al. 1981). Even during the early days of stranding when evaporation of volatile compounds was rigorous, the percentage of straight alkanes declined probably due to microbial degradation. After 9–16 months more or less persistent components were left. But here and there patches of rather fresh oil polluted the environment which came up from the deeper anoxic strata of the beach sediment, and it is feared that such contamination will continue for years (Vandermeulen 1977; see also p. 78)

Table 18. From the scientific investigation of a number of oil accidents a classification is derived in order to estimate what damage can be expected in case an area should be hit by oil drifting ashore from an oil accident. Vulnerability increases following a scale from 1 to 10 of an "oil spill vulnerability index" (Gundlach et al. 1978)

1	Exposed rocky cliffs	Under high wave energy, oil spill clean-up is usually unnecessary
2	Exposed rocky platforms	Wave action causes a rapid dissipation of oil, generally within weeks. In most cases clean-up is not necessary
3	Flat fine sand beaches	Due to close packing of the sediment, oil penetration is restricted. Oil usually forms a thin surface layer which can be efficiently scraped off. Clean-up should concentrate on the high tide mark, lower beach levels are rapidly cleared of oil by wave action
4	Medium to coarse-grained beaches	Oil forms thick oil-sediment layers and mixes down to 1 m deep with the sediment. Clean-up damages the beach and should concentrate on the high water level
5	Exposed tidal flats	Oil does not penetrate in compacted sediment surface, but biological damage results. Clean-up only if oil contamination is heavy
6	Mixed sand and gravel beaches	Oil penetration and burial occur rapidly, oil persists and has a long-term impact
7	Gravel beaches	Oil penetrates deeply and is buried. Removal of oiled gravel is likely to cause future erosion of the beach
8	Sheltered rocky coast	The lack of wave activity enables oil to adhere to rock surfaces and tidal pools. Severe biological damage. Clean-up operations may cause more damage than if the oil is left untreated
9	Sheltered tidal flats	Long-term biological damage. Removal of the oil nearly impossible without causing further damage. Clean-up only if the tidal flat is very heavily oiled
10	Salt marshes and mangroves	Long-term deleterious effects. Oil may continue to exist for 10 or more years

After the "Amoco Cadiz" accident in Brittany, only 8% of limpets *(Patella)* and periwinkles *(Littorina)* survived on heavily oiled rocky coasts, while barnacles *(Balanus)* were, even after 4 months, not much affected, because they live on a deeper level of the intertidal zone. On tidal flats, the lugworm *(Arenicola)* was much more resistant than the cockle *(Cardium = Cerastoderma)*. Experts were puzzled to find, on April 1, 1978, about 2 weeks after the accident, millions of dead heart urchins *(Echinocardium)* and hundreds of thousands of dead razor clam *(Ensis)* at the beach of St. Efflan, 95 km away from the wreck. These animals normally live in deeper water or at the low water mark. Apparently they could not have been damaged by oil covering the sea, but by toxic oil components in the water column or in the sediment (Chasse 1978); again it is interesting that some sensitive species suffered from oil, others did not.

Damage from oil accidents is most severe in narrow fjords and inlets without strong wave activity. Up to now no research reports are available about the long-term effects of the oil of the "Amoco Cadiz" on the fjords of Britanny. Immediate damage was specially severe close to the high water mark in the area which is preferred by the young bottom stages of lugworms and clams. In some salt marsh areas the oiled sediment had to be exchanged completely, which means a total devastation of the area. Salt marshes are known to be specially vulnerable (Table 18), and sometimes the effect of oil ist most severe after as long as 2 years. In Chedabucto Bay, Canada, the salt marsh recovered only 6–7 years after an oil spill (Vandermeulen and Ross 1977). Oil trapped in the clay sediment can exert its toxic effect for many years and prevent recolonization. It seems that terrestrial plants, including salt marsh vegetation, is much more sensitive to oil than algae, because oil components immediately penetrate their tissues (Cowell 1978).

On rocky shores, which have been cleaned with detergents after an oil spill, recolonization is comparable to the recolonization one observes on a nude rock. The first organisms which appeared on the Cornwall shore after the "Torrey Canyon" accident in 1967 were green algae of the genera *Ulva* and *Enteromorpha*. They covered the entire region because the limpets and periwinkles which usually eat away all algae seeds had been killed by the oil and by the subsequent cleaning procedures. In the autumn of 1967 brown seaweed *(Fucus)* settled and grew to such an extent that the barnacles, which up to this period had survived successfully, perished under the seaweed carpet. In 1968 limpets started to immigrate from the low water level to the higher zones and by their grazing activity prevented further settlement of algae seedlings. From 1970 to 1972 the carpet of brown seaweed became thinner and thinner and finally disappeared, because any new settlement was prevented by enormous numbers of limpets. When finally the limpets could not find enough food, their population declined to normal numbers, and barnacles had a chance to settle. Since 1975 they have dominated the region, as they did before the oil accident (A.I. and E.C. Southward 1978).

At Cornwall there was a small cove oiled from the "Torrey Canyon" accident, which was left without cleaning operations. Rock and organisms were covered with a 1–2 mm oil layer, but one month after the oiling some barnacles had managed to clean an opening through the oil, and took up filter feeding activity. Mussels *(Mytilus edulis)* seemed to survive, but limpets were gone. In 1968 about half of the barnacles were dead, some limpets had returned. After 2 years all oil had gone without any treatment, and there was never an algal bloom on the rocks.

If one cleans a rocky coast after an oil spill the result is dubious: one gets a carpet of slippery green algae. It can therefore be good advice to leave the oiled coast alone and to restrict cleaning operations to such coastal areas where tourists would be molested by the oil.

Sandy beaches sometimes have a good capacity of natural self-cleaning. When on May 12, 1976, the tanker "Monte Urquiola" stranded a few miles from La Coruña, Northern Spain, and 30,000 t of crude oil spoiled the coast, some sandy beaches were covered with a 30 mm layer of oil which persisted for weeks. The meiofauna, consisting of microscopical worms and crustacea living in the pores between the sand grains, was nearly completely killed. One year later, however, the beaches appeared clean, and an abundant meiofauna had recolonized the beaches (Giere 1979). As most tourist beaches of the German North Sea coast are of fine to medium sand and are wave-exposed, there is reasonable hope that waves will take the oil away after some period of time, and that the damage will at least not be catastrophic in this areas, if an oil accident should happen.

5.5 Combating Oil Pollution

Everyone agrees that every possibility for reducing the introduction of hydrocarbons into the ocean must be exhausted. In the final analysis, petroleum is too valuable a source of energy to waste by fouling the ocean with it. The increasing price of oil is stimulating ideas to effect the more efficient burning of hydrocarbons as an energy source and the recycling of waste oil. Ocean pollution by petroleum will obviously be halted when the source is exhausted. But with around 53×10^9 t, only a fraction has been used to date. The available reserves are greater by far.

The concrete question arises: what measures are to be taken so that an oil spill does not occur, be it a tanker collision, a harbor mishap, or a drilling rig accident? Better technical precautions, better developments in ship building, but above all better navigation are the most effective measures for preventing accidental oil pollution. When we hear that badly trained officers are working on poorly equipped and badly maintained tankers registered in countries with substandard maritime safety regulations, we have to conclude that there is much room for improvement in these areas (Krüger 1977).

After the "Amoco Cadiz" accident in March 1978 there were discussions in all coastal nations what could be done to prevent oil accidents. France made a regulation that tankers on the way to the English Channel should keep a greater distance to the coast of Brittany. The Federal Republic of Germany regulated that all tankers above 10,000 gross tons bound for German North Sea ports have to take a pilot at light ship "Deutsche Bucht" far away from the shore. All vessels above 300 gross tons carrying cargo have to take a pilot for the dangerous passage through the shallows of the Jade, Weser, or Elbe.

Such measures against the risk of oil pollution can be organized without world-wide conventions, on a national or regional scheme. In principle, the harbor captain of the port of Wilhelmshaven, if he is backed up by other authorities, could refuse to let substandard tankers berth at Wilhelmshaven oil pier. But in general, it is certainly reasonable that international agencies like IMCO work for better standards of construction, equipment, and crew of tankers (see Chap. 10).

But when oil is spilled, what is to be done? This question is taken seriously by all countries with a merchant marine fleet and by the scientific community. But the answer to the question has not yet been found. A precise definition of the purposes to be served by the measures taken to combat oil spills is part of the question.

If it could be determined that the wind was not going to drive it into a coastal area and the sea area were not frequented by marine birds, an oil spill would not need to be combated. Looked at soberly, an oil slick on the open sea is of little significance. Tar, however, will turn up later on a beach.

Little can be done for fish eggs and plankton in the water under the oil slick. They are harmed mainly by the toxic hydrocarbon components of the oil which are released into the water during the first few days after an oil spill.

If the oil slick moves toward a coast, the oil should be fought. Unfortunately the many different measures to fight oil on the sea have not been completely successful in rough weather on the open sea. One reason for this may be the lack of convincing technical concepts, but certainly another reason is the lack of funds to really build up effective oil-fighting units and oil recovery devices. Estimates from the U.S.A. are that for each 150 km of coast line 20 million US $ investments and 2 million US $ annual maintenance costs are necessary, which adds up to 250 million US $ for all U.S.A. coasts. This seems to be the amount of money spent for just one oil spill, the "Amoco Cadiz" accident, if preliminary estimates of clean-up operations and remuneration for damage are correct. However, a cost-benefit calculation is difficult because even with effective oil spill fighting precautions there is no 100% safety, and after some oil spills some oil will reach the coast on spite of all efforts. In 1977, the price to remove 1 t of oil from the sea surface, chemically or mechanically, was between 10 and 50 US $ (Fig. 52; Holmes 1977).

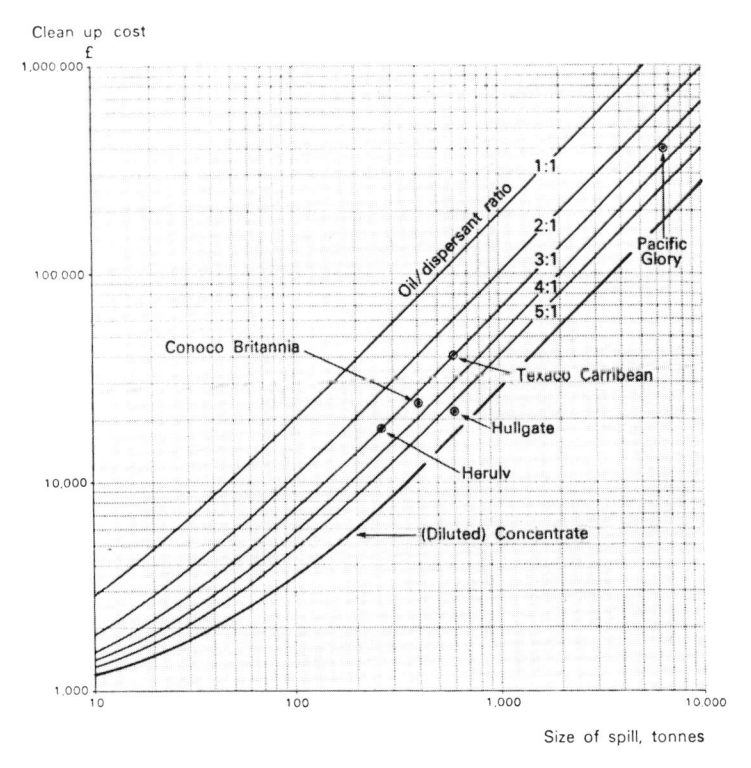

Fig. 52. Evaluation of the costs to remove oil from the surface of the sea following different oil spills, and a generalization regarding utilization of conventional dispersants (oil-dispersant ratio 1:1 to 5:1) or of concentrates diluted comparable to an oil-dispersant ratio 3:1. Expenses for transport to the site of the spill have not been included in the calculations (Department of the Environment 1976)

On the first day of the "Torrey Canyon" accident, in 1967, attempts were made to disperse the oil with dispersants. This reaction was understandable in view of the danger that the tanker's whole load might spread out along the coasts of the English Channel and possibly penetrate into the North Sea. So that the beaches would be fit for the approaching summer tourist season, detergents were used in large quantities to clean up the fouled coasts of Cornwall. In all, 10,000 t of dispersants and detergents were used to combat approximately 14,000 t of weathered oil. Biologically, more damage was done by the dispersants applied than would have been done by the oil. The dispersants applied in the "Torrey Canyon" accident were fairly poisonous. "BP 1002" was mainly used, a product with 12% nonionic detergents and 3% stabilizers in an aromatic solvent. The toxicity of "BP 1002" for subtidal organisms is 0.5–5 mg/l expressed as 24 h LC_{50}, the concentration which kills half of the experimental animals within 24 h. Toxicity for intertidal animals is 5–100 mg/l. In practice all animal life was killed where beaches had been cleaned with such types of dispersants (A.I. and E.C. Southward 1978).

As the consequence of such experience the Federal Republic of Germany decided not to use detergents in case of an oil spill, except in a few situations. According to the Helsinki Convention (see Chap. 10) dispersants are not allowed in the Baltic. In the U.S.A. they were only allowed to fight oil spills of more than 30 t on water more than 150 m deep and never close to a shore (Boesch et al. 1974).

In the meantime, however, the question of chemicals for fighting oil spills is again under discussion. Dispersants have been developed which are only about 1/1000th as toxic as "BP 1002", for example "BP 1100" and "Corexit 7664", and for some of them the 48 h LC_{50} is as low as 10 g/l. In Great Britain such dispersants are licenced for sea use which have a 48 h LC_{50} of above 3.3 g/l, and the toxicity of an oil-dispersant mixture must not be higher than the toxicity of oil alone (Table 19). In 1975, on the coasts of Great Britain there were stocks of dispersants available to fight an oil accident of 35,000 t.

Dispersants are surface-active substances which enhance the formation of small, 1–5 μm diameter droplets of oil suspended in the seawater, so that the oil slick disappears from the surface of the sea. The oil droplets, under the influence of dispersants, do not stick to the surface of particles or of the sediment. At the beginning of the use of dispersants for fighting oil spills, about 1 t of dispersant was necessary to disperse 1 t of oil. Modern concentrates can be used 1:30. Further, it is no longer necessary to provide good mixing of dispersant, oil, and water; modern dispersants are rather self-mixing so that some may be applied from airplane or helicopter. Important is the exact timing: even if it is an open question to what extent dispersed oil can disappear from the water column by evaporation, it seems reasonable to allow the light, volatile fractions of the oil to evaporate before dispersal; then the explosion risk is avoided during operations. On the other hand, spraying should be done before a water-in-oil emulsion, the "chocolate mousse" is formed, because this is much more stable and more difficult to disperse than an oil slick. One should keep in mind that dispersal of oil following an oil accident is done not only by chemicals, but to a fairly large degree, depending on the weather, by natural processes, too, and that under favorable conditions small oil spills may be dispersed by the stirring effect of a motor boat's propeller (Moss 1971).

Dispersion of oil in the water column therefore seems to be a practicable method to eliminate the oil from the sea surface. Some experiments (Fig. 42) indicate that the dispersal of oil by chemicals enhances the microbial breakdown of oil in seawater. Britsh oil spill experts summarize their experience as follows:

Norton and Franklin (1980, p. 19):

"Current arrangements for licensing ensure that the dispersants available for use in dispersing oil at sea or on beaches have as low a toxicity as technically possible. Nevertheless, due to the high toxicity of some oils, dispersions resulting from the use of chemical dispersants may be toxic to marine life. At sea, where conditions favour the rapid dispersal of a chemically-dispersed slick, the probability of causing significantly deleterious effects on marine organisms is low and probably restricted to the more sensitive stages such as fish eggs or larvae. Only in shallow waters and where the oil is relatively fresh is mortality likely to occur in adult fish and shellfish, and to other forms of marine life. Spraying of intertidal organisms is likely to lead to some increased mortality of exposed organisms under most conditions."

Department of the Environment (1976, p. 10):

"Clean-up on shore is likely to be less complete, more difficult, more expensive, more ecologically damaging and slower than treatment at sea. There is thus a clear case in favour of the latter. Mechanical recovery from the sea has many features of the ideal system, but unfortunately it is only feasible at the present time in calm conditions, needs ancillary equipment and trained operators and involves high capital costs. For the time being dispersants provide the only really effective and generally applicable method of dealing with oil at sea. They are most effective and can be used in smaller quantities when applied to fresh oil, before "weathering" takes place. Concentrate dispersants increase the effective spraying time of boats and are considerably more economical than ordinary dispersant."

However, to utilize chemicals as dispersants against oil on the sea surface is to fight the devil with Beelzebub, as the German saying goes. Up to now authorities in the Federal Republic of Germany are reluctant to organize large-scale oil spill fighting equipment based on chemical dispersants, and they are encouraged by public opinion which at this moment, 1981, is very much against any introduction of chemicals into the sea; certainly dispersants have to be ranged under pollutants. Experimental evidence regarding the toxicity of oil and oil-dispersant mixtures to marine life are not conclusive, and contrary to the British results from short-term toxicity experiments (Table 19) there are contradicting results from long-term studies (Fig. 53). It seems necessary to spend more scientific effort on the question, unbiased by industrial pressure or emotions, to have better arguments regarding the conflict of damage by dispersants in the open sea versus damage of oil on the shores.

Other chemicals used to fight oil spills are "herders" which counteract the spreading of the oil at the sea surface and have the effect that oil creeps together to form larger, water-repellent units, and "sorbents" which bind the oil by capillarity and adsorption. Polyurethane foam may take up oil 30 times its own weight, and can be reused after pressing the recovered oil out. Straw is a sorbent that takes up oil about five times its own weight. At the "Torrey Canyon" oil spill, in 1967 the French Government used pulverized blackboard chalk, to sink the already aged oil that approached the coasts of Brittany. By the use of 3000 t of chalk, approximately 30,000 t of oil were eliminated from the sea surface. However, is has been reported that in 1969 oil

Table 19. In the Fisheries Laboratory, Burnham-on-Crouch, England, a device has been developed to investigate the toxicity of oil and of oil-dispersant mixtures under realistic conditions, using special 18-l seawater experimental vessels, and shrimps *(Crangon vulgaris)* as experimental animals. The percentage of dead shrimps is identified after experiments of 100 min in 1 ml/l of fresh Kuwait crude oil or in a mixture of 1 ml/l oil plus 1 ml/l of the dispersant to be tested. Dispersant concentrates are diluted to 10% for the experiment. Beach tests were done by spraying limpets *(Patella vulgata)* with oil or with dispersants. An asterix means that the toxicity of the dispersant or oil-dispersant mixture is significantly greater than the toxicity of the oil control. In Great Britain such dispersants are licensed for use only in the case of oil spills which do not increase the toxicity which the fresh oil alone exerts (Norton and Franklin 1980)

Dispersant	Oil and dispersant premixed before addition to test tanks		Dispersant added to oil film before agitation		Beach test Limpets sprayed with	
	Oil alone	Oil and dispersant	Oil alone	Oil and dispersant	Oil	Dispersant
Conventional						
Agma EP 540	–	–	5	0	40	40
Ameroid Oil Spill Dispersant/LT	60	65	–	–	65	25
Applied 8-40	–	–	80	20	65	70
Atlan'tol 3211	–	–	70	90*	65	100*
Atlan'tol 3211/E	–	–	80	5	65	100*
BP 1100X	65	45	10	10	35	15
Corexit 8354	60	35	–	–	65	95*
Dasic Slickgone LT2	40	35	10	15	80	35
Emkem Spillwash LT	35	30	–	–	80	75
Emulsol LW	40	15	–	–	80	60
Finasol OSR-2	60	60	–	–	65	85*
Finasol OSR-3	–	–	70	15	65	90*
Gamlen OSR LT 126	60	45	–	–	65	85*
Kraken MC 563	65	35	–	–	80	50
Lankromul OSD	65	65	–	–	80	55
Petrocon Oil Spill Eliminator IV	–	–	5	15	40	95*
Rochem Oil Spill Remover (WSA)	–	–	10	10	50	55
Servo CD 2000	–	–	5	5	45	60
Shell Dispersant LTX	35	35	–	–	80	70
Teklene TC-48	–	–	40	45	45	20
Wellaid 311	–	–	65	95*	35	100*
X-3125	–	–	5	15	45	60
Concentrates						
Agma EP 559	–	–	5	15	40	75*
Atlan'tol AT7	–	–	55	80	65	95*
Atlant'tol AT7 floating	–	–	40	55	35	100*
Compound W1911	–	–	10	20	50	80*
Compound W1986	–	–	10	5	50	100*
Corexit 7664	–	–	10	65*	50	10
Corexit 9527	25	90*	30	75*	45	70*
Corexit 9600	40	70	55	70	95	90
Dasic Slickgone LTD	85	95	20	40	95	45

Table 19 (continued)

Dispersant	Oil and dispersant premixed before addition to test tanks		Dispersant added to oil film before agitation		Beach test Limpets sprayed with	
	Oil alone	Oil and dispersant	Oil alone	Oil and dispersant	Oil	Dispersant
Finasol OSR-5	35	95*	20	70*	80	90*
Finasol OSR-7	–	–	70	80	66	22
Gamlen Oil Dispersant LT	50	60	55	75	80	30
IMX 103	–	–	60	45	55	50
Leek A	–	–	45	80*	60	30
Leek B	–	–	30	65*	60	20
Nokomis-3 Conc.	50	70	20	25	80	65
Quell-Oil CI	60	75	55	80	95	55
Seawash	–	–	15	15	50	65
Shell Dispersant Concentrate	60	90	30	55	50	15
Spillaway	–	–	35	30	35	100*
Surflow OW1	50	70	55	95	80	40
Synperonic OSD 20	45	55	55	60	40	50
Synperonic OSD 41	–	–	80	85	30	20
Value 100	–	–	50	50	65	5

clumps floated back to the surface, and the method is not to be recommended, because it seems that microbial degradation of the oil at the sea bottom is minimal. Clumps of oil also floated back to the sea surface, when the dredging vessel "Geopotes II" tried to eliminate 100 t of oil by sweeping a mixture of water and sand over the drifting oil slick (Farn 1976). To burn an oil slick at the sea surface is possible only during the first half hour after the oil has streamed out of a wreck, because later the highly inflammable light components of the oil have evaporated. Later one may add chemicals to the oil slick to facilitate ignition, but this method has drawbacks, and air pollution by smoke is generated.

In any case the best method to fight oil at the sea surface is mechanical clean-up. There are plenty of devices on the market, and many of them have been efficient with oil spills on inland waters, in sheltered coastal regions, or on calm days. But every time when it has been necessary to clean oil from a wave-agitated sea surface, there were difficulties. At the "Amoco Cadiz" accident, the ship "Chandis" was at hand equipped for mechanical oil recovery: within 2 h, 80 t of oil were sucked up, but later rough sea made further operations impossible.

In principle, one of the following methods could be applied for mechanical oil recovery: (a) adhesion: oil adheres to surfaces, especially if they are oleophilic. Apparatus with rotating disks or drums or with continuous bands etc. have been invented, from which the oil is mechanically stripped off; (b) sill: a narrow surface layer of seawater and oil runs down over a sill; oil and water must subsequently be separated further; (c) cyclones: a circular water movement is created which has a depression in the middle; oil is pumped from the depression; (d) suction: the sea surface layer is sucked

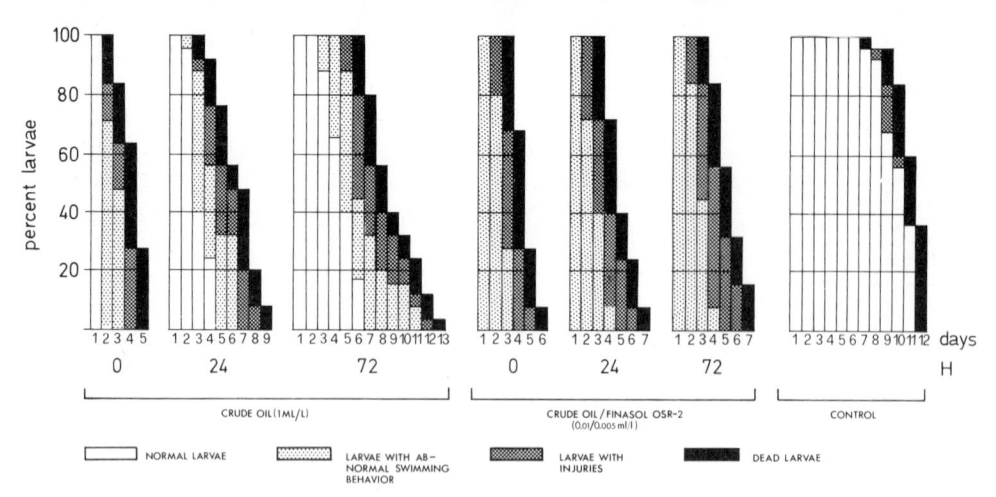

Fig. 53. Herring larvae *(Clupea harengus)* react in a sensitive manner to oil pollution. At first, they manifest abnormal movements when swimming (indicated in the graph by *light-dotted areas*). Later, the body appears deformed and the skin is damaged, in particular their fins *(dark-dotted areas)*. In the end, the larvae die *(black areas)*. If 1 mg/l of Venezuelan light crude oil without any additives is dispersed in seawater, the fresh dispersal (0) is much more toxic than a dispersal 24- or 72-h-old *(left diagram)*. If the relatively low toxic dispersant "Finasol OSR-2" is added to small concentrations of 0.01 mg/l of oil, then this mixture is much more toxic even than 1 mg/l of oil alone, and it retains its strongly toxic effect even after 72 h of aging. In controls, most larvae do not exibit abnormal behavior or damage over a period of 8 days *(right diagram)*. White *areas:* undamaged larvae with normal swimming behavior (Lindén 1975)

in, water and oil are subsequently separated. In principle, all these (and maybe other) possible methods work in connection with barriers or other devices to concentrate the oil at the sea surface before recovering it, and with separators to separate oil from surplus water recovered.

For the situation at the German North Sea coast, an oil-recovering device should meet the following standard: it should work up to significant wave heights of 3 m and wind speeds of 10 m/s (Beaufort 5-6), it should perform with oil quantities of 400 t per hour and it should be capable of coping with 20,000 t of oil within 3 days. The oil film will be between 1 and 3 mm, and usually one has to deal with light crude oil.

On paper, some constructions have been invented which claim to meet the qualification. For example Lühring Werft in Brake invented a twin hull 80 m vessel with both hulls linked at the rear end; both hulls can be swung around the link to open a triangle space between them. If the vessel moves downwind through an oil slick, the triangle provides an area of sheltered sea surface from which the oil can be recovered. However, the price is said to be about 40 million DM, and therefore, at the moment, only a small 34 m version of the so-called "Ölsau" is under construction, with 2.5 m draught and a capacity of 200 t of oil, and at a price of 5 million DM. Other oil-recovery vessels available at the German coast are much smaller, the Swedish-built catamaran "OSK 1" stationed at Cuxhaven has only a capacity to recover 4 t of oil per hour, and it cannot work at 1 m waves. The price was 2 million DM. For more than 10 million DM, in early 1981 the ex-offshore-supply-vessel "Ostertor", 56 m long, will be refitted to work as oil accident emergency vessel that can take over oil from a wrecked tanker, and can operate oil barriers and oil recovery equipment.

At the German North Sea coast, the Wadden-Sea area between the chain of low sandy islands and the dikes of the mainland, a region of tidal creeks, tidal flats, and salt marshes, is specially vulnerable to oil accidents. In case of an oil spill it would be a primary task to protect the Wadden-Sea by constructing oil barriers across the 1–3 km wide tidal channels between adjacent islands, in order to prevent oil from penetrating from the North Sea into the Wadden-Sea. Unfortunately oil barriers have a poor performance in streaming water: a tidal current of 1 knot or about 2 km/h is sufficient to carry the oil below any barrier of the types hitherto applied. Better systems should be invented.

At the "Amoco Cadiz" accident, 11 km of oil barriers were used, mostly inflatable constructions of the type "Sykore II", in order to protect oyster beds, ports, and estuaries. The results were poor, especially during the first weeks, and the experience is that an oil barrier is as good as the crew which maintains it, because oil barriers need constant supervision and repair. Further, an oil barrier is useless without a system to recover the entrapped oil and to avoid oil accumulations (Hann et al. 1978).

If all attempts to recover the oil slick from an oil accident or to prevent the oil from approaching a shore are unsuccessful, there is a final chance to remove the oil from the water surface immediately before it enters into the beach sediment or spoils a rocky coast, by laborious hand work, with buckets and shovels, by small tractor-driven vacuum trucks normally used in agriculture to pump and spread manure to the fields, by larger vaccum trucks or by skimmers. Cleaning of oiled beaches is done by bulldozers and other heavy gear, if the beach consists of fine hard sand, but is much more laborious in other coastal environments. On gravel and cobble beaches, trenches and pits are dug, in which the oily water collects after the beach has been treated by hosing with hot pressurized water jets (80°–140°C). Cleaning of harbor walls, rock surfaces, and individual cobbles and stones is done by hand, with hot water jet or steam, and with detergents. One man can clean, under good conditions, 500 m^2 per day. In some areas, oiled sediment and debris are transported away and either burnt or deposited. If beaches are important for tourist purposes, clean sand may be brought in to replace the oiled sediment.

Wherever an oil accident has polluted a coast, careful consideration is necessary to clean or not to clean the environment, to take the risk that cleaning operations may do more damage to nature, or to leave the situation as it is and to wait for nature to take care of the oil (Table 18). The Warren Spring Laboratory in the United Kingdom, which did considerable oil spill research, concludes (1972), that "from the point of view of protection of the environment, the best thing to do with oil pollution of the beach is to do nothing. With the climatic conditions and the types of sea around the British Islands, oil left alone will fairly rapidly become innocuous, or disappear altogether."

6 Radioactivity

6.1 Natural Background Radioactivity and Fall-out

In the marine environment, too, there is a natural basic exposure to radioactivity caused by radioactive potassium-40, rubidium-87 and the elements of the natural disintegration series from uranium to thorium. Added to these are radioactive hydrogen (^{3}H, tritium) and carbon-14, that develops through the influence of cosmic rays upon the atmosphere. The natural radioactivity of potassium-40 in the ocean is estimated at 5×10^{11} Ci [3].

In comparison to the natural background radioactivity, induced radioactivity caused by man in the ocean is still relatively low and estimated to be somewhat above 10^9 Ci (Preston et al. 1971). The radioactivity of tritium released up to now through nuclear testings amounts to 10^9 Ci, together with $0.2-0.6 \times 10^9$ Ci of radioactivity in form of other radioactive fission products and radioisotopes originating from neutron activation and nuclear fusion. Compared to this radioactive fall-out, the amounts of the radioactivity in the ocean caused by nuclear plants and reprocessing plants are relatively low: approximately 3×10^5 Ci of tritium and 3×10^5 Ci of fission products and isotopes (Table 20). Roughly the same figures apply to the radioactive wastes stored in the ocean.

Table 20. Average concentrations of radioisotopes in the surface water of the North Atlantic Ocean, that are responsible for natural radioactivity and radioactivity from fall-out of nuclear testings (Woodhead 1973)

Natural radioactivity	pCi/l	Fall-out radioactivity	pCi/l
^{40}K	320	^{3}H	48 (31–74)
^{87}Rb	2.9	^{137}Cs	0.21 (0.03–0.80)
^{234}U	1.3	^{90}Sr	0.13 (0.02–0.50)
^{238}U	1.2	^{14}C	0.02 (0.01–0.04)
^{3}H	0.6–3	^{239}Pu	0.0003–0.0012
^{14}C	0.16–0.18		

---

[3] The measuring unit for radioactivity, Curie (Ci) is still being applied: 1 Ci = 3.7×10^{10} disintegrations per second, 1 pCi = 1 Picocurie = 10^{-12} Ci.
In the future the measuring unit Becquerel will be used: 1 Bq = 1 disintegration per second. Another frequently applied measuring unit is disintegration per minute (dpm): 1 Ci = 222×10^{10} dpm

These global figures, however, do not refer to the regional distribution in the ocean. Over large areas of the world ocean, mixing with the water is a slow process (Figs. 54 and 55); concentration levels of tritium are still higher on the ocean surface than deep below.

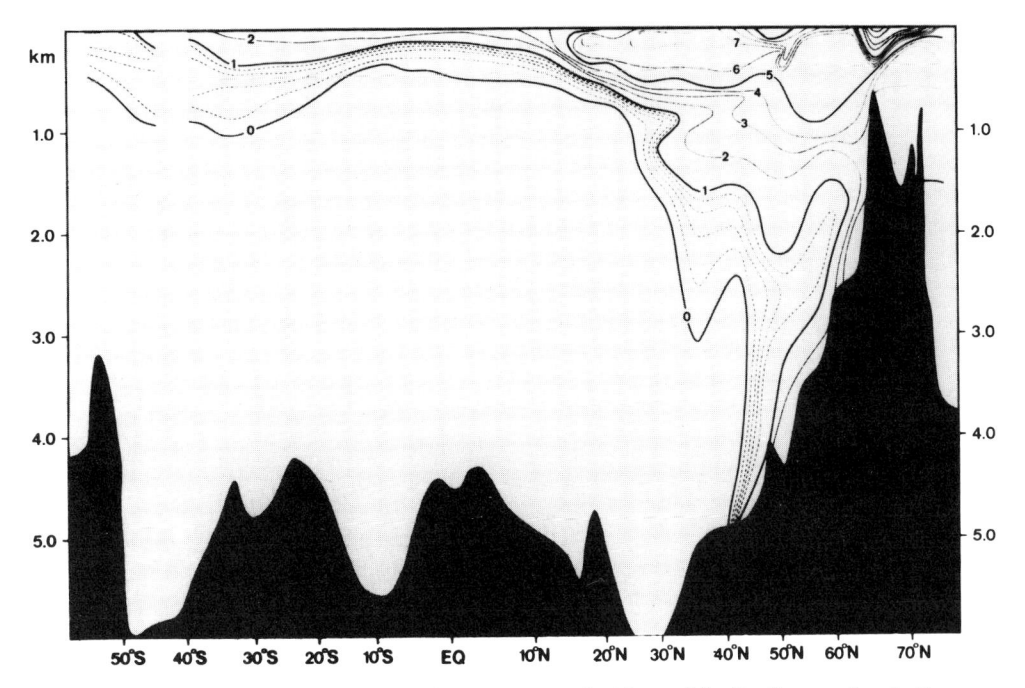

Fig. 54. Approximately 10 years after large amounts of tritium originating from nuclear testings entered the world oceans as fall-out, tritium can only be found in large areas of the Atlantic Ocean in the surface layers with a depth of less than 1000 m. But in the northern hemisphere very saline cold water sinks down, thereby also transporting tritium to the deeper water layers. Tritium can therefore be used as a tracer in finding out the water movements in the oceans. With modern analytic methods, it is possible to detect a concentration of less than 1 tritium atom in 10^{19} hydrogen atoms (unit: TU, tritium unit = $1:10^{18}$). (Results of H.G. Östlund, University of Miami, from Hammond 1977)

Radioactive fall-out was first produced toward the end of the Second World War through the nuclear bomb explosions of Alamogordo (New Mexico), of Hiroshima and of Nagasaki. In July 1948 two nuclear devices were detonated at the Bikini Atoll. Up to 1968 a further 470 nuclear weapon explosions took place, in which the amounts of emitted radioactive substances varied. The highest amount of radioactive fall-out was registered in 1963/1964 (Fig. 56). In 1963 the U.S.A., Great Britain, and the U.S.S.R. signed the Treaty Banning Nuclear Weapons Tests and since then fall-out radioactivity has been reduced by way of sedimentation, disintegration, and by mixing with the ocean water. French and Chinese nuclear testings between 1964 and 1968 have not produced sufficient radioactive fall-out to reverse this general trend.

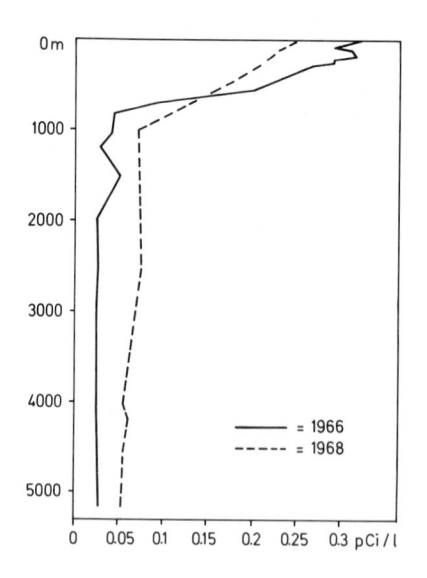

Fig. 55. In 1969/1970 the caesium-137-activity in the surface waters of the North Atlantic Ocean was generally above 0.1 pCi/l. It seems that the concentration levels are gradually decreasing: at 42°–43°N, 14°–15°W the figures for the water surface were at 0.3 pCi/l in 1966. In 1968, however, they were at 0.25 pCi/l. On the other hand, concentration levels are slowly rising at a depth below 1000 m, due to an increased mixing with surface water that contains caesium-137 (Kautsky 1973)

For nuclear weapons, enriched uranium and plutonium are employed. In nuclear fission or nuclear fusion more than 200 different radioactive fission products and isotopes are produced (Table 21), especially if the explosion takes place only a little above the ground or in the water. The radioactive material is transported partly in form of very fine dust right up into the stratosphere. Owing to atmospheric circulation it then falls onto the ground, mainly in the area between 45°N and 45°S, and most intensively in the northern hemisphere, since all nuclear explosions have taken place there. It seems as though the fall-out over the oceans is more intensive than over the land.

Many radioactive isotopes originating from nuclear explosions only have short half-lives; though they also can be detected in the ocean and in marine organisms right after the experiment, they are not to be found later in large quantities in world-wide fall-out. The radioactive isotopes strontium-90 and caesium-137 are primarily characteristic for this fall-out, both having half-lives of roughly 30 years. These isotopes only occur when produced artificially and are therefore useful in tracing world-wide radioactive contamination.

Transuranic elements are those elements with atomic numbers higher than that of uranium. In nature they only occur in uraninite ores, and then only in low amounts so that practically all transuranic elements that are now detectable in the environment originate from nuclear fission and fusion, meaning that they are man-made. The most important transuranic element is plutonium-239 with a half-life of 24,400 years. It is difficult to separate from plutonium-240 through analytic processes, this being the reason why both isotopes are usually mentioned together. Americium-241 is produced through the disintegration of plutonium-241 which has a short half-life. During nuclear reactions low amounts of plutonium-238, with a half-life of 86 years, are also produced.

During nuclear testing a very large amount of plutonium, in form of particles 1 μm in size, enters the atmosphere and then is washed into the ocean with precipitation.

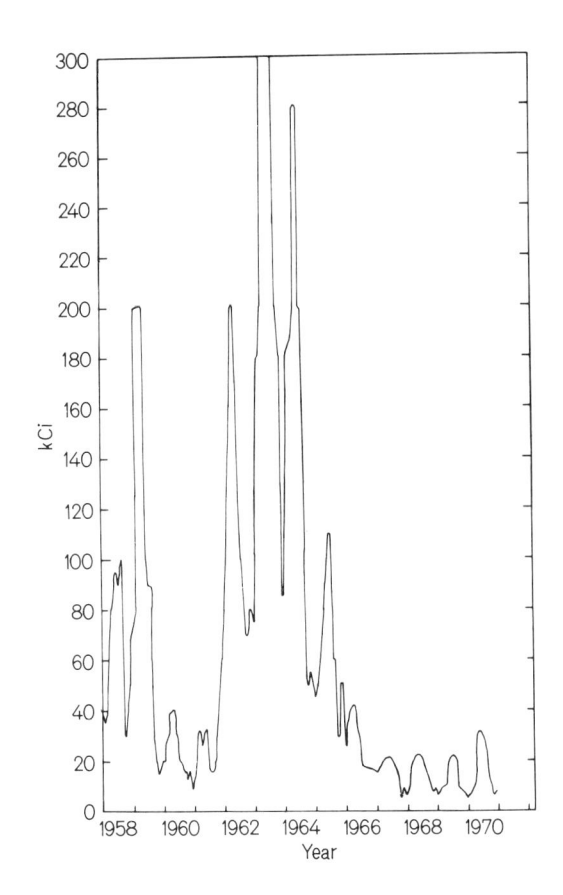

Fig. 56. The fall-out originating from nuclear weapon testings was especially intensive from 1958–1959 and from 1962–1965. The graph shows the monthly trend of strontium-90 sedimentation in the northern hemisphere. Data given in kCi (= 10^3 Ci). (Data of Volchok and Kleinmann, from Hodge et al. 1973)

As "hot-spots" these particles can expose their immediate surroundings to intense alpha-radiation. Chemically, plutonium is very noxious. Transuranic elements have properties similar to the natural elements polonium and uranium. Their water solubility is low, they are easily sedimented but can possibly also be remobilized from the sediment in form of an organic complex. What than happens to transuranic elements in the marine environment varies according to whether they occur dissolved or in the form of particles, and according to the different marine organisms they occur in. If plaice *(Pleuronectes platessa)* are kept in seawater containing 100,000 Ci/l of plutonium and fed with plutonium-contaminated worms for 2 months, almost no plutonium can later be detected in the fish's organs. The same experiment made with rays *(Raja clavata)*, however, shows that 0.2% of the plutonium dose is later to be found in the ray's liver (Pentreath 1978).

In 1971, 211,000 Ci of plutonium-239 and plutonium-240 (155,000 Ci of this in the northern hemisphere) as well as an additional 16,000 Ci of plutonium-238 had been introduced in the world oceans (National Academy of Science 1975a). From this, the plutonium-239 + 240-concentration of 0.0003–0.0012 pCi/l has developed in the surface waters of the North Atlantic and North Pacific Oceans (Woodhead 1973). Similar amounts have been measured in the German Bight and in the German coastal

Table 21. Important radioisotopes in the marine environment. The half-lives are stated in years. The various types of radiation are marked: α for alpha-emitters, β for beta-emitters, K for K-emitters and γ for gamma-emitters. The concentration factors given refer to wet weight (Rice and Wolfe 1971)

Symbol	Element	Half-life	Radiation type	Concentration factor in mollusc	fish
1. Natural radioactivity					
^{3}H	Tritium [a]	12.3	β		
^{14}C	Carbon-14 [a]	5760	β		
^{40}K	Potassium-40	1.3×10^9	β		
^{87}Rb	Rubidium-87	4.7×10^{10}	β		
^{210}Po	Polonium-210	0.38	α		
^{234}U	Uranium-234	2.5×10^5	α		
^{235}U	Uranium-235	7.1×10^8	α		
^{238}U	Uranium-238	4.5×10^9	α		
2. Transuranic elements					
^{238}Pu	Plutonium-238	86	α	200	5
^{239}Pu	Plutonium-239	24400	α	200	5
^{240}Pu	Plutonium-240	6600	α	200	5
^{241}Pu	Plutonium-241	13,2	α	200	5
^{241}Am	Americium-241	458	α, γ		
3. Fission products in nuclear reactions					
^{85}Kr	Krypton-85	10.6	β	1	1
^{89}Sr	Strontium-89	0.14	β	1	0.2
^{90}Sr	Strontium-90	28.0	β	1	0.2
^{90}Y	Yttrium-90	0.007	β	15	10
^{91}Y	Yttrium-91	0.16	β	15	10
^{95}Nb	Niobium-95	0.10	β-γ	5	1
^{95}Zr	Zirconium-95	0.18	β-γ	5	1
^{103}Ru	Ruthenium-103	0.11	β-γ	10	1
^{106}Ru	Ruthenium-106	1.0	β-γ	10	1
^{131}J	Iodine-131	0.02	β-γ	50	10
^{137}Cs	Caesium-137 [b]	30.0	β-γ	10	10
^{144}Ce	Cerium-144	0.78	β-γ	400	3
4. Activation products					
^{32}P	Phosphorus-32	0.04	β	6000	3300
^{51}Cr	Chromium-51	0.08	K-γ	400	100
^{54}Mn	Manganese-54	0.86	K-γ	10000	200
^{55}Fe	Iron-55	2.7	K	10000	1500
^{57}Co	Cobalt-57	0.74	K-γ	500	80
^{60}Co	Cobalt-60	5.3	β-γ	500	80
^{65}Zn	Zinc-65	0.67	K-β-γ	10000	1000
^{110}Ag	Silver-110	0.69	β-γ	10000	–
^{134}Cs	Caesium-134	2.1	β-γ	10	10

[a] Is also produced from induced radioactivity
[b] Caesium-137 is a β-emitter, the γ-emission usually detected during classification originates from the disintegration product barium-137

waters (Murray and Kautsky 1977). In the sediment, plutonium can accumulate extensively (15,000-fold) so that concentrations of up to 1–10 pCi/kg of dry sediment develop in German coastal areas.

Concentration levels in organisms vary enormously: in large algae: 0.3–0.6 pCi/kg (wet weight), in invertebrates: 0.2–3.5 pCi/kg (wet weight), in fish however, below 0.01 pCi/kg (wet weight) in the muscle tissue and slightly higher in the liver.

The specific activity of plutonium-239 is 0.0614 Ci/g, so that 16 g of plutonium correspond to 1 Ci. If, up to 1971 then, a total of 2.1×10^5 Ci of plutonium have been delivered into the world oceans, this is equivalent to 3 t of plutonium. According to some information, the nuclear weapons stored in the arsenals of the Nuclear Superpowers contain 10^7 Ci of plutonium. It appears superfluous to speculate about the intensity of radioactive fall-out, in the form of plutonium and other fission-products like strontium-90 and caesium-137, that would result if this potential of nuclear weapons were put into use. It can, however, be concluded that these nuclear weapons constitute the greatest danger for the world and consequently also for the world oceans. In Western Europe alone 7000 tactical nuclear war heads are stored, altogether mention is made of 50,000 nuclear devices.

Up to now accidents involving nuclear weapons were of little importance. On 21 January 1968 a U.S. atomic bomber crashed into the ocean in the vicinity of Thule in Greenland: high concentrations of plutonium were measured as far as 15 km away from where the bomber had crashed and even in 1974, 25–30 Ci corresponding to roughly 0.5 kg of plutonium, could be detected in the sediments of the crash site, meaning that it had therefore not spread far over the ocean (Aarkrog 1977).

6.2 Reprocessing Plants, Plutonium Plants, and Nuclear Reactors

Where highly enriched nuclear fuels that can also be applied for nuclear weapons are being produced, the environment may be exposed to a high dosage of radiation.

From 1944 to 1971 the U.S. plutonium plant Hanford on the Columbia River was using the river water as coolant, with the result that, per month, 25,000 Ci in form of 60 different radioactive substances, among others zinc-65 and chromium-51, that are extensively accumulated in organisms, reached the Pacific Ocean 360 miles away. In the meantime circulatory cooling has been abolished and the radioactive contamination of the coastal waters has stopped. European plants, however, do not only utilize water for cooling purposes, they also release radioactive wastes into the ocean. This is for instance the case in Windscale by the Irish Sea and (on a smaller scale) in Dounreay in Northern Scotland, as well as in the French plant La Hague by the English Channel. In these plants spent-fuel elements for nuclear reactors are regenerated, by chemically separating the undesired substances produced in the fuel elements, while the nuclear plant is in operation, from the uranium. In 1974 Windscale was allowed to dispose of 300,000 Ci of beta-emitters into the Irish Sea; 207,000 Ci equivalent to 69% of the allowed amount were thus disposed of (Hetherington 1976).

During 1971 the following amounts of radioactive isotopes were released from Windscale into the sea together with waste water (Woodhead 1973):

^{106}Ru	36,400 Ci	^{144}Ce	17,200 Ci
^{137}Cs	35,800	^{90}Sr	12,300
^{3}H	31,500	^{91}Y i.e.	4,400
^{95}Zr	18,000	^{103}Ru	830
^{95}Nb	17,300	^{89}Sr	390

From 1969 to 1976 this sums up to an overall 500,000 Ci of caesium-137 and over 100,000 Ci of caesium-134. Added to these are alpha-emitters like plutonium-239. The permissible amount for these was raised to 6000 Ci per year in 1970 and in 1974, 4560 Ci equivalent to 76% of the amount permitted were released into the sea.

The emissions from Windscale are the reason why in the northwestern part of the North Sea 5–10 pCi/l of caesium-137 are being measured, far more than the 0.2–0.3 pCi/l basic level in the surface waters originating from radioactive fall-out (Fig. 57). The emissions from La Hague are the reason why in the southeastern part of the North Sea 0.7–1.0 pCi/l of caesium-137 are being measured. Since 1975, however, only very low quantities of caesium-137 are being discharged into the water and since 1977 the figures measured in the North Sea have decreased accordingly (Deutsches Hydrographisches Institut 1978).

According to calculations of the International Atomic Energy Agency (IAEA) concentrations of up to 900 pCi/l of caesium-137 in the ocean water can be tolerated. Therefore the caesium-137 concentrations measured in the North Sea are not injurious to health. However, the exposure is of course higher directly near the processing plants: in the area of Windscale 100 pCi/l, and directly in the emission zone 1000–2000 pCi/l. The concentration levels in the organisms are also accordingly high (Table 24). Obviously plutonium is not that easily spread over wide areas by water masses. Figures from the North Sea are not too different from those of the Atlantic Ocean (Murray and Kautsky 1977).

In the future increasingly more spent-fuel elements will be produced and new reprocessing facilities are being planned in many countries. Here it should be possible, to keep the concentration levels of radioactivity in the waste water lower than at Windscale and La Hague, by way of improved cleansing processes. More reprocessing, however, means more transporting, also across the sea. It is estimated that in 1990, 50 t of plutonium will constantly be en route between nuclear reactors and reprocessing facilities. This figure may be wrong, just like all the other forecasts about the further development of nuclear energy, but it is certain that accidents will occur during transportation whereby plutonium will enter the environment.

During operation, too, nuclear reactors emit radioactivity into the surroundings in the form of tritium, krypton-85, carbon-14, caesium-137 and caesium-134, strontium-90, iodine-131 and cobalt-60. The amounts of radioactivity in waste water is, however, not very high (Table 22). The nuclear power plant Unterweser at Esenshamm, Germany, has permission to emit 950 Ci of tritium and 2 Ci of dissolved fission- and activation-products into the river Weser, per year.

The emissions from nuclear-powered merchant vessels are also low. The German freighter "Otto Hahn" was allowed to emit 2 Ci of radioactive isotopes per month into the seawater, but from 1974–1976 the overall emission only amounted to 2–3 Ci per year. Concerning atomic-powered war-ships, of which there are some hundred,

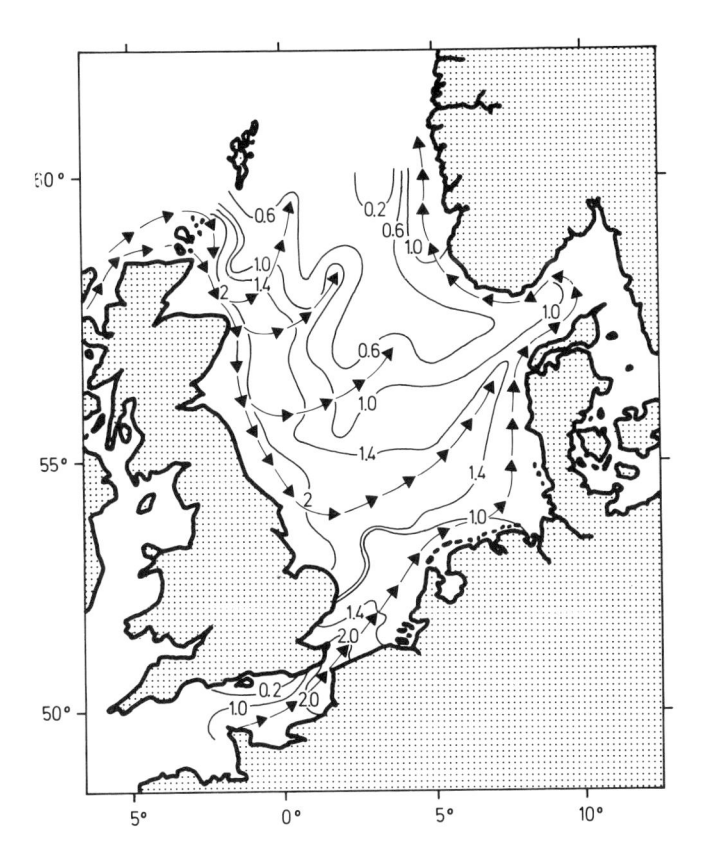

Fig. 57. From the French processing plant La Hague at Cherbourg and from the British processing plant Windscale on the Irish Sea, caesium-137 is emitted by waste water. Owing to very sensitive measuring methods, the increased caesium-137 activity can be detected all over the North Sea. One singular emission of 3000–4000 Ci of caesium-137 during February/March 1971 made it possible to directly trace the drifting of water masses marked with caesium-137: the transportation velocity is 1 nautical mile per day. From the distribution of caesium-137 in the North Sea (example February/March 1975, see map) general conclusions about the water movements there can also be made. Flowing relatively close along the British eastern coast, water with a high caesium-137 activity from Windscale at the Irish Sea flows from the north into the North Sea. At 57°N, 50% of this labelled water is diverted toward the east and enters the Norwegian Sea, whereas the rest is transported further south and only begins to flow toward the northeast at 53°–55°N, clearly separated from the water of the English Channel that runs along the coast. The transportation velocity is 0.6–0.7 nautical miles per day. Data given in pCi/l (Murray and Kautsky 1977; Kautsky 1977)

the situation is not that clear. When starting off, atomic submarines emit radioactivity due to the expansion in the coolant cycle, and in addition the synthetic resin inside the ion exchangers, applied for demineralizing the coolant, emits radioactive isotopes. I cannot investigate the data stating that, from atomic-powered warships 1 million Ci originating from the exchanger-resin, 5000 Ci of fluid wastes and an additional 3400 Ci lost by leakage, enter the ocean per year (Weish and Gruber 1979); nor can I investigate whether it is true that sardines and mackerels swallow the radioactive

Table 22. Amounts of permitted and actually emitted radioactivity in waste water from nuclear reactors in Great Britain during 1974 (Hetherington 1976)

	Tritium		Other elements	
	Permitted Ci	Emitted Ci	Permitted Ci	Emitted Ci
Amersham	400	244	72	25
Berkeley	1500	60	200	24
Oldbury	2000	40	100	33
Bradwell	1500	120	200	90
Dungeness	2000	240	200	70
Hinkley Point	2000	40	200	126
Sizewell	3000	240	200	16
Wylfa	4000	120	65	1
Hunterston	1200	72	200	70

resin particles and transport them over large areas of the ocean (Miyake 1971). It is also not possible to discuss in detail how dangerous accidents in nuclear power plants on the coast, in reactors used for ships and in atomic-powered satellites can be.

Radioactive isotopes are applied in many ways, for instance in medicine, in ecological and physiological tracer-techniques, in material testing, for fluorescent colors and for many other purposes. Radioactivity is thereby regularly emitted into the environment, but the amounts that enter the ocean seem insignificant compared to the amount of fall-out. What may be significant is the radioactivity emitted from isotope batteries, the disintegration heat of which is used as a power source. On a smaller scale, such atomic batteries are, for instance, to be found in pace-makers. Larger ones are used as power source for ocean buoys, for mobile drilling barges (with strontium-90-titanite) and in certain satellites (with plutonium-238). In 1964, the U.S. atomic-satellite SNAP-9A, containing 17,000 Ci of plutonium-238, entered the Earth atmosphere 46 km over the southern part of the Indian Ocean. By 1966, 88% of this plutonium radioactivity could still be detected in the atmosphere. By mid-1970, 95% had reached the Earth's surface in form of fall-out and could be detected everywhere between 70°N and 44°S. The result was that the plutonium-238 concentration originating from nuclear testings was doubled (National Academy of Science 1975a).

6.3 Radioactive Waste

The production and application of radioactive material generates radioactive wastes, the storage of which is problematic; many countries dump the waste into the ocean. Between 1946 and 1970 the U.S. Atomic Energy Commission authorized the dumping of 14,000 Ci into the Pacific Ocean and 80,000 Ci into the Atlantic Ocean, until in 1972 this was forbidden by legislation. Between 1951 and 1966 Great Britain dumped 45,000 Ci of radioactive waste into the Atlantic Ocean. Since 1967 the Nuclear Energy Agency of the OECD had been regulating the dumping of radioactive waste of eight European countries (Table 23). After the enforcement of the London

Table 23. Quantities of radioactive wastes, originating mostly from nuclear research centers in Belgium, France, Federal Republic of Germany, Italy, the Netherlands, Sweden, Switzerland, and the United Kingdom, which have been dumped into the Atlantic under the supervision of the Nuclear Energy Agency of the OECD. Most dumping campaigns were with the British freighter "Topaz". 1967–1969 the dumping area was at 42°–43°N, 14°–15°E, 1971–1975 a 70-mile circle around 46°15'N, 17°25'W, 1976 a rectangle at 46°N, 16°00'–17°30'E, all approximately 1000 km away from the European coast, on 4500 m of water in the deep sea (Olivier 1978)

Year	Dumped weight t	Alpha activity Ci	Beta-gamma activity Ci
1967	10,840	250	7,600
1969	9,180	500	22,000
1971	3,970	630	11,200
1972	4,130	680	21,600
1973	4,350	740	12,600
1974	2,270	420	100,000 (mostly tritium)
1975	4,460	780	60,500 (30,000 tritium)
1976	6,770	880	53,500 (21,000 tritium)
total	45,970	4,880	289,000

Convention (see Chap. 10) the OECD has stopped the dumping of radioactive waste; since then this is settled by the participating states at their own responsibility and the Nuclear Energy Agency has only a controlling function.

The U.S.S.R oppose this practice of dumping radioactive waste into the ocean, and both Sweden and the U.S.A. hold nothing in favor of the "diluting-philosophy" and prefer depositing radioactive waste on land which often is also cheaper. Especially in Great Britain one is, however, of the opinion, that the best way to get rid of such wastes, which are somewhat too active to deal with by simple burial and have half-lives too long to permit storage for decay, is to dump them into the sea.

Incidentally, the waste dumped into the sea does not consist of high-level radioactive material; the London Convention forbids this. A considerable amount of the waste is made up of isotopes that have half-lives of less than 1 year, so that within 10 years the radioactivity of these materials will have disappeared almost entirely (Kautsky 1972). In addition the waste is mixed with cement and bitumen and packed into steel drums. It is expected to take some years until the iron coating is corroded to such an extent that the nucleus of cement and bitumen is fully exposed to the influence of the seawater, and that it will then take many more years until the substances inside the nucleus, too, could be leached out. By that time their radioactivity will have almost completely disappeared.

In the meantime the International Atomic Energy Agency has defined (IAEA 1975) what is to be regarded as high-level radioactive waste: this is the case when more than 1 million Ci tritium, more than 1000 Ci beta-, and gamma-emitters, more than 100 Ci strontium-90 + caesium-137 or more than 10 Ci alpha-emitters with half-lives of over 50 years, are included in 1 t of waste. These figures were based on the assumption, that up to 10^{10} Ci of alpha-emitters, up to 10^{13} Ci beta-, and gamma-emitters and up to 10^{15} Ci of tritium could annually be discharged into the Atlantic Ocean without causing any damage, the safety factor being higher than 10,000. In the meantime, however, one has become more sceptic and in early 1978 the standards were lowered considerably, for example down to 10^{11} Ci for beta-, and gamma-emitters (Olivier 1978).

Plans to deposit high-level radioactive wastes in the ocean, too, have always existed. The idea of taking advantage of zones, where geophysical processes cause the seabed to be gradually transported further down, seemed tempting. Unfortunately the speed of transportation is low and the zones that come into question are geologically unstable. One has to think 1 million years ahead if one wants to create secure final deposits for radioactive substances with long half-lives. Strontium-90 and caesium-137 which are mainly contained in radioactive waste, have half-lives of only roughly 30 years, but the half-life of plutonium-239 is 24,000 years, that of techneticum-99 is over 200,000 years and the half-life of neptunium-237 is over 2 million years.

Simple dumping into the ocean is out of the question, for the deep sea too, is linked to the productive surface waters of the ocean by organism-migration and ocean currents; when a deep-sea fish enters the intermediate water layers it can very quickly transfer radioactivity into the higher water layers, too. The latest concepts therefore do not intend utilizing the deep sea as such, but rather regard the sea bed as a possible site for permanent storage (Fig. 58).

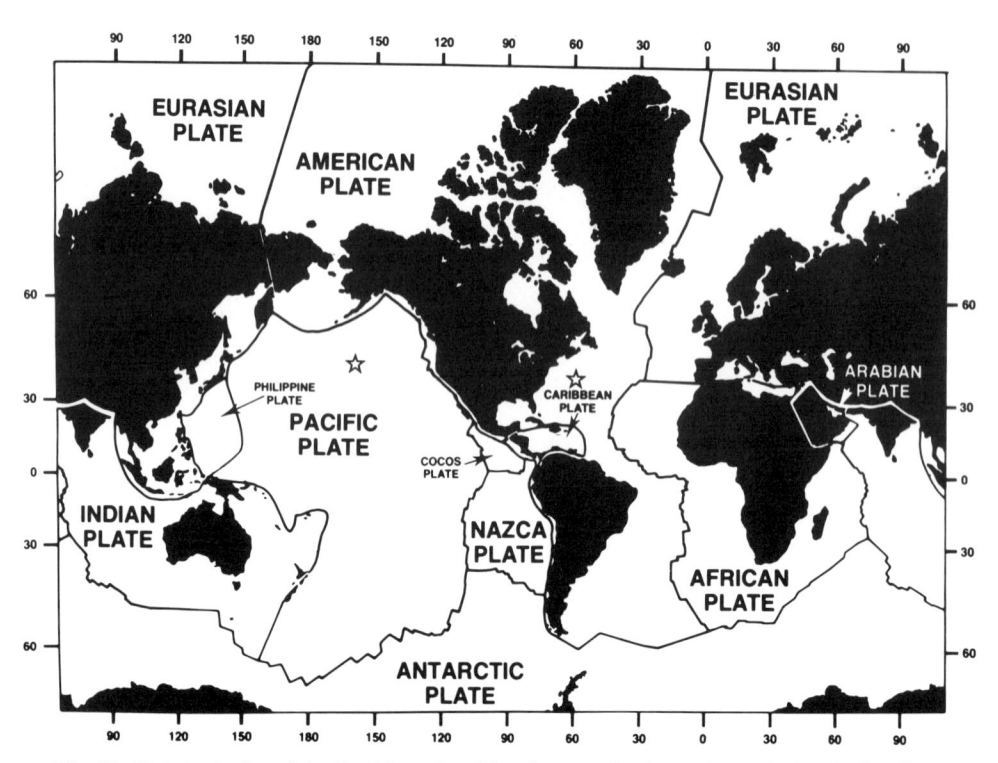

Fig. 58. Plate tectonics of the Earth's surface. The plates are slowly moving against each other, but are themselves relatively stable. Central areas of various plates *(stars)* were examined with regard to the possibility of permanently storing high-level radioactive substances there (Hollister 1977)

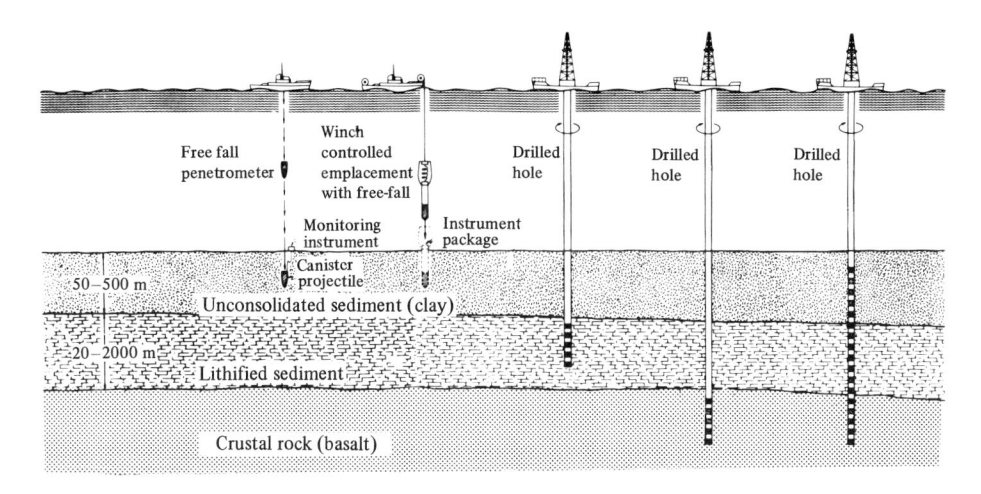

Fig. 59. Concepts for the permanent storage of high-level radioactive material in the geological layers below the deep-sea bed (Silva 1977)

The large geotectonic deep-sea plates are the most stable formations of the lithosphere, they develop in the central oceanic mountain ridges and gradually drift toward the continents, thereby becoming increasingly covered with sediment. It is very possible that there are no changes over millions of years and that the absorption and diffusion properties of the deep-sea clay would prevent radioactive substances from seeping through when after 1000 years they should leak out of their encasement of glass and ceramic. In the U.S.A. a team of scientists has been examining this problem since 1973 (Hollister 1977) and up to now one has not come upon facts that speak against permanent sub-sea bed storage, except from the technical point of view: burying drums containing high-level radioactive material deep enough into the deep-sea clay (and this at a depth of 5000 m) is not unproblematic (Fig. 59); accidents will happen and therefore this idea cannot be put into practice at the moment.

In 1976 the U.S.A. alone produced 284,000 m^3 of high-level radioactive waste originating from the reprocessing of nuclear fuels for nuclear energy-powered submarines, and from the plutonium production, corresponding to 270 million Ci of strontium-90 and roughly the same amount of caesium-137. In the meantime U.S. nuclear reactors in operation have quite likely again produced just as much nuclear waste which will have to be disposed of in the future (Krugmann and Hippel 1977).

Countries like Great Britain, Belgium, The Netherlands, Switzerland, and Japan have no adequate geological formations for disposal. In the U.S.A. and in the Federal Republic of Germany, where adequate geological formations exist and could be utilized for storage on the continents, the federal states, the communities, and the affected citizens are increasingly opposing such plans and effectively demonstrating with the slogan: "Not in my garden!" Therefore one has to consider the possibility that in the future, due to such understandable personal reasons, attention will increasingly be directed toward international territories, which also includes the deep sea.

6.4 Effects of Radioactivity

The distribution of natural radioactivity in the ocean varies: the ocean water has a radioactivity level of approximately 320 pCi/l, the sand has a level of 5000–10,000 pCi/kg and the mud between 20,000–30,000 pCi/kg. Algae in coastal areas, like *Fucus* and *Porphyra*, have a radioactivity of 5000–15,000 pCi/kg, mollusc and fish of 1000–3000 pCi/kg, both referring to wet weight. When the level of radioactivity is raised artificially, either by fall-out or due to wastes from recycling plants, the levels of radioactivity in the sediment and in the organisms rise accordingly (Table 24).

Table 24. Radioactivity of organisms from near the recycling plant Windscale on the Irish Sea, 1974. The data is given in pCi/kg referring to wet weight (Hetherington 1976)

	^{106}Ru	^{134}Cs	^{137}Cs	^{90}Sr	$^{239+240}$Pu	^{241}Am
Plaice *Pleuronectes platessa*	600	2,500	11,200	40	5.4	11
Dab *Limanda limanda*	1,800	3,700	15,600	–	–	–
Mussel *Mytilus edulis*	183,000	2,000	7,800	–	1,090	5,000
Red algae (laver-bread) *Porphyra*	340,000	–	15,000	600	4,400	14,000

It would, however, be misleading to generalize the effects of radioactivity on all organisms, for in the ocean radioactivity is tied to completely different elements and isotopes. Tritium reacts like a hydrogen atom and can be a component of the water and of all organic compounds. Like potassium, caesium too is easily soluble in seawater. Strontium reacts in a similar way to calcium and is primarily included in bone and other calcareous skeleton substances, Iodine, zinc, iron, manganese, and cobalt are important essential trace elements (see Chap. 7.3), they are considerably accumulated by ocean organisms, since they are necessary for certain enzyme or vitamin reactions; the radioactive isotopes are absorbed together with elements that are not radioactive. Some other elements of no physiological usefulness are also accumulated; among them are silver-110 and chromium-51, which, in addition to the radioactive toxicity also have a general toxic effect. What in the end happens to the various radioactive substances after contact with clay particles suspended in the water, or with the sediment, also varies from element to element.

To begin with, the toxicity caused by radioactivity is a unit, that, similar to the toxicity of other poisonous substances, can be analyzed by defining the lethal dose. Hereby the concomitant circumstances are also important, for the effect of radioactivity also strongly depends upon temperature and salinity of the environment and adaptability of the organisms. Just as heavy metals, like mercury and cadmium (see Chap. 8) can possibly have a harmful effect on marine life, even in low, natural trace concentrations, natural radioactivity too, may be disadvantageous to marine life.

Radioactive radiation does not only directly have a toxic effect on physiological processes, it also affects the genes in the chromosomes or breaks up the chromosomes, so that they either remain in the form of fragments or coalesce in an unnatural order: genetic changes and deformations are the consequence. In embryos of the scorpionfish *(Scorpaena porcus)* 6%–14% of all the chromosome sets are already fragmented as a result of natural radioactivity and other natural influences, but if the embryos are put into seawater with an artificial strontium-90 activity of 100,000 pCi/l, the amount of chromosome fragmentation rises noticeably. But this is a concentration level 1 million times higher than the strontium-90 concentration (approx. 0.1 pCi/l) in the ocean originating from nuclear testings. In experiments with carbon-14 as a radiation source, a dose ten times higher was needed to bring about the same effect of chromosome fragmentation (Tsytsugina et al. 1973). There are also reports stating that additional artificial radioactivity of just 100 pCi/l is harmful to the development of fish eggs, whereas in other experiments much higher radiation doses have not had any harmful effects at all. So no final judgment can be given about the minimum level of harmful radioactivity in the ocean water (Ravera 1978).

Radioactivity is emitted from the various isotopes in different qualities: alpha-radiation consists of positively charged helium atomic nuclei, that only penetrate 0.06 mm deep into normal tissue. Beta-radiation penetrates 20 mm deep and consists of negatively charged electrons, the K-radiation consists of electrons of the inner atomic shell. The photons of gamma-radiation penetrate practically every tissue unhindered. Radioactivity does not only affect marine organisms from the outside, it is also emitted by isotopes that are incorporated in the body tissue of the organisms themselves. This is the reason why organisms of different forms and sizes are differently affected by radiation. Referring to human beings these complicated standards have already been examined long ago and have been given the unit termed "rem"[4], a radiation equivalent applicable to human beings. First attempts are being made to set up a differentiated rating applicable to marine organisms (Woodhead 1973).

A pelagic fish can be regarded as a cylinder with a diameter of 10 cm and a length of 50 cm. To begin with, it is exposed to the natural radioactivity contained in its own body, from potassium-40 and polonium-210, of 3–4 μrad/h [5].

Added to this is a slight radiation exposure of 0.1 μrad/h from potassium-40 contained in the seawater. Cosmic radiation is only of importance to organisms living near the water surface. For the model fish, cosmic radiation would here amount to 4 μrad/h; at 20 m below the water surface it would already be reduced to 0.5 μrad/h and at 100 m below the surface it can be dismissed. Instead radioisotopes in the sediments become more important the more one approaches the sea bed. For a fish living near the surface of the sediment the radiation dose would be 1.5–16 μrad/h of gamma-radiation and 1.6–21 μrad/h of beta-radiation. If the fish were buried into the sediment, the dose would be twice as high.

Altogether the exposure to natural radiation amounts to 5–21 μrad/h for fish living near the sediment at a depth of 20 m. For the fall-out caused by nuclear weapon

---

[4] 1 rem = 0.01 J/kg, equivalent dose from a mixture of different emitters of radioactivity, applied to human beings; 1 mrem = 10^{-3} rem

[5] 1 rad = 0.01 J/kg, unit of the absorbed radiation dose; 1 μrad = 10^{-6} rad

explosions an additional radiation exposure of 0.1–1.8 μrad/h has to be considered. For smaller animals the radioactive exposure is considerably higher and assuming the model calculation is correct, natural radiation and fall-out radiation are of about equal dimension as regards the effect on plankton; fish eggs are just as small. Incidentally such radiolaria that have a skeleton of strontium sulfate may be exposed to a higher radiation dose.

The radioactivity of marine sediments, too, can vary. On beaches, where due to the sorting effect of wind and water, placer-type accumulations develop from especially heavy, dark mineral granules, uranium-238 and thorium-232 can accumulate together with zirconium. On the beach of Norderney Island on the German North Sea coast, a local dose of up to 500 mrem per year was measured in such places. Higher doses were measured in the monazite sands of India and Brasil (Bonka 1980).

The discussion about the level of the radiation dose a human being can endure, continues. At first 500 mrem per year were considered acceptable by the International Commission on Radiological Protection, later 5 rem per generation (that means 167 mrem per year), or a dosage that is as high as the natural radiation exposure, was considered. Now, a maximum of 30 mrem per year are considered acceptable, as stated in the Radiation Protection Regulation of the Federal Republic of Germany of 1976. This is a low amount compared to the radiation doses used for medical reasons, especially for X-ray treatment, for which in the Federal Republic of Germany an average of 50 mrem a year, and in the U.S.A. even 100 mrem a year may be guessed. This too, is little compared to natural radiation levels. For a human being living at an altitude of 3000 m on a granite ground the natural radiation dose is made up of 100 mrem/year of cosmic radiation, 90 mrem/year of radiation from the mineral sub-soil and 20 mrem/year from potassium-40 in his own body; added together this amounts to 210 mrem per year. For a human being living at sea level on sediment ground the result is more favorable: 30 mrem/year of cosmic radiation, 23 mrem/year of radiation from the sub-soil and 20 mrem/year from potassium-40, together 73 mrem per year. The exposure to natural radiation of the population of the Federal Republic of Germany can be assessed at an annual average of roughly 110 mrem.

It does not look as though the increase of man-made radioactivity in the water of the North Sea up to now, contributes essentially to the permissible radiation dose.

While freshwater fish caught in rivers near nuclear plants may contain 1–200 pCi/kg of artificial radioisotopes in their flesh, and while on the other hand in fish from the North Sea the natural radioactivity, mainly based on potassium-40, amounts to 1000–2500 pCi/kg, the artificial radioactivity in plaice and sole amounts to 2 pCi/kg of strontium-90 and 60 pCi/kg of caesium-137. There is less in the North Atlantic redfish *(Sebastes marinus)* with 0.2 pCi/kg of strontium-90 and 10 pCi/kg of caesium-137 (Deutsche Forschungsgemeinschaft 1979). For the fish eater this seems not to be hazardous to his health.

In the vicinity of Windscale on the Irish Sea, the additional radioactivity originating from the reprocessing plant is regularly being checked, as percentage of a dose of 5 rem per generation which is regarded as permissible. For a person who stands on contaminated mud for a longer period this amounts to 7% of the permissible dose, for a fisherman who regularly eats fish from the contaminated area (Table 24) up to 14%, and for an inhabitant of South Wales who regularly eats "laver-bread" (prepared from *Porphyra* red algae; Table 24) from near Windscale, the dose is up to 22% of the permissible dose.

For 26,000 people in Wales "laver-bread" is a speciality; it is prepared from the alga *Porphyra* which extensively accumulates ruthenium including ruthenium-106. In the meantime, however, the last fishermen of *Porphyra* from Windscale have retired, so that no more contaminated *Porphyra* is being used (Hetherington 1976).

Applying the stricter guiding principles of the Federal Republic of Germany for our calculations, the resulting percentage figures are much higher.

In a book about the contamination of the ocean, no well-based discussion can be given concerning the points of controversy such as: the maximum additional radiation dose that man can still endure, the genetic consequences and whether the risk of cancer will increase if the radiation dose in man's environment is raised to twice the amount of natural radiation. It is, however, legitimate to ascertain that momentarily the radiation caused by man in the world oceans has only risen by a fraction of the natural radiation, and that there are only very few problematic areas.

The fact that this phenomenon is identifiable and can even be put down in figures, is because radioactivity can, even in very low quantities, be measured about 1 million times more accurately than radioactive and other elements can be analyzed with the best chemo-analytic methods. Even though it is encouraging that man's activities have, up to now, only led to a minimal radioactive contamination of the world oceans, one should still consider the potential of radioactivity stored in the nuclear weapon arsenals of the nuclear super-powers, and what quantities of radioactivity, from nuclear plants no more in operation, are waiting for a secure, permanent storage. Oceanology will have to carefully register the further development.

7 General Problems of Harmful Substances in the Sea

7.1 Toxicity

Many environmental chemicals have a toxic effect on marine organisms. They reduce the number of survivors, influence metabolism, photosynthesis or breeding efficiency, and alter behavioral patterns. To date, too little has been known about sublethal toxic effects and long-term effects. It is also uncertain to what degree laboratory tests apply to effects in natural environments. The toxic effect of a heavy metal, for example, may be quite different, depending on the composition in which that element occurs.

In the presence of chelating or complexing substances, the toxicity of heavy metals is often less than when metal ions have a direct effect. Metalorgano complexes are not absorbed by organisms from seawater to the same degree, and since natural seawater contains different amounts of complexing substances, different samples of seawater may result in different toxicity reactions, after the addition of the same amounts of heavy metals. However, detergents may facilitate the entry of heavy metals and other poisonous substances into organisms and increase the toxic effect.

Certain metalorgano compounds are much more poisonous than the metal is in ion form. These are applied in plant insecticides and fungicides (Fig. 60). Under microbial influence in marine environments, lead salts are changed into methyllead which in experiments has a much more poisonous effect on diatoms than lead nitrate (Wong et al. 1975). Methylmercury is not only produced via bacteria from metallic mercury, but also from various anorganic mercury compounds. Certain bacteria can also make metallic mercury from methylmercury (Spangler et al. 1973). Under anaerobic culture conditions, some bacteria are not harmed by mercury because the mercury introduced is changed directly into nontoxic mercury sulfide (Gillespie and Scott 1971; Vosjan and van der Hoek 1972).

No wonder, then, that natural bacteria populations in experiments react unpredictably to increased amounts of heavy metals. Within a few days, a selection takes place. Sensitive species, which can only take 0.1 mg/kg of cadmium in the agar nutrition substratum, die off at higher toxic concentrations. In 2–3 days, however, species that tolerate cadmium increase and even thrive on concentrations of 5 mg/kg of cadmium (Thormann 1975). In corresponding experiments, lead-resisting bacteria can survive 200 mg/kg of lead in the nutrient substratum.

Under natural conditions, not just one, but a whole variety of harmful substances has an effect on organisms. But to date, little is known about whether the toxic effect of various harmful substances increases or reduces the effect of a certain poison or whether a synergistic, hyperproportionally harmful effect might also occur. The small number of findings are contradictory.

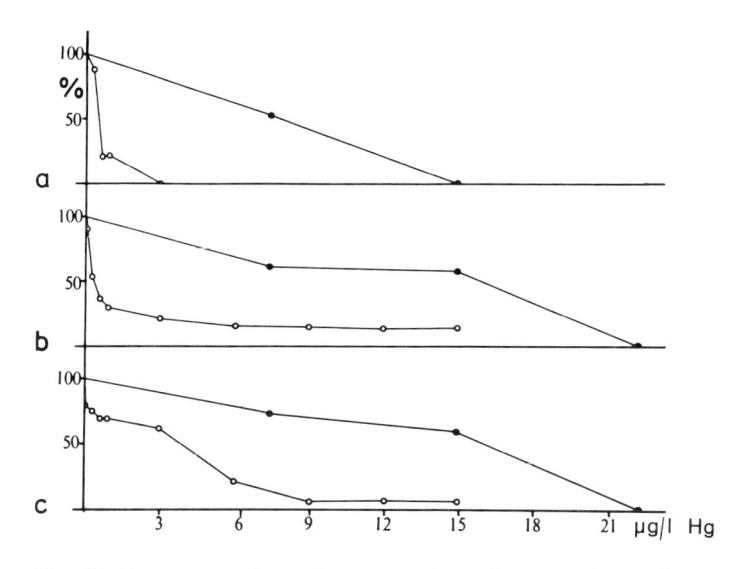

Fig. 60. Even extremely small concentrations of mercury in a culture medium influence the growth of phytoplankton algae. Considerable differences between anorganic mercury (mercurychloride, *closed circles*) and mercury in an organic compound (phenylmercuryacetate, *open circles*) emerge. Compared with the control experiments without mercury (100%), the 3 diagrams show at what percentage the cell concentration is reduced by the influence of the poison in *a Chlamydomonas* sp. (control: 2.2 million cells/ml), *b Chlorella* sp. (control: 6.5 million cells/ml), and *c Phaeodactylum tricornutum* (control: 27.4 million cells/ml) (Nuzzi 1972)

Other ecological factors must also be taken into consideration. As a rule, an organism living under optimum conditions is more resistant than an animal under stress (Fig. 61). For a species to continue living in an environment, the tolerance at the most sensitive stage is the decisive factor. Marine organisms are often particularly sensitive when hatching out of an egg or during metamorphosis. Full-grown *Echinometra mathaei* sea urchins manifested 50% mortality (96 h LC_{50}) in a copper concentration of 0.3 mg/l in seawater. But even at concentrations of 0.02 mg/l, the formation of the calcareous skeleton is disturbed in developing larvae (Heslinga 1976). Finally, the sensitivity of organisms vis-à-vis toxic substances also depends on whether or not they have previously been able to adapt to toxic concentrations or whether they are coming into direct contact with poisonous substances for the first time in an experiment (Table 25; Fig. 62).

Experiments are also influenced by the fact that they have to be conducted in small containers where it is possible that poisonous substances are harbored on the walls of the container instead of being effective in the experimental water. Then too, the organisms themselves often react in an uncontrollable manner in the experiment. Sometimes they store up poisonous substances so intensively that the poison in the experimental medium itself diminishes. Under certain conditions, plankton algae grow even better in a culture experiment if lead nitrate is added (Fig. 63). Sometimes, organisms transform poisons into a form that reacts differently than the lab scientist expects. For example, mercurychloride in a watery solution is not very volatile.

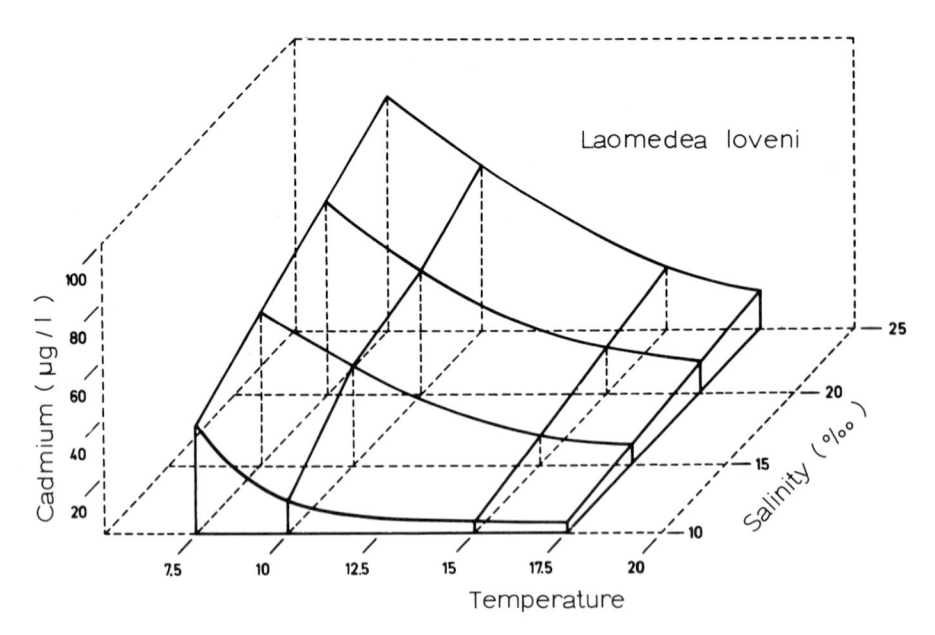

Fig. 61. Experimental animals react differently to poisons, depending on the ecological conditions under which they have been kept. The diagram shows the cadmium concentration at which half the number of polyps in a colony of hydroid polyps *(Laomedea loveni)* is reduced, after 7 experimental days (7 d LC_{50}). At below 20 µg/l, cadmium is most poisonous at high temperatures and low salinity (Theede et al. 1979b)

Table 25. Different populations of an animal species can be immune to the poisonous effect of heavy metals to different degrees. Depending on the geological formation and mining activity in the vicinity, the sediment in the estuaries of southwest England has a different heavy metal content (see Fig. 67). As a measuring unit for the sensitivity of the *Nereis diversicolor* polychaet vis-à-vis copper and zinc, the concentration of poison is indicated at which half of the experimental animals died in the course of a 4-day experiment (96 h LC_{50}). Animals from Restronguet Creek, which has a high heavy metal content, are more resistant than animals from the Avon Estuary which is not influenced by heavy metals (Bryan 1974)

Origin of the animals	96 h LC_{50}	
	Copper	Zinc
Restronguet Creek	2.3 mg/l	94 mg/l
Avon Estuary	0.5 mg/l	5 mg/l

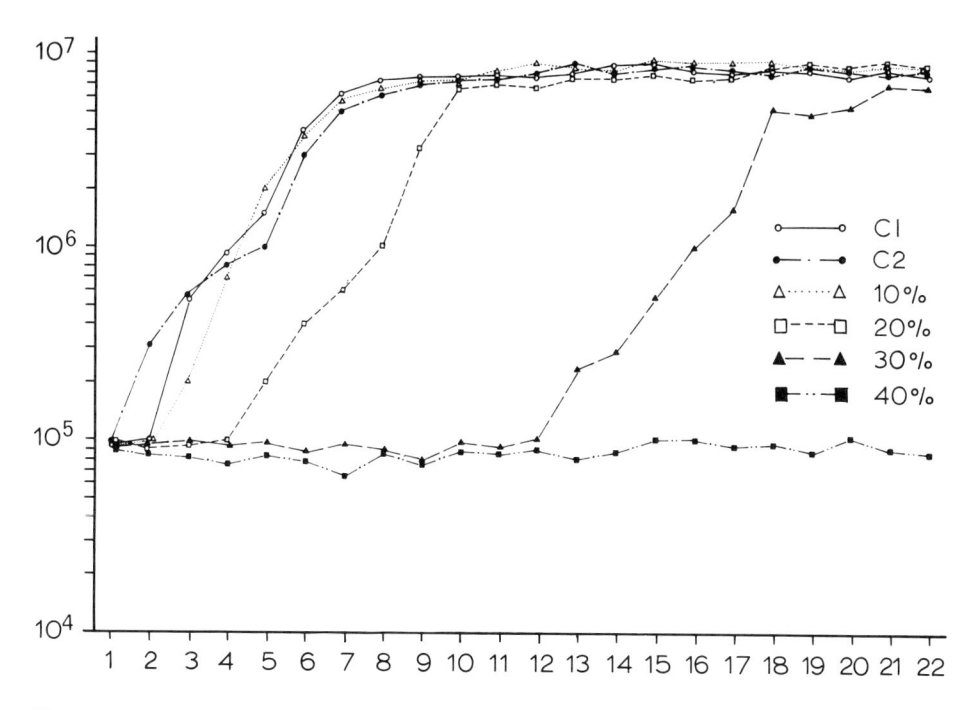

Fig. 62. If effluents from a cellulose plant are added with 20%–40% to the culture environment of the diatom *Skeletonema costatum,* no increase in the number of cells takes place. This suggests definite growth interference. But 12 days later, the algal culture begins to grow, even after 30% of effluents has been added. It attains the same cell count after 25 days as unpolluted cultures do after 9 days. Obviously, this alga is capable of adapting to conditions, apparently by pursuing other metabolic avenues than it does under normal conditions (Stockner and Anita 1976)

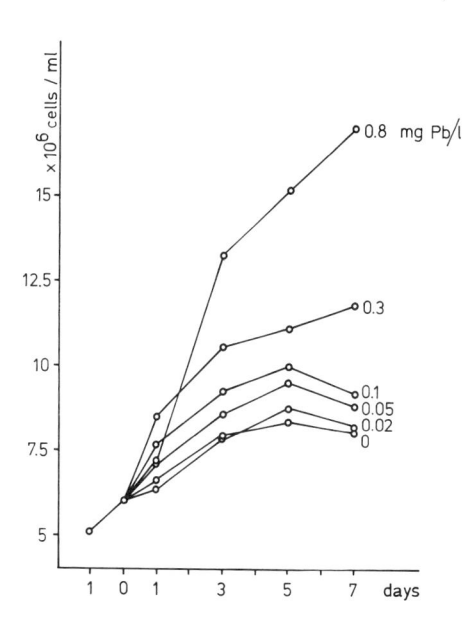

Fig. 63. Under certain conditions of the culture experiment, plankton algae may grow better with than without a toxic heavy metal in the culture medium. The graph shows the increase of *Phaeodactylum* diatom cells in a culture, to which 1 day after the start lead nitrate in various concentrations has been added. Presumably the toxicity of lead is so small for these rather tolerant algae that the fertilizing effect of the nitrate dominates the toxic effect. (According to unpublished data of Schulz-Baldes from Gerlach 1976)

The concentration remains practically the same, even when air is blown through the experimental container. If, however, an algae culture is living in experimental water, the rapid disappearance of the mercury is in proportion to the concentration of the algae culture. The mercury cannot be found in the experimental water, nor on the walls of the container, nor in the algae themselves. The explanation is that the algae have transformed the mercurychloride into mercury-organic compounds which then escape from the system with the aquarium's air supply. This experiment also shows how careful one must be when interpreting the results of toxic experiments with heavy metals (Ben-Bassat and Mayer 1975).

From all these examples and arguments it is understandable that toxicity experiments lead to very different results, if one applies one and the same substance to one and the same species of organism, but under differing environmental and experimental conditions. On the other hand, toxicity experiments with one and the same substance, even when conducted with the same experimental setup, come to different results when different experimental organisms have been tested. According to toxicity experiments, mercurychloride can have a toxicity threshold between 0.5 and 100 $\mu g/l$ (Table 26).

It does not seem sound to go on with toxicity experiments including more and more organism species, except when trying to find organisms which are more sensitive than species known, and which could set up new standards for toxicity evaluation. It would be more important if future work could point to fundamental differences, if they can be demonstrated, between freshwater and seawater experiments. The higher concentration of salts in seawater, not only of sodium chloride, could be the reason for reduced toxicity of some elements. On the other hand, it could be that organisms from the high seas are more sensitive to stress from pollution than freshwater organisms, because the high sea is an extremely stable environment where even small-scale fluctuations of environmental factors are the exception. In this respect, marine organisms from nearshore waters and from estuaries should like freshwater organisms be adapted to cope with changing environmental conditions and stress, and the assumption is made that they may be more resistant to some pollutants, too, compared with sensitive organisms from the high seas. The dilemma is that many organisms from the high seas are so sensitive that even without the addition of pollutants it is impossible to keep them for longer periods in culture. But without submitting them to experiments, science has difficulties in assessing the effect of pollution to such highly sensitive species.

Anyway, if it should be possible to deduct from experiments with standard freshwater test organisms, or standard terrestrial test organisms the possible effects a certain environmental chemical will have upon the marine environment, this would economize an enormous amount of laboratory toxicity testing procedures with marine organisms.

Table 26. Toxicity of mercurychloride: the mercury concentrations at which 50% of the organisms die off (LC$_{50}$) in experiments of different duration, or other effects occur. (From different authors quoted by Stebbing 1976)

LC$_{50}$ (μg/l)	Experiment, duration	Organism	Reference
0.5	Growth rate: 11-day experiment	*Campanularia* (hydroid polyp)	Stebbing (1976)
1.6	Same experiment, 6 days	*Campanularia*	Stebbing (1976)
1.3	Morphological changes	*Eirene* (hydroid polyp)	Karbe (1972)
1−3	48 h LC$_{50}$ of larvae	*Ostrea* (oyster)	Connor (1972)
1.6	Cell size, inhibition of reproduction rate	*Isochrysis* (plankton algae)	Davies (1974)
1.8	Metabolism and mortality of larvae	*Uca* (fiddler crab)	De Coursey and Vernberg (1972)
2.5	Inhibition of reproduction rate	*Cristigera* (cilicate)	Gray and Ventilla (1973)
3	Inhibition of reproduction rate	*Chlamydomonas* (plankton algae)	Nuzzi (1972)
3	Inhibition of reproduction rate	*Chlorella* (plankton algae)	Kamp-Nielsen (1971)
4.8	48 h LC$_{50}$ of larvae	*Mercenaria* (clam)	Calabrese and Nelson (1974)
5.6	48 h LC$_{50}$ of larvae	*Crassostrea* (oyster)	Calabrese et al. (1973)
10	48 h LC$_{50}$ of larvae	*Crangon* (shrimp)	Connor (1972)
10	Stimulation of rate of respiration	*Congeria* (mussel)	Dorn (1974)
14	48 h LC$_{50}$ of larvae	*Carcinus* (crab)	Connor (1974)
27	Morphological changes in larvae	*Crassostrea* (oyster)	Woelke (1965)
33−100	48 h LC$_{50}$ of larvae	*Homarus* (lobster)	Connor (1972)
50	25 h LC$_{50}$	*Acartia* (copepod)	Corner and Sparrow (1956)
50	Inhibition of growth	*Scenedesmus* (plankton algae)	Matida et al. (1971)
50−64	96 h LC$_{50}$	*Petrolisthes* (crab)	Roesijadi et al. (1974)
75	48 h LC$_{50}$	*Pandalus* (prawn)	Portmann (1968)
100	Inhibition of rate of growth	*Chaetoceros, Cyclotella, Phaeodactylum* (plankton algae)	Hannan and Patouillet (1972)
170	Inhibition of growth	*Chlamydomonas* (plankton algae)	Ben-Bassat et al. (1972)

7.2 Accumulation[6]

Some heavy metals and other trace elements, radioactive elements and chlorinated hydrocarbons are accumulated so intensely by organisms that considerable concentrations result in the tissues of the organisms from the extremest dissolutions in seawater. Such accumulations have previously been noted and been considered a curiosity. It was learned that tunicates accumulate vanadium, and later that a number of trace elements are essential for biochemical processes, especially for the synthesis of vitamins and enzymes. Among these are copper, zinc, cobalt, and manganese. It is evident that organisms needing these substances must be equipped to accumulate them from the strongest dissolutions.

During earlier stages of marine pollution research, it became clear that an accumulation of such elements and organic compounds results, too, which according to current knowledge, have no positive metabolic function; on the contrary, they can manifest harmful effects. Now, following important first-stage results, research is beginning to explain the different mechanisms of uptake which are evidently different from substance to substance, to localize the point of storage, and to understand the mechanisms of elimination. Here too, it will not be possible to make sure judgments for a number of years. To date, interest has centered only on those substances which seem to be of direct importance to those fish that man uses as a source of nutrition.

In the smallest concentration it occurs in seawater, DDT is difficult to analyze quantitatively. In experiments, however, radioactively marked DDT can be used because even the smallest doses of radioactivity can easily be measured and the traces of radioactively marked DDT can be tracked. If only 0.01 $\mu g/l$ of DDT is present in aquarium water, marine worms *(Lanice)* absorb enough DDT to attain a more than 200-fold wet weight concentration. Higher concentrations are attained if the experiments run longer (Ernst 1972), until finally an equilibrium concentration between concentration in the seawater and concentration in the tissues may be reached. There is a wide range of concentration factors as the result of many experiments using many different organisms and experimental conditions (Table 27).

Methodological errors from past experiments may be responsible for the divergence of existing data. But certain regular features are also implied, and it is possible that there is a direct connection between the solubility of a hydrocarbon substance and the concentration factor (Fig. 64). It is also known that accumulation is stronger in tissues that are rich in fat than in those that contain little fat.

The accumulation of chlorinated hydrocarbons takes place not only from the surrounding seawater, but through food as well. Under experimental conditions, soles can be fed pieces of meat that have been prepared with DDT. If a fish has been fed 17 μg of DDT in a 4-week period, a DDT content of 0.5–0.8 μg shows up in its muscles, while its brain and liver show a higher concentration. During the experiment, these fish stored up approximately 60% of the total amount of DDT fed to them (Fig. 65). Likewise, 58%–93% of the PCB's fed to them through food was found in the body tissue of the polychaete *Nereis virens* after a 3-week period of digestion (Goerke and Ernst 1977).

---

[6] In this text the term accumulation includes bioconcentration (from concentration in seawater) and biomagnification (from concentration in food)

Table 27. Organism absorb DDT from small concentrations in water and accumulate until a balance between uptake and elimination is attained. As a concentration factor, one can determine the relationship between the concentration in seawater and the concentration in the organism. (Data from various authors, from Ernst 1975a,b)

Organism tested	Duration of experiment (in days)	Concentration of DDT in experimental water (μg/l)	Concentration factor (relative to wet weight)
Lagodon, Micropogon (brackish water fish)	14	0.1–1.0	10,000–38,000
Mya (clam)	5	0.1	8,800
Mercenaria (clam)	5	0.1	1,260
Nereis (polychaete)	5	0.3	2,033
	5	0.75	1,653
	5	3.0	1,400
Lanice (polychaete)	31	2.24	2,300
	3	0.11	273
	3	0.06	208
	3	0.01	233
Penaeus (prawn)	13	0.14	1,500
Euphausia (plankton shrimp)	0.1	0.03	1,200
		0.02	1,100
		0.01	1,100
			(relative to dry weight)
Cyclotella (diatom)		0.7	37,000
Skeletonema (diatom)		0.7	32,000
Amphidinium (peridinea)		0.7	4,300

Approximately 60% of the DDT contained in their body is given off when the affected soles are kept in water that is free of DDT for two months. Accumulation and elimination of DDT and other chlorinated hydrocarbons is, then, a complex interrelationship between harmful substances in water, in food, and in the organisms. This explains, too, why animals living close together in the same region may have different spectra of chlorinated hydrocarbon concentrations in their tissues: they all have different ways of feeding and they all come in different ways into contact with the chlorinated hydrocarbons either in seawater, or adsorbed to suspended particles in the seawater, or incorporated in their food (Fig. 66). In general predators, the higher they are in the feeding hierarchy, show these higher concentrations of DDT and PCB's.

But even with one organism, for example the mussel (Table 28), to date no one has been able to establish with sufficient clarity the relationship between accumulation under laboratory environments versus natural environments, which are much more complicated.

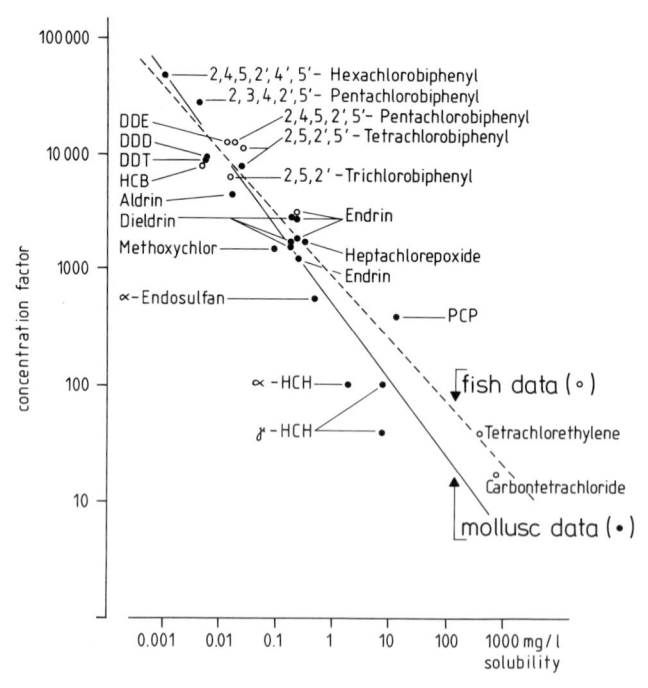

Fig. 64. Concentration factors, referring to wet weight, of various chlorinated hydrocarbons in experiments with marine molluscs and fish. Data from various research are compared with the water solubility of the applied compound. Strongly water-soluble compounds are accumulated to a lesser degree than less water-soluble chlorinated hydrocarbons. Sometimes it might be difficult to identify low solubilities, and it could be better to correlate concentration factors with partition coefficients which describe the partition process between the aqueous phase and the organic compartment (Ernst 1980)

In principle, the accumulation of trace elements, for example heavy metals in organic tissues, presents general problems similar to the accumulation of chlorinated hydrocarbons. Uptake is partly from seawater, partly from food, and partly from sediment or contact with suspended particles, but it is not clear which percentage to the total uptake is delivered by these three sources, which, of course, are interrelated (Fig. 67). As with chlorinated hydrocarbons, it is still not permitted to conclude directly from trace element concentrations in organism, for example in mussel tissue, to trace element concentrations in seawater, but it may be that this goal can be achieved with better scientific efforts. A correlation between concentrations in tissues and in seawater can be found either by identifying uptake rates in the initial phase, when an organism is exposed to higher than normal trace element concentrations (Fig. 68), or by identifying equilibrium concentrations which finally are reached in tissues (Fig. 69).

Each trace element has a specific concentration factor, and elements with different valences may have different concentration factors. For example, 4-valent selenium is more highly accumulated in mussels than 6-valent selenium. Each trace element, too, has a specific elimination characteristic, described as the biological half-life time.

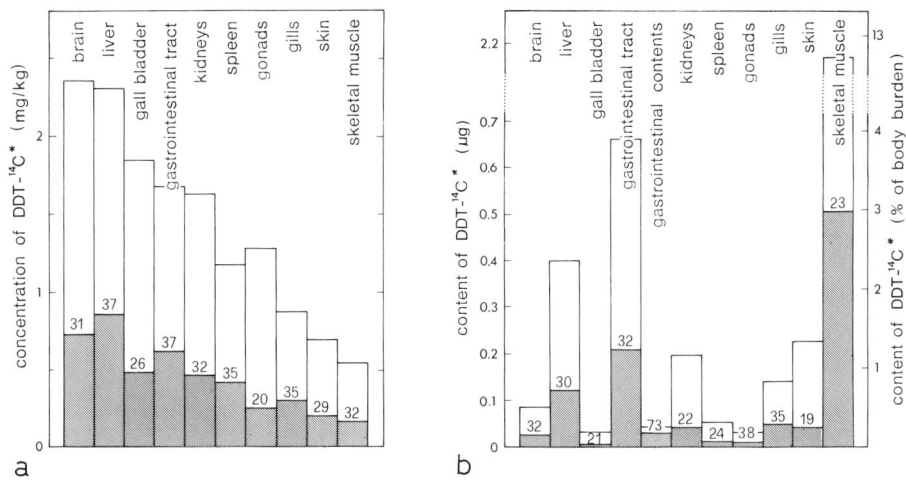

Fig. 65 a,b. DDT is stored up in different quantities in the various organs of a fish. The diagram show the result of experiments with sole *(Solea solea)* which had been fed 17 µg of radioactively marked DDT. The DDT concentrations (on wet weight basis) in the various organs are indicated in **a**. The absolute DDT amounts in fish that were analyzed after being fed DDT for 3 days are indicated in the *white bars* of **b**. The *black bars* indicate how much DDT was still in tissues after the fish were kept in a DDT-free environment for 2 months and were able to eliminate DDT into the aquarium water. The *figures* indicate the percentage of DDT that was not eliminated in this period (Ernst and Goerke 1974)

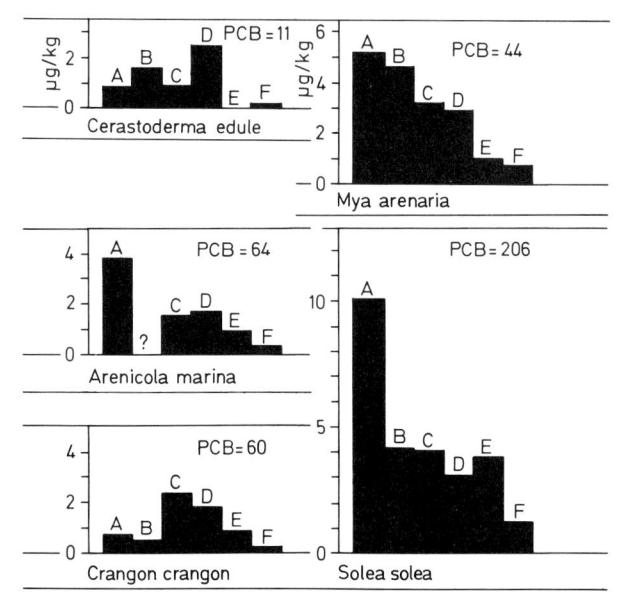

Fig. 66. Even in one and the same marine locality different organisms have different patterns of chlorinated hydrocarbons (per wet weight) in tissues. Different sources of food and different metabolic pathways are reflected. The graph shows patterns for cockle, lugworm, shrimp, clam, and sole from a subtidal area in the Weser Estuary, North Sea, several miles away from the coast and away from the main shipping route, and not by a point source polluted. PCB's are represented by *figures,* the other compounds by *bars* with the meaning: *A* DDD, *B* dieldrin, *C* alpha-HCH, *D* lindan, *E* DDE, *F* alpha-endosulfan (Goerke et al. 1979)

Table 28. Analysis of PCB's and DDT in seawater and in tissues of mussels (*Mytilus* sp., on wet weight basis) in different marine locations. The concentration factor calculated varies up to one order of magnitude from analysis to analysis. For this reason, it is at present not possible to extrapolate directly figures for the concentration of harmful substances in the seawater from the concentration of harmful substances in the mussels (Risebrough et al. 1976a)

Marine area and year	Seawater concentration (ng/l)	Concentration in mussel (mg/kg)	Concentration factor
1. PCB's			
France:			
Valras Beach, 1974	0.94	0.25	270,000
Saintes Maries, 1974	0.95	0.45	470,000
Carro, 1974	0.68	0.20	300,000
Marseille, 1974	1.60	1.10	690,000
California:			
Los Angeles, 1972	1.20	0.23	190,000
San Francisco, 1975	0.98	0.068	69,000
2. DDT or DDE			
France:			
Valras Beach, 1974	0.54	0.051	90,000
Saintes Maries, 1974	0.35	0.013	40,000
Carro, 1974	0.25	0.052	210,000
Marseille, 1974	1.30	0.900	690,000
California:			
Los Angeles, 1972	5.10	1.600	310,000
San Francisco, 1975	0.11	0.005	45,000

That is the time which passes until the trace element concentration in a tissue decreases to 50%, while the organism is kept in uncontaminated water. Figures of biological half-life time in mussels are: cadmium 307–1254 days (Fowler and Benayoun 1976), mercury 377 days (Fowler et al. 1978), selenium 63–81 days (Fowler and Benayoun 1976), cobalt 57–72 days, zinc 48–76 days, antimony 14–20 days (Walz 1979). Unfortunately, not only do various organisms differ as to concentration factors and biological half-life times of one and the same trace element, but the various tissues within one organism have different characteristics. A good amount of general knowledge, and a strict standardization are therefore necessary before it can be attempted to utilize the trace element concentration, or the chlorinated hydrocarbon concentration within a plant or an animal to tell something about contamination of the seawater during the past period. Such a programm has been suggested under the name "Mussel Watch", making use of the fact that mussels (family Mytilidae) are distributed in nearly all the world oceans, and that they are sessile animals which cannot emigrate, but have to cope with all qualities of water that pass by. Therefore they should integrate all kinds of contamination events (Fig. 26).

By order of the U.S. Environmental Protection Agency the Scripps Institution of Oceanography in La Jolla, California, has, since 1976, collected mussels or oysters at 107 localities of both U.S. coasts once a year. The soft body of the molluscs is

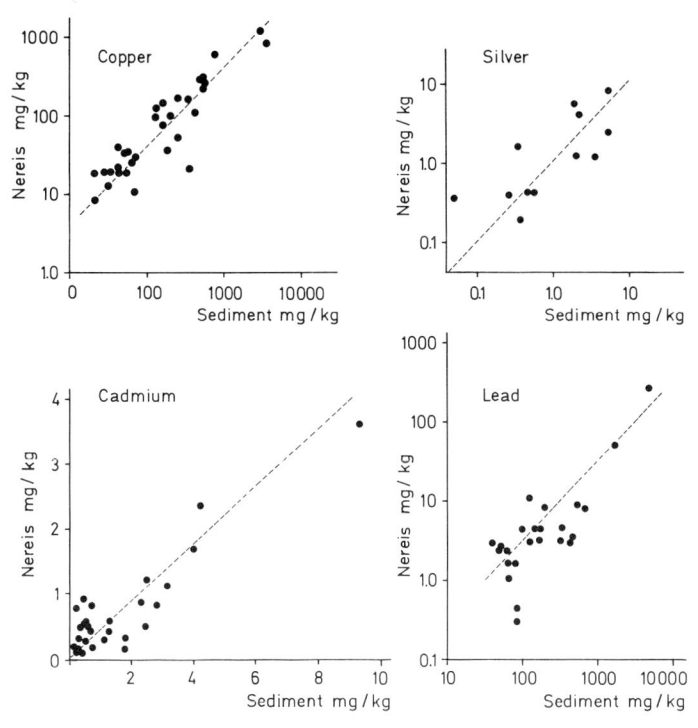

Fig. 67. In southwest England, sediment with widely differing amounts of heavy metals can be found, depending on geological-mineralogical conditions and waste water relationships of the mines (see Table 25). The heavy metal concentrations in the *Nereis diversicolor* polychaet differ corresponding to the different locations in which they are found. Simple relationships exist between the heavy metal concentrations in the sediment and body tissue. Concentrations are indicated in mg/kg dry weight (Bryan 1974)

analyzed for heavy metals, chlorinated hydrocarbons, hydrocarbons, and radioactive elements. During the pilot phase the cost to obtain each sample was 1000 US $, each series of heavy metal analyses amounted to 50 US $, each series of organic pollutant analysis to 1000 US $. From the results it is evident that most compounds are rather uniformly distributed over the coastal regions, an expression of large-region or global marine pollution. But some regional influences can also be detected (Goldberg et al. 1978).

Uptake of trace metals from seawater follows several different ways which are partly combined. The mechanisms can best be studied in experiments with a heavy metal which is of rather low toxicity and can be added to the seawater or the culture medium in rather high concentrations without impairing the organisms too much. For this reason, many experiments have been done with lead.

If algae cultures are contaminated with lead, an amount dependent on the lead concentration is at first absorbed by the surface layer of the algal cells. This does not seem to be an active process, but is caused by the physicochemical characteristics of the cell surface. The addition of complexing agents (EDTA) hinders the absorption

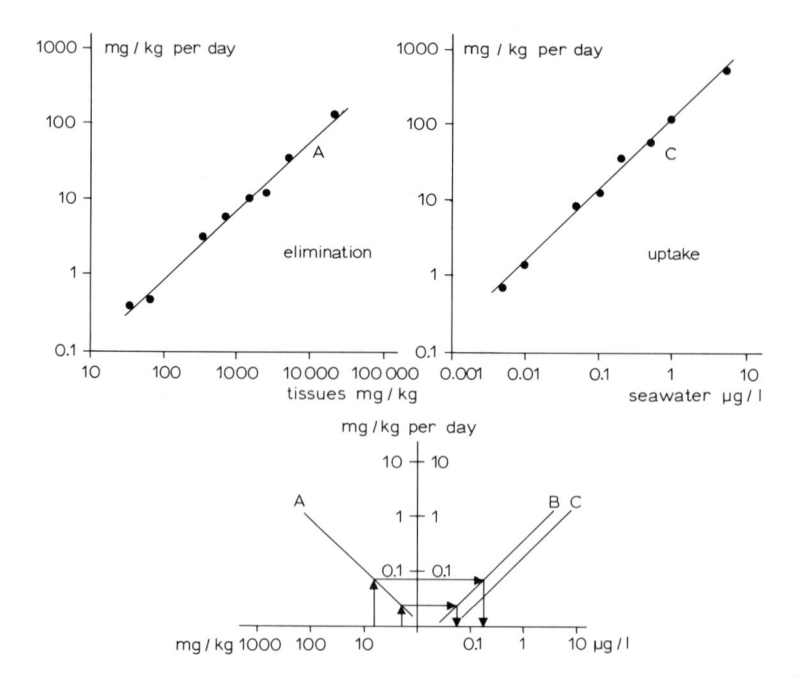

Fig. 68. If *Mytilus edulis* mussels are kept in seawater to which lead nitrate has been added, simple relationships between the rate of uptake (relative to dry weight of mussel soft body) and the lead concentration in the seawater result. If the mussels are afterwards kept in unpolluted seawater, they give off lead. Simple relationships between the elimination rate and the lead content of the mussel's soft body again result. The experiments on which the data rest were conducted over a 40-day period.
If lead uptake and elimination rates *(upper figures)* are compared with one another, the straight lines *A* and *C* in the lower figure result. Mussels, however, do not absorb lead from seawater alone, but also from the algae they consume. This has been taken into account in line *B*. With the rates of uptake and elimination, every value of lead concentration in seawater corresponds to a certain value of lead concentration in the tissue. *Arrows* point to the lead concentration in mussels from Helgoland in the North Sea (2.2 mg/kg) and from the estuary of the Weser River (6.4 mg/kg). The corresponding concentrations of lead in the seawater were not measured. By and large, however, they correspond to the analyses of various other coastal regions (0.05–0.2 µg/l) (Schulz-Baldes 1974)

of lead, and considerable amounts of lead can still be washed off the algae with EDTA within the first two days (Schulz-Baldes and Lewin 1976).
Similar results are arrived at if either living or dead cultures of the alga *Chlamydomonas* are brought into contact with a medium containing 0.02 µg/l of mercury. In the first 10 h of the experiment, up to 360 mg/kg of mercury (on dry weight basis) is absorbed by the algae (Lock 1975). What role the absorption of trace elements plays in the surface layer of animal bodies is largely unknown. However, water fleas *(Daphnia)* absorb from a medium contaminated with 0.25–1 µg/l of mercury the metal within the first hours while later the concentrations in the soft body do not change significantly.

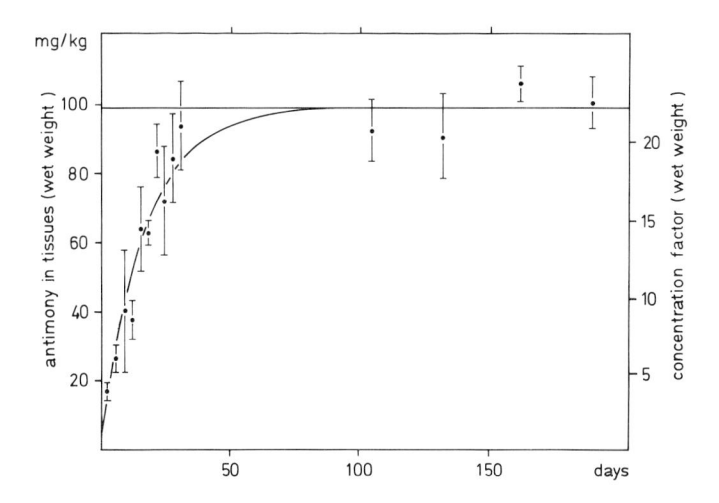

Fig. 69. In accumulation experiments with 20 mm-long *Mytilus edulis* mussels, a balance between the antimony concentrations in experimentally contaminated seawater (5 mg/l) and in the soft body of the mussel occurs within a month. The relationships can only be described with approximate correctness by means of an exponential function. In fact, the increase of the concentration in the soft body during the first 30 days of the experiment follows almost a straight line (Walz 1979)

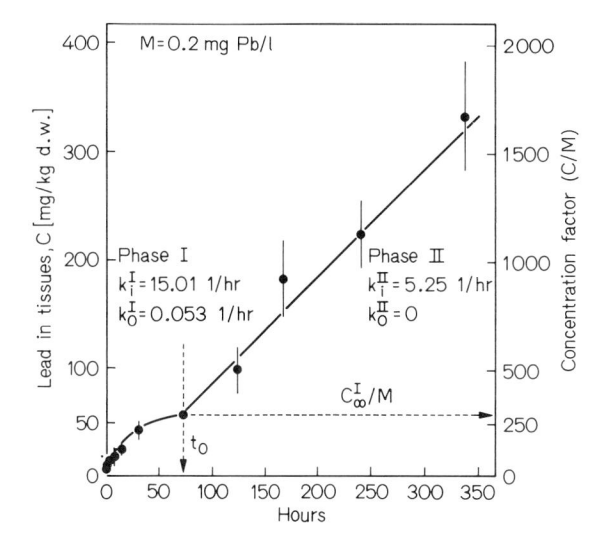

Fig. 70. Uptake of lead from a concentration of 0.2 mg/l in seawater by *Mytilus edulis* mussels. The uptake in the whole soft body is represented in accordance with the conception of a two-compartment phase model (Schulz-Baldes 1978)

Also in larger animals, the rate of uptake is different in the initial hours of the experiment than later. It seems improbable, however, that simple surface absorption could play a significant role, as surfaces are too small in relation to volume or body weight. The attempt has been made to describe this phenomenon as a two-compartment system in which the accumulation occurs exponentially in the initial phase (as in a single-compartment model of the type: exchange with a medium of constant concentration) while the accumulation proceeds linearly in the second phase (Fig. 70).

The difference might be found in the diversely rapid uptake of heavy metals into the various tissues of an animal, in the different concentration factors achieved in various tissues, and in different elimination characteristics.

In experiments with lead, the lead concentration in the blood of *Mytilus edulis* mussels stabilizes quickly; it depends on the lead concentration in the aquarium water and is approximately 2–3 times higher. By means of ultrafiltration, the lead also reaches the pericardial liquid. The lead is reabsorbed by the kidneys from the primary urine. The excretion cells of the kidneys absorb the lead and store it in granula enclosed by membranes (Fig. 71). The excretion cells partially constrict and the apical bodies with the lead granula are eliminated via the pore of the kidney. In an EM picture using the method of X-ray microanalysis, it becomes clear that lead and phosphorus occur together. The presumption is justified that lead in the granula is stored up in an indissoluble complex with phosphates (Schulz-Baldes 1978).

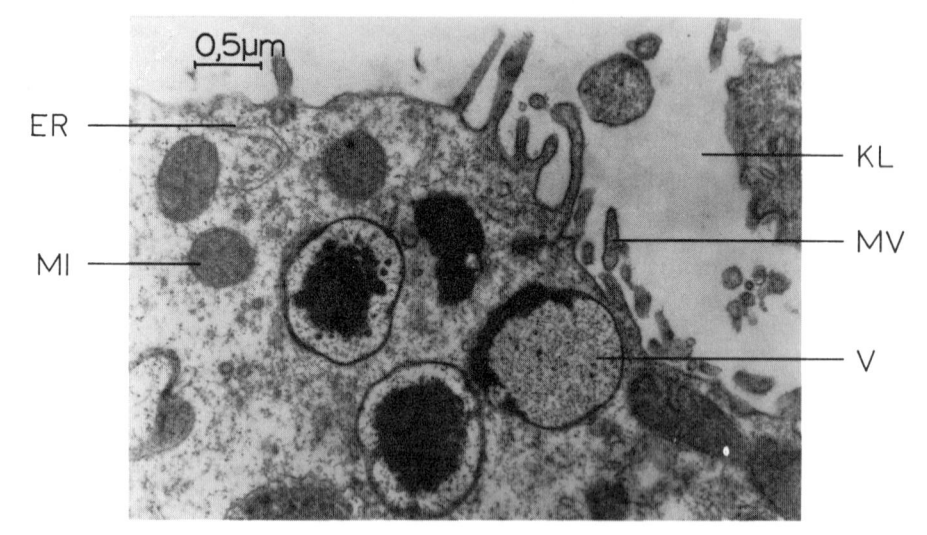

Fig. 71. Electronmicroscope photograph of excretion cells in the kidney of a *Mytilus edulis* mussel. *ER* endoplasmatic reticulum, *MV* microvilli, *V* vesiculum with granules of lead, *MI* mitochondrium, *KL* kidney lumen (Schulz-Baldes 1978)

It can be assumed that a linear increase with time of lead concentrations in the entire body of a mussel then becomes apparent when the excretion capacity of the mussel kidney no longer keeps pace with uptake of lead so that a storing occurs in the kidney.

Organisms need at least 11 essential trace elements: Fe, Cu, Zn, Co, Mn, Cr, Mo, V, Se, Ni, and Sn. These are important for enzyme metabolism. It is obvious that accumulation mechanisms were developed by organisms in the course of evolution to enrich these essential elements from the extreme dissolution in seawater. If along with them, nonessential and extremely toxic elements like mercury, cadmium, and lead are also absorbed from seawater, then it can be assumed that the same mechanisms are activated as for the essential two-valence trace elements. It is possible that mercury, cadmium, and lead also behave like zinc or copper during ultrafiltration from the blood to primary urine and during reabsorption in the kidneys. To be sure, zinc and copper are necessary, but they are also poisonous in higher concentrations. To date, no biochemical reactions are known in which mercury, cadmium, and lead play a positive role. For this reason, these elements, even in the smallest concentrations, are held to be basically toxic, that is, they have a negative effect on physiological processes. That organisms continue to survive in the sea despite this is due to the fact that they tolerate the toxicity of these trace elements, not only at the low concentrations that occur naturally in seawater, but also in the higher concentrations which organisms inevitably accumulate in their tissues. By and large, uptake mechanisms cause more trace elements to be absorbed than the organism needs (Bryan 1976b). Excretion processes see to it that trace elements are eliminated, too; if excretion is not sufficient, toxic trace elements may be transformed into a nontoxic compound and stored away in liver or kidney, or even in hair, feathers, or in the calcareous shell of molluscs. As long as excretion and storage mechanisms are effective, sensitive tissues like the brain are protected from accumulating too high concentrations of toxic elements which might be harmful to physiological processes.

In detail, physiological and biochemical strategies for uptake, storage, and elimination of toxic heavy metals may differ widely. For example, predatory fish of the same 40–50 cm length living under very similar environmental conditions at a depth of 2500 m have very different mercury concentrations in their flesh: *Aldrovandia macrochir* a mere 0.07–0.08 mg/kg; *Antimora rostrata,* on the other hand, 0.5–0.7 mg/kg, ten times as much on wet weight basis (Barber et al. 1972). These figures could indicate different growth rates and, therefore, different ages for fish of the same size, or different physiological adaptations to mercury uptake.

Larger predators among marine animals absorb considerable amounts of heavy metal through the fish they eat. Cases are well known in sharks and other large edible fish, for example (see Chap. 8.2). Seals that absorb highly toxic methylmercury through the fish they eat also have a long life. Seals accumulate mercury in their liver, but no more than 15% methylmercury has been found in them. The greater part is present in less toxic form and has apparently been detoxified. Their brain, which is sensitive to mercury, is protected from high concentrations by the mercury's being stored up in their liver. Only when the mercury concentration in their liver rises above 200 mg/kg, relative to wet weight, does their protective mechanism fail. Then the brain also attains high concentrations (Roberts et al. 1976).

In experiments with rats and quails, scientists were able to prove that selenium diminishes the toxic effect of mercury. In swordfish and seals, the concentration of selenium increases with that of the mercury (Table 33 and Fig. 72). According to findings in California sea lions *(Zalophus californianus),* there are indications that the proportion of the atom equivalents of mercury, selenium, and bromine, not the absolute quantities plays a role; stillbirths are more common if the proportion is not 1:1:1 (Martin et al. 1976). In the connective tissue of the liver of dolphins, concentrations of pure mercury selenide were found, apparently a product of detoxification of the methylmercury absorbed through their diet of fish (Martoja and Viale 1977).

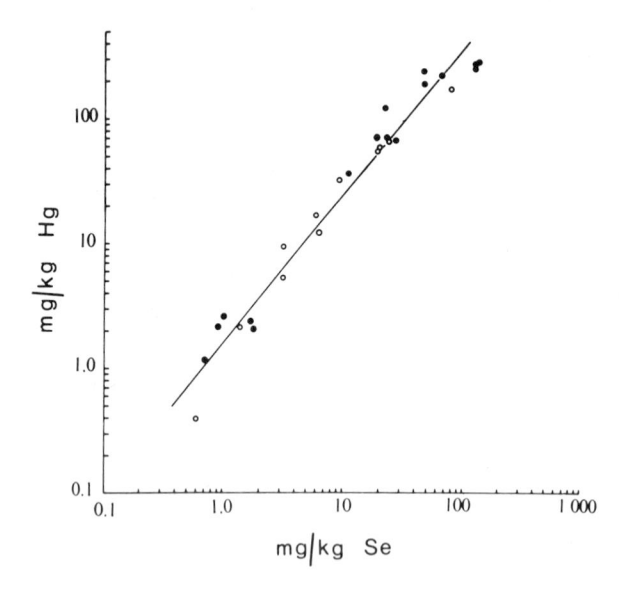

Fig. 72. Seals and whales manifest extremely high concentrations of trace elements. In the livers of 5 Dutch seals *(Phoca vitulina)*, the following concentrations were analyzed: 257–326 mg/kg of mercury; 46–134 mg/kg of selenium; 25–34 mg/kg of zinc; 0.2–1.7 mg/kg of arsenic; 0.05–0.3 mg/kg of cadmium; and less than 0.01 mg/kg of antimony (on wet weight basis). There is a correlation between the concentrations of mercury and selenium both in the liver of Dutch seals *(closed dots)* and the liver of dolphins and porpoises of various origins *(open dots)* (Koeman et al. 1973a)

These are the first, tentative findings indicating that there are regulatory mechanisms which permit marine animals to live with high mercury concentrations. For other trace elements, other biochemical reactions come into consideration. Cadmium, for example, is bound as the protein compound metallothionin.

7.3 Geochemical Processes

Only such substances accumulate in the tissues of organisms which are to a certain degree, persistent. Heavy metals and other trace elements are persistent as elements. For practical purposes, even radioisotopes which have long half-lives can be regarded as persistent, even if they disintegrate with time. The same assumption can be made for a number of chlorinated or, in another way, halogenated hydrocarbons: they are not absolutely, but to a certain degree, resistant to chemical, photochemical, and microbial impact.

While erosion is predominant on land, sedimentation is the dominating factor in the sea. Harmful substances in seawater, when they sedimentate and become incorporated in the bottom sediment, disappear from the biosphere. They must, however, be so deeply buried, and the overlying sediment must be so thick, that neither the burrowing activity of bottom fauna nor occasional erosion due to storm impact bring the deep sediment back into contact with the biosphere. In superficial layers of the sediment, such circumstances are guaranteed only in deep anoxic regions, where water depth safeguards against water dynamics, and lack of oxygen does not allow burrowing macrofauna to exist (Fig. 81). Otherwise, burying animals live as deep as the deep sea. If PCB's have been found below 15 cm sediment depth (Table 29), this must not mean that sedimentation was more than 15 cm during the decades when PCB's were introduced into the biosphere and into the oceans. It may be that a burrowing animal

Table 29. Sediment cores normally have a higher concentration of harmful substances in the strata close to the surface; however, the picture is often complicated by the burrowing activity of bottom animals which mixes up the layers of sediment. The figures refer to PCB's in µg/kg of dry sediment from the Central North Sea and the Skagerrak (Eder 1976)

Sediment depth	Central North Sea, 1972				Skagerrak, 1972–73	
	50 m depth muddy sand			45 m depth fine sand		420 m depth clay	505 m depth clay	
1	33	51	36	14	17	35	41	37
3	29	22	19	10	15	211	39	19
5	28	45	22	10	58	18	34	22
7	42	23	16	19	30	111	23	16
9	14	26	36	17	25	21	37	19
11	0	17	8	18	8	16	0	19
13	0	4	8	18	39	51	0	19
15	17	0	13			93	0	26
17	11	0	12			340		
19						27		

transported sediment from the surface down to 15 cm depth right yesterday. In shallow waters the situation is even more complex. When higher concentrations of heavy metals are found in the upper 20–40 cm of the sediment in Helgoland Bight, at 22 m water depth (Fig. 73), this must not be an indication that heavy metal concentrations increased during the past centruy. Bioturbation of the sediment by burrowing macrofauna could have mixed any recent addition of heavy metals to the sediment with the deeper layers, and whenever a heavy storm hits Helgoland Bight, large-scale erosion occurs, with subsequent sedimentation of suspended sand, silt, and clay. In this way, layers of several centimeters may be sedimented within a day (Gadow and Reineck 1969).

In principle, wherever there is a reasonable sedimentation rate, a gradual burying of pollutants occurs, even when the surface layer again and again is disturbed by bottom-living animals. Gradually, as new sediment layers are added to the surface, the biosphere, that is the layer inhabited by organisms, moves up and leaves the deeper layers less and less disturbed. Therefore, the knowledge of sedimentation rates is of high importance when assessing the possible fate of pollutants in the marine environment (Table 30).

Geologically, it can be assumed that the trace element concentration in seawater was not constant in the past. If mountains came into being as a result of movements of the Earth's crust, then erosion was also heavy and led to large amounts of trace elements entering the oceans. High volcanic activity during such periods contributed, too, to the heavy metal concentrations in the biosphere. The concentrations were certainly higher then than in times when the greater part of the Earth was covered by water and hardly any erosion could take place on land. In these periods, sediment and trace elements were transported away from the biosphere into relatively deep sediment layers.

While most persistent hydrocarbons owe their origin to man, heavy metals and other trace elements owe their origin to the Earth's crust. They reach the biosphere by

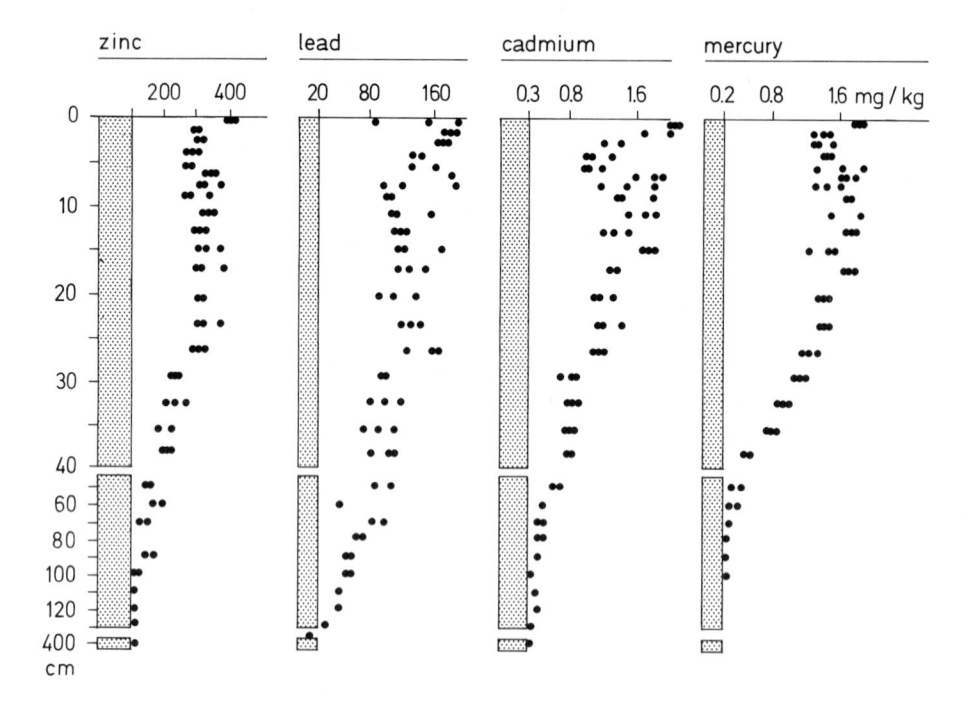

Fig. 73. Concentrations of heavy metals in a sediment core from 22 m water depth in the Bay of Helgoland, North Sea. The natural rate of sedimentation in this area is approximately 50 cm per century. The *shadowed vertical bars* represent as background the heavy metal concentration before man and his civilization affected it. At a core depth of 125 cm, corresponding approximately to a period of 200–300 years ago, fairly high concentrations of some heavy metals begin to stand out. There are especially high amounts in the upper 20–40 cm. It has been concluded, from the chemical form of the heavy metals that atmospheric contribution played the most important role and that a 3.5-fold increase of the lead concentration and a 2.5-fold increase of the cadmium concentration has taken place (Patchineelam and Förstner 1977). These conclusions are not beyond doubt as there are plenty of burrowing macrofauna living in the silty mud of Helgoland Bight, and as sometimes erosion and subsequent resedimentation of the suspended sediment occurs and disturbs the stratification. Data represent mg/kg of dry clay fraction of the sediment with grain sizes less than 2 μm (Förstner and Reineck 1974)

Table 30. Examples of sedimentation rates in various marine areas (Seibold 1974)

Wadden-Sea	20 cm/year
Coral reefs	1 cm/year
Oxygen-deficient basins off coast of California	2–3 m/millenium
Black Sea	40 cm/millenium
Bay in the Adriatic	25 cm/millenium
Continental slope	10 cm/millenium
Globigerina deep sea clay	5 cm/millenium
Red deep sea clay	1 cm/millenium
Red deep sea clay in Pacific	1 mm/millenium

means of volcanic activity and the weathering of stones. Naturally, they are present in seawater. It was not until particular questions arose about ocean pollution that attention was directed to the geochemical processes. But the few years of research in this field have not yet produced sufficient information to determine the relationship of quantities supplied by erosion and precipitation and by sedimentation. To estimate whether increased amounts of man-made input will be matched by increased amounts of sedimentation, so that increased concentrations do not take place in the ocean, seems an important task.

Knowledge is quite insufficient on the fate of pollutants at the water-sediment interface, and what exchanges may occur between the overlying water, the pore water in the oxidized surface layer and the pore water in the anoxic deeper layer of the sediment. Quite insufficient, too, is knowledge on the fate of pollutants at the freshwater-seawater border in rivers and estuaries where the situation is complicated by the formation of a turbidity cloud (Fig. 5). Probably a reasonable fraction of the contaminants which originally are dissolved in river water, or are bound to colloidal material, will go into the flakes which form at the brackish water border, and accumulate in the brackish water region. It may be that only a fraction of the pollutant load of a river (Table 8) really reaches the open sea, but that a large fraction accumulates with mud and clay in the estuarine region. As a matter of fact, fine material which sedimentates off the mouth of a polluted river contains tremendous concentrations of heavy metals, more than upriver and more than farther in the sea (Fig. 74; see Chap. 3.2).

Experimentally it is easy to demonstrate changes in the chemical form of heavy metals when introduced into the marine environment. If one adds 1 μg/l of inorganic mercury to seawater in a large, 100 m^3 tank, 12–24 h later one finds about 70% of the mercury in a nonreactive form (see Fig. 80), probably bound to particles which exist in seawater as suspension, and which will become incorporated into the sediment after sedimentation (Topping and Davies 1980). Plankton organisms are helping in the process to eliminate heavy metals from the water column by sedimentation to the sea bottom (Table 31).

A remarkable correlation exists between cadmium, phosphate, and nitrate concentrations in seawater (Bruland et al. 1978). In deep seawater high nutrient concentrations are combined with a cadmium concentration of about 100 ng/l. In the biologically active surface zone of the sea, however, cadmium concentrations are as much depleted as phosphate and nitrate concentrations, and figures as low as 4 ng/l are measured (Fig. 75). All three elements are, on the other hand, concentrated in plankton and other living organisms. The correlation between cadmium and phosphorus in plankton samples is not as close as in seawater, but can nevertheless be demonstrated. Mean cadmium concentration in plankton collected off Baja California was 5.8 mg/ kg on dry weight basis. One can imagine that with dead plankton phosphorus as well as cadmium is transported to the sea bottom. When under special conditions phosphate ores are formed on the sea bottom, they automatically include reasonable concentrations of cadmium. In this way it can be explained why synthetic fertilizer contains more cadmium which then is transported into agricultural crops, than one would like considering the health risk (see Chap. 8.6).

As regards air-sea interaction, interesting phenomena have been discovered in recent years, but they have not yet been conclusively explained. Air-borne dust samples

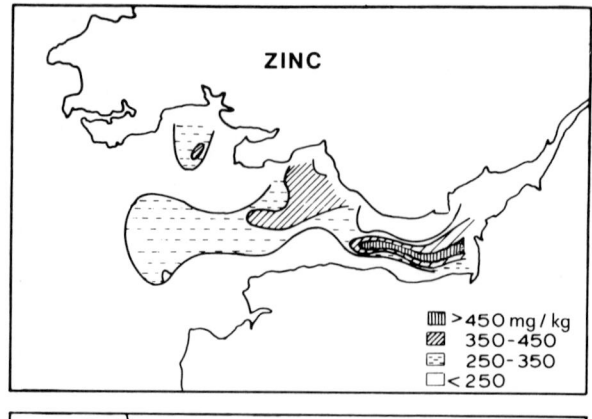

Fig. 74. The estuary of the Severn River and the Bristol Channel receive large amounts of heavy metals from the industrial area around Bristol and Cardiff as well as from the mining areas of South Wales. This is reflected in the concentrations of zinc and lead in the fine-grained fraction of sediment (under 61 μm). Data in mg/kg dry weight (Chester and Stoner 1975)

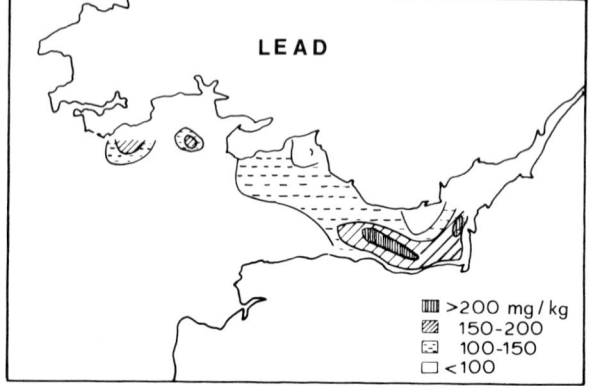

Table 31. Concentration of trace elements in plankton larger than 0.08 mm on which the krill shrimp *Meganyctiphanes norvegica* feeds, in krill eggs, moulted exuvia, and fecal pellets. The analyses were conducted in the Mediterranean near Monaco. The considerable amounts of heavy metals stored up in the fecal pellets which sink down and sedimentate on the bottom of the sea are impressive. Data in mg/kg dry weight. For conversion to mg/kg wet weight, apply the relevant numbers indicated in the following table (Fowler 1977)

Material	Dry weight %	Cd	Cr	Cu	Hg	Pb	Sb	Se	Zn
Plankton	10.7	2.1	4.9	39	0.05	11	0.22	2.7	483
Meganyctiphanes whole animals	4.7	0.7	0.8	48	0.35	1.1	0.07	4.4	62
Eggs	10.0	0.6	7.9	17		8.9			348
Exuvia	4.6	2.1	5.3	35	0.17	22	0.80	1.9	146
Fecal pellets	4.4	9.6	38	226	0.34	34	71	6.6	950

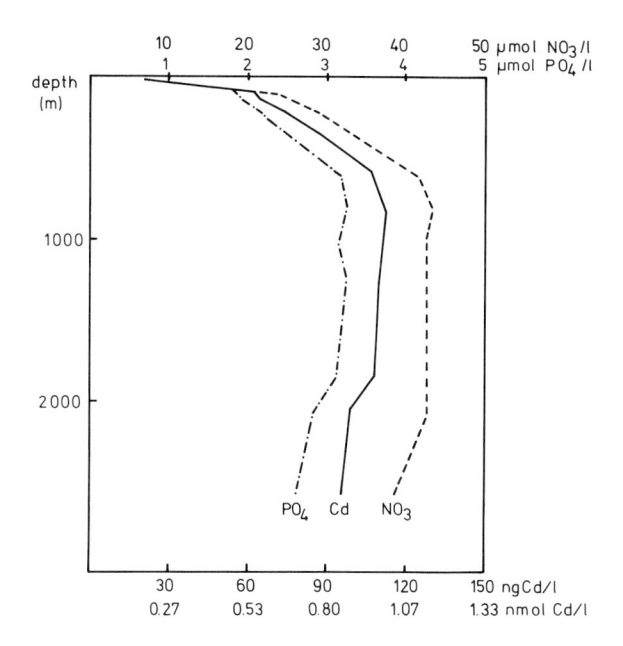

Fig. 75. Depth profiles for cadmium, phosphate and nitrate at a station off Central California in April, 1977 (Bruland et al. 1978)

over the Antarctic and over oceans far away from any source of pollution have been analyzed. If the composition of these air samples is compared with that of the common material of the Earth's crust, zinc and copper are 60–100 times, antimony and lead 1300–2500 times, selenium even 18,000 times more strongly concentrated (Zoller et al. 1974; Fig. 76). Air-borne dust, then, cannot be dust that is kicked up by sandstorms. Because the elements mentioned above are particularly volatile ones, it has been presumed that they enter the atmosphere primarily while high temperature processes are taking place. With lead, this is easy to explain because considerable quantities are added to automobile gasoline in the form of tetraethyllead (see Chap. 8.7). But how many trace elements enter the atmosphere and ocean water by incineration by man on the one hand and by volcanic activity on the other hand is not accurately known (Table 36).

It could even be that heavy metals in atmospheric dust are not on their way from land to sea, but have emerged from the sea itself. There are some facts which point to the hypothesis that heavy metal concentrations in the air above the sea and in seawater are interlinked, and that this holds true for other compounds, too, like hydrocarbons and chlorinated hydrocarbons. From tidal flats, microbially produced tetraethyllead is emitted to the atmosphere (Harrison and Laxen 1978). Copper found in the air above the sea originates from seawater (Catell and Scott 1978), and in general, all air samples collected a few meters above the sea surface have a good chance to be contaminated from the sea; therefore they are not conclusive for calculations of the atmospheric input of pollutants into the sea (Peirson et al. 1974).

Heavy metals and chlorinated hydrocarbons are concentrated in the microlayer at the air-sea interface, a film only a fraction of a millimeter thick. When the sea is agitated and waves break, air bubbles are formed which burst when they reach the sea surface, so a small area of the film becomes air-borne. For mass calculations,

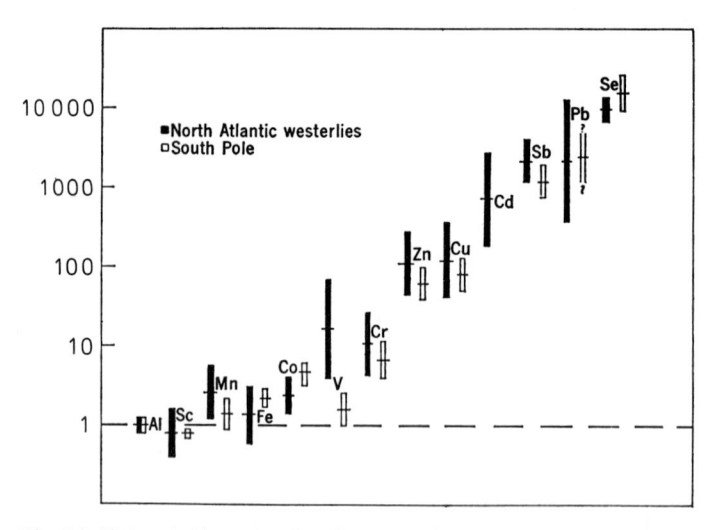

Fig. 76. If dust is blown by the wind away from deserts and other land regions, the dust should have more or less the same composition as normal Earth crust material. However, dust sampled over the Antarctic and over unpolluted areas of the North Atlantic have a quite different elementary composition, with selenium, lead, antimony, and cadmium specially enriched. A man-made source of air pollution is not likely: there is no indication from snow samples that lead, cadmium, copper, zinc, and silver have been precipitated in 1974 in larger quantities than in 1914 (Boutron and Lorius 1979). The similarities of dust from the South Pole and from the North Atlantic is striking. The graph shows the concentration factor against normal Earth crust material (Duce et al. 1975)

the amount of heavy metals and chlorinated hydrocarbons in the microlayer is, in spite of high concentrations, insignificant. For exchange dynamics, however, they may play an important role, and high concentrations of cadmium and DDT and PCB's in this microlayer explain data from oceanic birds and insects.

The finding that open sea petrels (Tubinares) show high concentrations of cadmium, was at first surprising. Specimens of five different species which breed on the island of St. Kilda, in the Atlantic 80 km west of the Outer Hebrides, had up to 49 mg/kg cadmium (on dry weight basis) in their liver and up to 240 mg/kg of cadmium in their kidneys. Petrels live almost exclusively on the open sea and for this reason it is unlikely that they come into contact with cadmium pollution during their lifetime. Petrels feed on material which they collect directly and often in flight off the surface of the water. Thus, it is no surprise that they have extremely high concentrations of cadmium and equally high ones (up to 66 mg/kg wet weight) of PCB's and DDT (Risebrough 1971). Petrels presumably have a detoxification mechanism against cadmium poisoning, probably cadmium-binding proteins (Bull et al. 1977). The sea skater *Halobates,* a hemiptera which is the only open sea insect that skates on the surface of the ocean in fairly warm regions, has been shown to have regular dry weight cadmium concentrations of 33 mg/kg and occasionally up to 300 mg/kg. These bugs feed on plankton animals that have come up from below and are stranded on the surface (Bull et al. 1977). It seems that the cadmium concentration in seaskaters is correlated with the system of oceanic currents (Fig. 77).

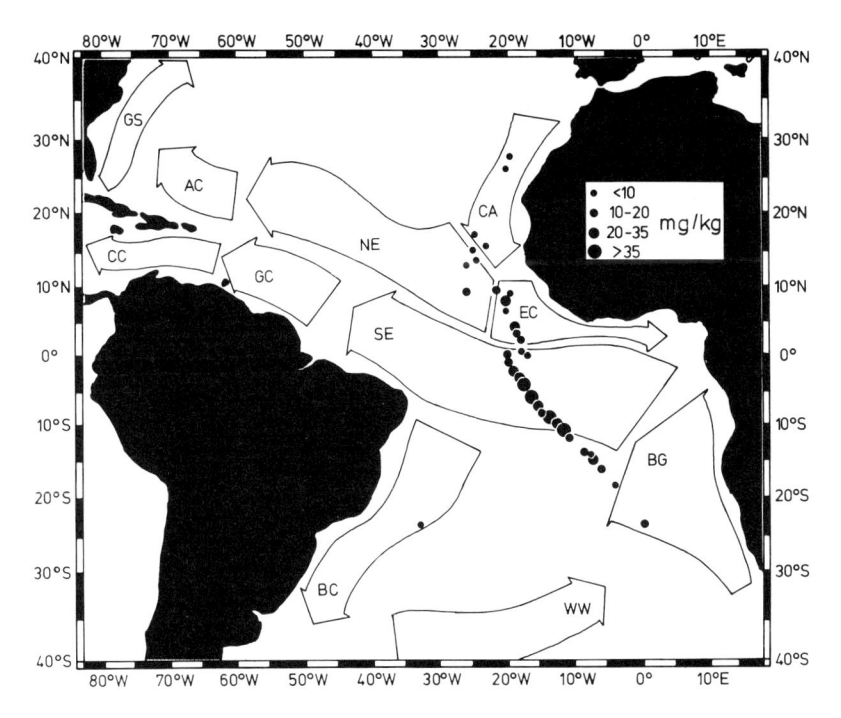

Fig. 77. Cadmium concentrations in the sea-skater *Halobates* living at the sea surface follow distinct patterns and are highest in the equatorial region, which receives nutrient-rich water from upwelling regions. Letters refer to ocean currents; data explaining size of points are mg/kg dry weight (Schulz-Baldes and Cheng 1980)

7.4 Global Considerations

In recent decades, mankind has had to learn to think in global terms. We sublimate the nightmare knowledge that available nuclear weapons are capable of destroying the world. In the meantime everyone has come to know that raw material, food, and energy sources are limited. Pictures transmitted by television from a space capsule showed us how vulnerable the biosphere is, the film of air and water which alone guarantees life on Earth. Calculations are being undertaken to determine the extent to which pollution of the atmosphere by dust and carbon dioxide is altering the influence of the sun's rays and how strongly the production of heat due to fossil combustion and atomic energy is influencing the world's climate. Thought has to be given to the question whether the emission of toxic chemicals has already reached the point that might have considerable global consequences.

The oceans cover 71% of the Earth's surface. They are the mirror of the health of the planet Earth. In recent years, the role of providing a diagnosis on the Earth's state of health has fallen to marine science.

Ocean pollution due to domestic effluents, to the introduction of industrial effluents, and the marine dumping of toxic substances have primarily regional effects. Tar floating on the surface of the ocean and plastic refuse, however, assume global proportions.

How much time do we have until we are faced with serious problems of world-wide ocean pollution?

The oceans of the world have a surface of approximately 360 million km² or 3.6×10^{14} m² (Table 32). Based on an average depth of approximately 3700 m, the oceans of the world contain a total of approximately 1.4×10^{18} m³ of seawater. The surface layer of 100 m is biologically the most important; it contains around 3.6×10^{16} m³ of seawater.

Table 32. The size of the world's oceans and of their depth strata. (According to Seibold 1974)

Ocean regions	Depth range	Million km²	Percent of Earth surface
Continental shelf	0–200	27.1	5.3
Continental slope	200–1000	16.0	3.1
	1000–2000	15.8	3.1
Continental foot	2000–3000	30.8	6.1
	3000–4000	75.8	14.8
Deep sea plains	4000–5000	114.7	22.6
	5000–6000	76.8	15.0
Deep sea trenches	6000–7000	4.5	0.9
	7000–11000	0.5	0.1
Total of all oceans		362.0	71.0

	Atlantic	Indian	Pacific
Medium water depth	4188 m	3767 m	3872 m
Area (million km²)			
Continental shelf	6.1	2.6	2.7
Continental slope	6.6	3.5	8.6
	5.4	4.2	2.7
Deep sea plains	68.1	62.8	147.5
Deep sea trenches	0.4	0.3	4.8
Total (without adjoining seas)	86.6	73.4	166.2

According to our present-day knowledge, a harmful reaction might even emanate from harmful substances if the concentration amounts to a mere 0.001 µg/l = 0.001 mg/m³. If this substance is to have a global effect, a pervasive distribution of 1.4 million t in the water of all the world's oceans would have to be present. If this concentration is to have an initial effect solely in the 100-m-deep surface layer, 36,000 t are sufficient. Much smaller quantities could be sufficient to change the system of chemical signaling in the sea, the rather unknown action of pheromones (see Chap. 5.5) and of other substances which signal a partner, food, or home to aquatic organisms. The eel *(Anguilla vulgaris)* is able to perceive the odor of betaphenylethylalcohol, a component of attar of roses, at a dilution of $1 : 2.8 \times 10^{18}$ (Teichmann 1959). Only 0.5 t of this substance, mixed with the volume of the world oceans, would be sufficient for the eel to smell it.

These figures are only meant to serve as a reference point to imagine the parameters that have to be considered. They do not take into account the elimination of harmful substances from the sea by means of sedimentation and they take absolute persistence for granted. Actually, the world-wide distribution will seldom be equal.

8 Global Contamination of the Oceans by Heavy Metals

8.1 How Much Mercury May Be Tolerated in Fish for Human Consumption?

Mercury is dangerous for man, as was demonstrated with the casualties of Minamata (see Chap. 3.2). The United Nations World Health Organization (WHO 1976a) summarized what was known about mercury poisoning and concluded that toxic effects of mercury can be expected to become manifest in the most sensitive groups of human adults when over longer periods of time a weekly intake of 1.3–2.9 mg methylmercury takes place, equivalent 0.02–0.05 mg/kg body weight. Such mercury intake results in blood concentrations of 0.2–0.5 mg/l and in hair concentrations of 50–125 mg/kg. Victims of the "Minamata Disease" had consumed, with polluted fish, about 14 mg of mercury per week, and analyses of their blood and hair showed respectively high mercury concentrations (Table 7). Concentrations of less than 0.02 mg/l in blood and less than 6 mg/kg in hair, on the other hand, have to be considered as normal for human beings (Fig. 78).

As early as 1972 a joint committee of experts from WHO and from the United Nations Food and Agriculture Organization (FAO) determined the quantity of mercury which may be tolerated in human food: 0.2 mg of methylmercury or 0.3 mg of total mercury per week for a person, or 0.005 mg total mercury per kg body weight. This limit is only about ten times lower than the threshold of methylmercury intake which may result in symptoms of mercury poisoning. But to issue a lower intake figure than 0.3 mg per person and week would be wishful thinking because mercury is not only present in contaminated food stuff but occurs in traces everywhere in our environment, in water, rock, sediment, earth, and in all organisms.

An average citizen in the Federal Republic of Germany has a weekly intake of 0.2 mg of mercury, that is two thirds the amount health organizations tolerate at all. The 0.2 mg are composed of 0.04 mg from grain, 0.03 mg from alcoholic beverages, 0.02 mg from milk and butter, and among other food stuff, 0.016 mg from the 200 g (wet weight) fish which the statistical average citizen consumes weekly. The average mercury concentration in this statistical edible fish is 0.08 mg/kg, mostly present as methylmercury (Ernährungsbericht 1976). In the U.S.A., the figures are similar. On the average, only 14 g fish are eaten per day, about 100 g per week. Concentrated in this fish, the statistical U.S. citizen takes up 0.014 mg of mercury. This quantity, as in the Federal Republic of Germany, is only a small fraction of the total mercury intake with food, which in the U.S.A., too, is not above 0.2 mg per person and week except for families eating very much fish (Finch 1973).

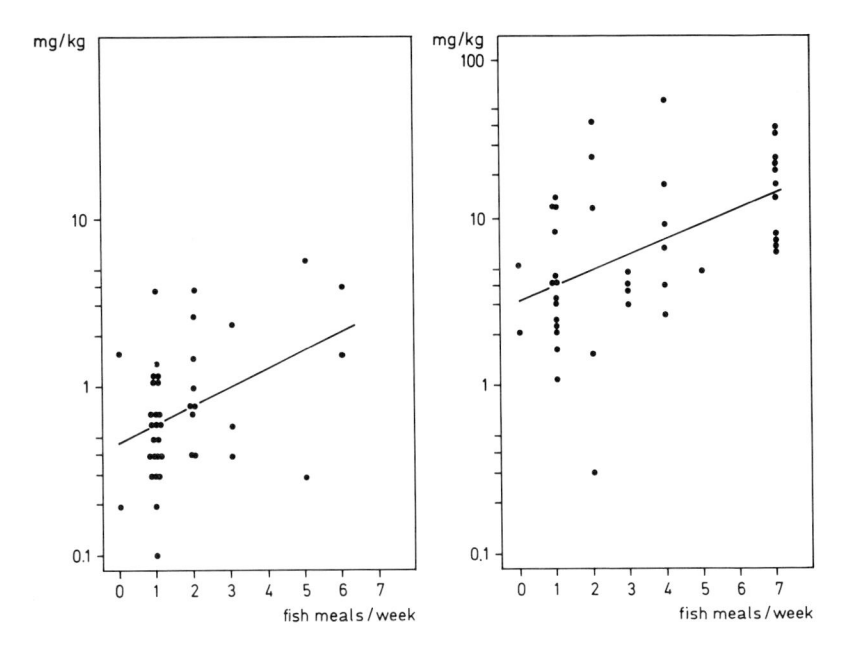

Fig. 78. In Finland test persons were analyzed for mercury in their hair, and they were asked how often they had had fish meals during the preceding 3 months. Mercury concentrations in hair from persons living close to Lake Mouhijärvi *(left diagram)* are all below 6 mg/kg. Lake Mouhihärvi is not industrially polluted, and fish living in this lake have 0.05 mg/kg mercury in their meat. In the region of Kotka and Munapirtti a river runs into the sea which comes from an area with paper mills and chlorine industry which years ago had emitted wastes containing mercury. At the period of the investigation (1971) fish from the region of Kotka and Munapirtti still had mercury concentrations of 2–3 mg/kg. In this region even persons who had not eaten fish at all have 2–6 mg/kg mercury in their hair *(right diagram)*, and those with three fish meals per week come into the range of 6 mg/kg. But even persons who had eaten fish day for day do not come to the range of 60 mg/kg where toxic symptoms may occur (Sumari et al. 1972)

When it became known how dangerous mercury is for man and to what extent high mercury concentrations in fish may accumulate from polluted waters, governments had to react with legislation. When it became known from Japan that Minamata Bay fish had mercury concentrations of 50 mg/kg, the Swedish Health Ministry determined 0.5 mg/kg to be the limit of mercury concentration tolerable in fish for human consumption. Such regulation would provide a safety margin of 100 which is common practice in environmental regulations. But then two facts became known: first it was discovered that fish in Lake Vänern in Sweden exceeded this limit, and therefore the limit was elevated to 1 mg/kg. Then, at a public hearing, it became known that information from Japan related to dry weight, whilst Swedish regulations related to wet weight of fish (Ackefors et al. 1970; Ui 1971). This means that polluted fish (about 80% water content) from Minamata Bay had a mercury concentration of only 10 mg/kg, on wet weight basis. The safety margin was down from 1:100 to bare 1:10. The calculation is not quite correct because the statistical Japanese, and specially the victims of the Minamata area, eat more fish then the average Swede.

1 mg/kg [7] is the maximum mercury concentration allowed in edible fish in Sweden, Japan, Switzerland, and, according to the Regulation on Maximum Amounts of Mercury in Fish, Crustacea and Molluscs from 1975, in the Federal Republic of Germany. In Italy the rule is 0.7 mg/kg, in the USA and Canada 0.5 mg/kg, but in Norway 1.5 mg/kg. The variety of limits is a mirror of the difficulties which governments encounter in setting reasonable standards.

Governments want to be sure that mercury from industrial and other marine pollution does not find its way to the dinner of the citizen. In fact the regulations of all countries mentioned, from 0.5 to 1.5 mg/kg, safeguard against a drastic pollution catastrophe like Minamata, if properly monitored. The question, however, remains to be discussed what level of mercury concentration can be achieved and can be tolerated in such fish which live in nearshore and estuarine regions polluted by rivers and other diffuse sources, and what to do with high natural mercury concentrations in some fish. The question becomes more complex, too, by the fact that one has to consider the risks of people who eat considerable quantities of fish.

In the U.S.A. the statistical average citizen eats 14 g of fish per day, but 1% of the population eat 77 g/day, and 0.1% of the population eat as much as 165 g/day, or 1155 g of fish per week (Finch 1973). Eating fish with 1 mg/kg mercury, such a person would take in 1.3 mg mercury per week, 1.1 mg from fish and 0.2 mg from other food. This exceeds about four times the quantity tolerated for the average person; when fish with 0.5 mg/kg of mercury is eaten, the mercury quantity taken up is double the tolerance set for an average citizen. 0.5 mg/kg of mercury quite regularly occur in fish from such estuaries and coastal seas which are generally polluted, like the Irish Sea, the mouth of Thames and Rhine, of Weser and Elbe (see Chap. 3.2 and 4.2). In negotiations about the Paris Convention (see Chap. 10) regarding marine pollution from land, efforts are under consideration to fight mercury pollution with the aim that marine fish should not have more than 0.3 mg/kg of mercury in their meat. But at present this goal has not been achieved. If one sets the tolerance limit for mercury in fish at 0.3 mg/kg, some fish could not be sold any longer, if one sets the limit at 1 mg/kg, there might be a risk for extraordinary fish eaters.

It is good to know that all the species of edible sea fish which are landed in great quantities in the German fishery ports have only small concentrations of mercury in their meat (Meyer 1972; Priebe 1976). Concentrations in cod *(Gadus morhua)*, coalfish *(Pollachius virens)*, haddock *(Melanogrammus aeglefinus)*, herring *(Clupea harengus)*, and mackerel *(Scomber scrombrus)* are generally 0.05–0.1 mg/kg, very rarely up to 0.4 mg/kg, and only very exceptionally, in very old larger fishes, up to 1 mg/kg. This fish reach catchable size in 5–8 years. Flatfish (Pleuronectidae) have somewhat higher concentrations of 0.1–0.2 mg/kg, perhaps because they come into contact with the sediment of the sea bottom.

Again somewhat higher, up to 0.25 mg/kg are mercury concentrations in catfish *(Anarhichas)* and redfish *(Sebastes)*, both of which grow slowly. A wide-ranging analysis of North Sea fish has shown that mercury concentrations are, in general, not higher than in fish from the open Atlantic: cod 0.13 mg/kg (variation 0.03–0.26),

---

7 If not stated otherwise, all data on concentrations of mercury in organic tissues, given in this chapter, are per wet weight

herring 0.06 mg/kg (variation 0.02–0.26). It is good to know that apparently the North Sea is not polluted by mercury, apart from regional pollution in coastal areas (Fig. 76; Ices 1974; see Chap. 3.2 and 4.2).

8.2 Mercury in Large Fish, Seal, and Open Sea Marine Birds

Tuna fish *(Thunnus thynnus)* live mainly in the open seas and avoid the areas where rivers flow into an ocean. Therefore, they do not come into contact with ocean areas that are contaminated by effluents. Some quantities of canned tuna fish are seized in various countries, however, because they contain more than 0.5 mg/kg of mercury. This question is under public discussion. To be sure, of 56 brands of canned tuna fish tested by the Bundesforschungsanstalt für Fischerei in Hamburg, Federal Republic of Germany, half contained less than 0.2 mg/kg of mercury (28 samples), an additional 20 were between 0.2 and 0.4 mg/kg, 7 between 0.4 and 0.5 mg/kg, and only one contained 0.9 mg/kg (Mundt and Feldt 1971). But other tests again and again show amounts around 1 mg/kg in tuna fish from various sources, primarily in older fish that are over 2 m long (Peterson et al. 1973). The concentration is even higher in swordfish *(Xiphias gladius)*. Five fish cought in the Straits of Gibraltar contained between 1 and 2 mg/kg (Establier 1972). Indeed, a case of mercury poisoning did occur in the U.S.A., following the eating of swordfish meat (Kahn 1971), but the affected person had eaten a lot of swordfish meat over a fairly long period of time as part of a diet to lose weight.

The marlin, a relative of the swordfish, seems to have the highest concentrations among fish (Table 33). Sharks, too, are heavily contaminated by mercury, and in 1972 topes *(Galeorhinus)* longer than 112 cm were declared unfit for human consumption in Australia because on the average they contain more than 0.5 mg/kg of mercury. Fault was found with 81% of the catch (Fig. 79).

In the Federal Republic of Germany, the Regulation on Maximum Amounts of Mercury in Fish, Crustacea and Molluscs (see Chap. 8.1) insures that no edible fish containing more than 1 mg/kg of mercury shall reach the market. Tuna, swordfish, and

Table 33. According to research to date, the highest concentrations of mercury in any fish have been found in the marlin *(Makaira indica)* which is related to the swordfish. These fish weigh up to 500 kg. Sportsmen fish for them in the waters of northern Australia. Japanese fishermen brought around 44,000 of them into Japan in 1970. The figures are in mg/kg wet weight; range of variation in parentheses (Mackay et al. 1975)

Trace element	Musculature	Liver
Zinc	8.6 (5.8–14.6)	47.5 (4–375)
Mercury	7.3 (0.5–16.5)	10.4 (0.3–63)
Selenium	2.2 (0.4–4.3)	5.4 (1.4–13.5)
Cadmium	0.9 (0.05–0.4)	9.2 (0.2–83)
Arsenic	0.6 (0.1–1.6)	1.0 (0.1–2.7)
Lead	0.6 (0.1–0.9)	0.7 (0.4–1.1)
Copper	0.4 (0.3–1.2)	4.6 (0.5–22)

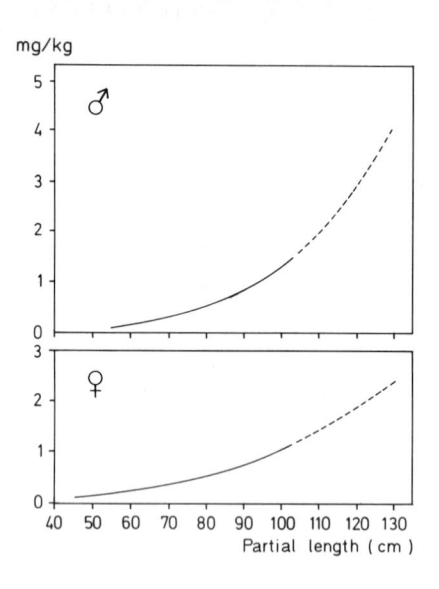

Fig. 79. The mercury content in the meat of south Australian tope sharks *(Galeorhinus)* increases exponentially with body size. *Partial lengths* in the graph are lengths of the dressed fish from front end of pectoral fin to tail incisure. 63 cm partial length corresponds to 92 cm total length and an age of 4 years, 83 cm partial length corresponds to 122 cm total length and an age of 8 years, 102 cm partial length corresponds to 150 cm total length and an age of 16 years (Walker 1976)

shark meat with higher concentrations are not allowed to be imported, even though canned tuna accounts for only 4.5% of the fish consumption in the Federal Republic of Germany.

Halibut *(Hippoglossus hippoglossus)* reach a length of 4 m and a weight of 300 kg. It is assumed that they are then 50 years old. The older and, consequently, the bigger halibut are, the higher the mercury concentration in the meat (Priebe 1976). By calculation, it emerges that a halibut with a dressed weight of over 115 kg has a mercury concentration of over 1 mg/kg. However, the variations in individual fish are great. In general, it can only be said that approximately half of all halibut weighing more than 60 kg exceed the maximum permissible amount of mercury. These fish are decleared unfit for human consumption and are either ground into fishmeal or exported to countries that do not have the corresponding maximum permissible tolerances for mercury. Only 22 t of halibut over 60 kg were landed into the fishery port of Bremerhaven in 1975, accounting for approximately 18% of the halibut catch. Every fish is individually analyzed for mercury, which is quite costly. This poses the question for the fishing fleet whether or not it is worth bringing in these large halibut. The situation is similar for large sharks, for example the Greenland shark *(Somniosus microcephalus,* 3–4 m in length, 300–500 kg in weight); 60% of the analyses show more mercury than 1 mg/kg. It is interesting that the Greenland shark from the waters of eastern Greenland, with an average mercury content of 0.95 mg/kg, contain somewhat less mercury than fish from Iceland, the Faeroes, and the Hebrides, which average 1.45 mg/kg. This might point to a higher mercury content in the water above the Middle Atlantic Ridge and its volcanic influence (see Chap. 8.3). All porbeagle *(Lamna nasus)* landed in Bremerhaven fishery port also had above 1 mg/kg of mercury.

There are stories about Greenland sledge dogs which get convulsions and die when fed exclusively with the meat of the Greenland shark. It is not known whether mercury can be responsible for these symptoms.

High mercury concentrations are, of course, encountered in fish-feeding sea birds and in seals, because these animals reach great ages. Apparently they take up more mercury with their food than they are able to eliminate, and therefore accumulate more mercury in their tissues from year to year. Examples of high mercury concentrations in sea birds and seals from polluted coastal areas have been provided in Chapter 3.2. However, high mercury concentrations have been found also in animals which live far away from any source of pollution on the high seas.

It seemed ironic: mercury was found, in 1970, in the liver extract an American firm intended to market as highly natural and unpolluted, because they had used the liver of *Callorhinus ursinus* fur seals from the Pribiloff Islands in the North Pacific. The livers of the adult animals, however, contain 3–19 mg/kg of mercury. That is less than the amount found in the livers of fur seals off the coast of Washington State (up to 172 mg/kg), but more than that found in most marine animal life (Anas 1974).

High concentrations of mercury are regularly found in marine bird life: 1–5 mg/kg and occasionally up to 20 mg/kg in goosander *(Mergus merganser)*, guillemot *(Uria aalge)*, kittiwakes *(Rissa tridactyla)*, fulmars *(Fulmaris glacialis)*, skuas *(Stercorarius)*, and eiderducks *(Somateria mollissima)*. A Finnish women became ill with mercury poisoning after eating fairly large amounts of goosander eggs which she had collected on a deserted skerry far from any source of pollution (Nuorteva 1971).

The regulations regarding maximum permissible concentrations of mercury in fish and other seafood are not very sound (Gerlach 1978). Delicatesses should be eaten in small quantities. Instead of regulations it would be better to make the public aware that some marine food, even if it comes from unpolluted areas, may have high mercury concentrations, but that these do not matter much as long as one eats small quantities only of such food like swordfish or shark steak or gull eggs. Since Canada has had a regulation forbidding the sale of fish with more than 0.5 mg/kg of mercury, swordfish has been smuggled into the country in large quantities. Since 1979 swordfish is legal again in Canada in spite of its mercury.

It would be interesting to know how eskimos living in Greenland and depending in their diet to a large proportion on sea food manage regarding the mercury problem.

8.3 Mercury in the Water of the World's Oceans

Even up to few years ago, chemists agreed that concentrations of mercury from 30–50 ng/l were present, with a variation of 10–140 ng/l, in uncontaminated ocean water. If, however, analyses are made taking every precaution known, lower concentrations show: for the Northeast Atlantic, an average of 13 ng/l with a variation of 3–20 ng/l; for the Northwest Atlantic, 7 ng/l with a variation of 6–110 ng/l (Fitzgerald and Lyons 1975). An average concentration of only 6.9 ng/l was also found in the North Sea with a variation in the individual measurements of between 3.4 and 22 ng/l (Baker 1977). Aside from the higher amounts in the Irish Sea and in the immediate vicinity of river mouths and points of effluent introduction, only very small differences have been noted regionally (Fig. 80). In the course of these analyses

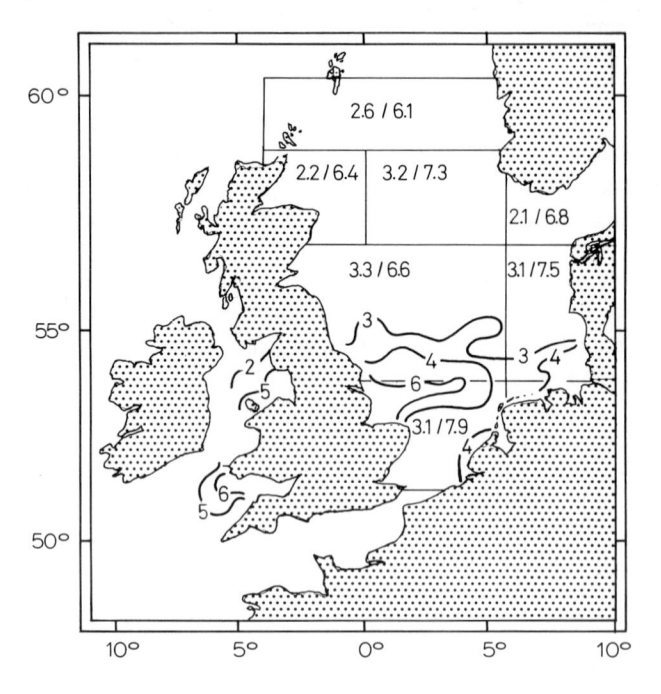

Fig. 80. Mercury concentration in the surface water of the North Sea and in British waters. The average concentration from all samples is 6.9 ng/l. For the various sea areas, average concentrations are indicated as well for reactive mercury (as reduction by zinc chloride; *first figure*), and for total mercury (subsequent oxidation with ammonia persulfate; *second figure*). In addition, some isolines of the distribution of reactive concentrations of mercury are included indicating a higher concentration of mercury in the Thames Estuary, the Bay of Liverpool, the Humber Estuary, and, to a lesser extent, the Bristol Channel, as well as the estuaries of the Rhine, Weser, and Elbe Rivers. Figures in ng/l (Baker 1977)

it became clear, however, that mercury is present in seawater in different forms. Depending on the intensity of the acid treatment, more or less high results are attained on total mercury. Besides this, it is clear that a considerable portion of mercury in the water samples is not in solution but bound to suspended particles. When there are many suspended particles in the water after a storm, the measured amounts are higher than after quiet periods when suspended particles settle to the bottom as sediment.

8.4 Sources of Mercury in the Sea

On the whole, 0.1–0.5 mg/kg of mercury are contained in unpolluted rock, sand, and earth material. Where seawater comes into contact with sediment in coastal areas, or where rivers carry their load of suspended particles to the sea, the mercury concentration is increased by natural processes. Global estimates (Table 36) suggest that 3500 t of mercury are annually released by the weathering of rocks and reach the oceans of the world. Mercury is also transported from land to the sea via the atmosphere which is said to contain around 12,000 t of mercury as dust particles (Williston 1968). But within this figure, it is not known what part of it comes from human activity and what part from natural processes. In regions where minerals containing mercury occur naturally, the air has a 20-fold higher concentration of mercury than elsewhere (Saha 1972). Active volcanoes also emit mercury into the atmosphere and mercury is washed out by rain and snow. Presumably, 50,000 t per year thereby reach the surface of the sea (Jernelöv 1975). Incidentally, mercury also seems to be transported to the ocean from the fairly deep strata of the Earth's crust, namely from where material from the Earth's crust wells up and leads to sea floor spreading and continental drift. While mercury analyses of around a thousand seawater samples generally showed values between 2 and 40 ng/l, concentrations of up to 1090 ng/l were measured in some seawater samples from a depth of 3200 m above the Middle Atlantic Ridge (36°N, 33°W) (Carr et al. 1974). What consequences this input can have is still totally unknown. An amount of 25,000 to 150,000 t of mercury annually is given as an estimate for the entire input into the ocean from degassing of the lithosphere (Table 36).

There are only a few mercury mines in the world. In all, mining produces 9000 t of mercury per year. From the types of use, it is clear that the major portion of it can sooner or later wind up in the environment, that is in the ocean as well. Mercury was used in the U.S.A. in 1969 for the following purposes: 26% for chlorine fabrication by electrolysis (see Chap. 3.2); 23% for electric systems and lighttubes; 12% for paints hindering the growth of marine organisms in ships' hulls (see Chap. 4.4); 6.5% for instruments; 3.5% as a catalyst in chemical processes (see Chap. 3.2); 3.5% as a filling material by dentists; 3.5% as a fungicide in agriculture; and the remaining 22% for various purposes, among others construction and new expansion of chlorine-alkaline plants (Saha 1972). Figures released in 1973 by the Bayerische Landesgewerbeanstalt in Nürnberg show that the situation in the Federal Republic of Germany in 1971 was similar: 660 t of mercury were imported; 250 t of this amount were used for new chlorine-alkaline plants. Of the remaining 410 t, 25% were consumed by chlorine-alkaline electrolysis and had to be replaced; 13% were used in the production of various chemicals; 12% for the production of pesticides; 11% was used as a catalyst in the chemical industry; 8% in the electrotechnical industry; 7% for instrument manufacture; 7% as a poisonous ingredient in paints; 5% by dentists; and an additional 12% served various other purposes. Every fluorescent candle contains 50 mg of mercury, and it is estimated that from this source alone, waste or broken fluorescent candles, about 2 t of mercury per year become incorporated in scrap (Hütter 1978).

Mostly in the range of around 1 mg/kg coal and petroleum contain mercury in highly varied quantities. When coal is burned, over 90% of the mercury reaches the atmosphere. A 660-MW power plant emits around 2.5 kg of mercury per day (Table 34). It is no wonder then that samples of air dust collected by a plane over San Francisco contained up to 50 ng/m^3, 10–100 times more than the 1–10 ng/m^3 of mercury in air dust over unpolluted areas (Williston 1968). Rain and snow in the vicinity of big cities contain 0.2–2 µg/kg of mercury (Keckes and Miettinen 1972), in Sweden, by contrast, 0.05–0.07 µg/kg (Ackefors et al. 1970), if analytical techniques of these

Table 34. A 660-MW, coal-fired power plant gives off mercury into the environment, even if the coal used contains only 0.3 mg/kg of mercury. The exhaust gas contains 0.03 mg/m^3 of mercury (Billings and Matson 1972)

Material	Average amount of material transformed (in t/day)	Average concentration of mercury (in mg/kg)	Average amount of mercury (in g/day)
Coal	7,750	0.3	2,580
Ash on the grate	330	0.2	66
Screened light ash	1,300	0.2	260
Light ash given off into the atmosphere	2	0.2	below 1
Exhaust gas	81,000 m^3/day	0.033 mg/m^3	2,500

older results are reliable. The total input of mercury to the world oceans from burning of fossil fuel is estimated to be 2000 t per year (Table 36).

8.5 Has Man Increased the Concentration of Mercury in the World's Oceans?

When the Minamata catastrophe became known, and the facts on mercury pollution from chlorine plants and paper mills (Chap. 3.2), the immediate conclusion was: high mercury concentrations found in tuna and seal (see Chap. 8.2) are also a consequence of marine pollution by mercury. Some findings seemed to support this theory; in the meantime, however, their conclusions are no longer beyond doubt.

Recent layers of sediment in the anoxic Santa Barbara Basin off California, U.S.A. (Fig. 81) have double as much mercury as sediment layers deposited before 1920, and much higher concentrations than layers deposited 3500 years ago, at 1500 B.C. However: the industrial region of Los Angeles is not far away. The increasing mercury concentrations could reflect industrial pollution on a regional scale, and forbid world-wide extrapolations.

Another objection may come from the possibility that bubbles of methane could form in the anoxic sediment layers and migrate through the sediment in an upward direction, disturbing the sediment structure and taking along traces of mercury.

In the 1970's, some scientists believed that they could demonstrate an increase of mercury concentrations in Greenland glacier ice cores, and that ice formed from snow before 1952 had only half as much mercury as ice which formed more recently. However, recent analyses with more modern techniques lead to other results (Appelquist et al. 1978): concentrations of mercury in the ice are only 2–19 ng/kg, one order of magnitude lower than previous data, and there were no significant differences in concentrations in ice from the past decades.

Other authors tried, by analyzing museum material, to prove that mercury concentrations in the marine environment 100 years ago were about the same as they are now,

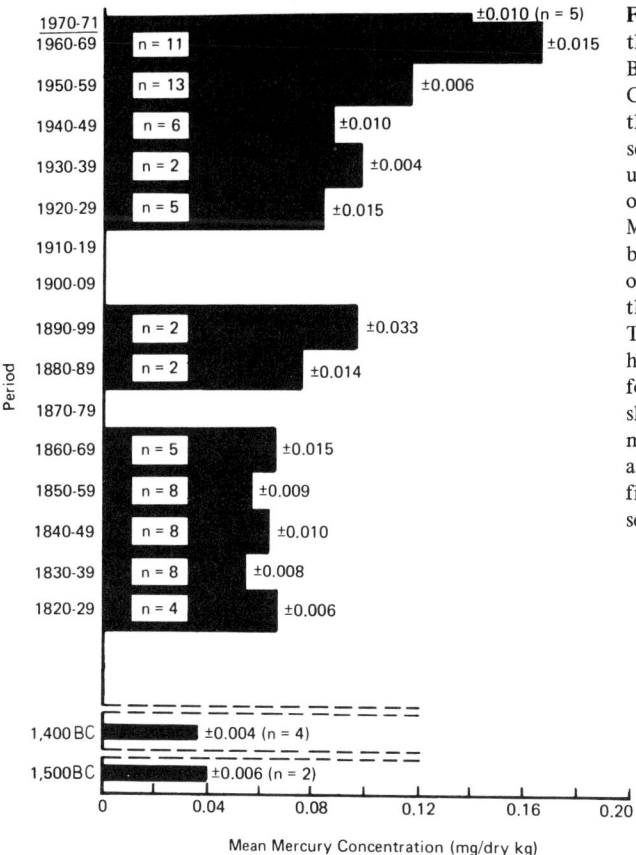

Fig. 81. There is no oxygen in the deep water of the Santa Barbara Basin off the coast of California, U.S.A. Consequently, there is no bottom fauna, and sedimentation takes place undisturbed. The annual layers of sedimentation can be traced. Material that was deposited before 1920 as sediment contains only half as much mercury as that which sedimented after 1920. The concentration is especially high in the layers which have formed since 1960. The graph shows the concentration of mercury in a core sampled from a water depth of 580 m. The figures are in mg/kg of dry sediment (Young et al. 1973)

which means that there has not been any significant increase of mercury concentrations on a world-wide scale, away from regional pollution sources.

There are several studies on the mercury content of fish which were caught decades ago and that have been preserved in museums. The results show that even 70–90 years ago the mercury content was just as high then as it is in tuna fish caught today (Table 35; Hammond 1971). However, these analyses must be judged with caution because mercury is a volatile element. If the preserving agent was changed relatively often, changes in the amount of mercury in the fish may have resulted. The fact that metal identification plates were formerly used in museums and that through them traces of mercury may have entered the tissues of the analyzed fish cannot be excluded from consideration.

It seems that mercury deposited by birds in their feathers is rather conservative. Therefore, if one finds rather constant concentrations in feathers of Greenland guillemots since 1840, this may be a significant evidence that mercury concentrations in seawater and in fish from the open North Atlantic have not altered since then (Fig. 82).

The best argument for the theory that mercury concentrations in the water of the world oceans cannot have changed much under man's impact is provided by a simple

Table 35. It must be feared that the concentration of mercury has changed in museum specimens during their preservation, but it is nevertheless noteworthy that the meat of tunafish caught 60–100 years ago shows similar concentrations of mercury as do specimens caught recently. The figures are in mg/kg wet weight (Miller et al. 1973)

Year caught	Species, locality	Mercury concentration in mg/kg
1878	Bonito *(Katsuwonus pelamis)*, Massachusetts, 2 samples	0.27
		0.64
1800	White Tuna *(Thunnus alalunga)*, California	0.27
1886	Tunafish *(Thunnus thynnus)*, Massachusetts	0.38
1890	Bonito, California	0.45
1901	Bonito, Hawaii	0.42
1909	Bonito, Phillipines	0.26

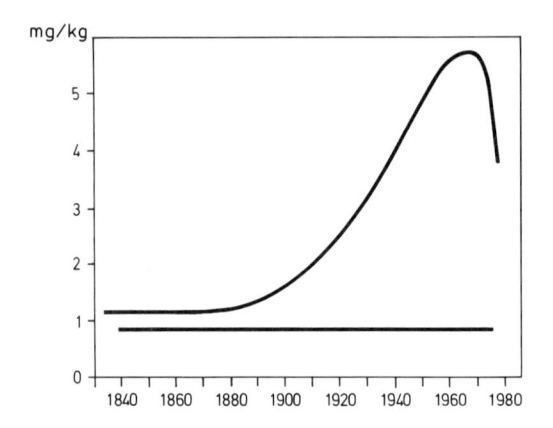

Fig. 82. Mercury concentrations in the feathers of the guillemot *(Uria aalge)* shot in Greenland waters since 1840 and preserved in museums did not increase. On the other hand, mercury concentrations in feathers of guillemots shot in the Baltic Sea increased during the twentieth century to a peak between 1960 and 1970; fortunately they decreased recently (Somer 1978)

calculation (Table 36). If mercury concentration in the oceans should be 7 ng/l, an amount of 10 million t of mercury is contained in the 1.4×10^{21} l of ocean water, Man is responsible for a yearly amount of 9000 t of mercury from mining and 2000 t from burning of fossil fuel etc. If 10,000 t per year should finally reach the sea from man's activities, this is 0.1% of the total amount of mercury in seawater, less than the input from weathering and from volcanism. Therefore the probability seems good that a yearly addition of 0.1% to the oceans will not result in measurable increases of the mercury concentrations in seawater. From a simple arithmetical point of view, it would take 1000 years to double the present mercury concentration in seawater. Whether such a doubling really would happen is a point of discussion: sedimentation and other geochemical processes could influence the figure, and one should have in mind that estimates of mercury released to the biosphere from degassing of Earth crust material are about 10 times higher than estimates of man-made input.

But what would happen if mercury concentrations in oceanic waters really should rise to double the present figure? There is at least a possibility that even the present natural concentration of mercury in seawater (below 10 ng/l) is not conducive, but harmful to life in the sea, because laboratory experiments have shown that certain

Table 36. Geochemical and industrial data for some trace elements, according to the UN Joint Group of Experts on the Scientific Aspects of Marine Pollution (GESAMP 1976) which looks upon the following metals and trace elements as potentially dangerous for man and the marine environment: antimony, arsenic, beryllium, cadmium, chromium, cobalt, copper, lead, manganese, mercury, nickel, selenium, silver, vanadium, and zinc. Data given for mercury (see Chap. 8.3), for lead (see Chap. 8.7) and for cadmium (see Fig. 75) have been updated according to more recent information. According to Bewers and Yeats (1977) the riverborne input to the global oceans is 7.6 x 10^3 t/y for cadmium, 127 x 10^3 t/y for copper, and 180 x 10^3 t/y for zinc. Sea water concentrations of copper and zinc probably are lower than figured

	Unit	Pb	Hg	Cd	Sb	Cr	Se	As	Cu	Zn
Earth crust concentration	mg/kg	15	0.06	0.2	0.2	?	0.09	2	45	40
Seawater concentration	µg/l	0.002	0.007	0.1	0.3	0.3	0.5	2	2	3
Total amount in world oceans	10^6 t	2.8	10	140	420	420	700	2800	2800	4200
Input by erosion	10^3 t/y	150	3.5	0.5	1.3	236	7.2	72	325	720
Input through burning of fossil fuel and cement fabrication	10^3 t/y	34	2.0	0.2	?	1.5	1.1	8.2	2.1	37
Mine production	10^3 t/y	3500	9.0	15.0	70	3000	1.2	30	7500	5000
Annual mine production in percent of total amount in world oceans	%	125	0.1	0.01	0.015	0.7	0.0002	0.001	0.3	0.1

mercury compounds are toxic even at concentrations of only 100 ng/l = 0.1 µg/l. They inhibit the photosynthetic efficiency of the diatom *Nitzschia delicatissima* (Harris et al. 1970) or impair the growth of plankton algae (Fig. 60). This toxic concentration is merely about 10 times higher than the natural one in seawater. Because especially effective organic complexes of mercury have been used in the experiments, a direct comparison to the toxicity of mercury in seawater, however, is not permitted. But there are interesting observations from nature, too. In upwelling regions, seawater from the deep sea comes in contact with the surface layer of the sea, and a good supply with nutrients and energy from the sun stimulate blooms of planktic algae. However, sensitive dinoflagellate algae cannot develop before more resistant diatom algae have conditioned the seawater by excreting organic compounds. It is a hypothesis that such organic compounds have a chelating effect on ions of heavy metals, and decrease their toxicity. If this hypothesis is correct, the natural concentrations of some heavy metals in clean, unpolluted seawater are toxic to marine life, and certainly double concentrations would be more toxic. However, up to date it is not clear which of the different highly toxic elements in seawater could be responsible for the toxic effect of fresh upwelling water; besides mercury, it could be copper, cadmium, lead, or another heavy metal.

A doubling of actual natural mercury concentrations in seawater would certainly result in about double mercury concentrations in all marine organisms, including fish, seals, and marine birds. It seems that some of the long-lived large fish-eating species like seals and tuna at present have mercury concentrations in some tissues which are close to the threshold of toxic effects (see Chap. 8.2). There is at least a possibility that these species could not cope with double concentrations. Certainly, too, under present health standards, tuna and some other large fish would be rendered unfit for human consumption if mercury concentrations should be double the present value (see Cap. 8.1).

Therefore it is good to know that according to the present state of knowledge man does not alter the mercury concentrations in the biosphere on a world-wide scale. However, the concentration of mercury in the oceans has surely changed in the course of the Earth's history and was higher at periods of strong erosion and volcanic activity. Whether or not that had consequences for the organisms of earlier periods is not known. Maybe they could adapt to live with higher mercury concentrations in their tissues, and the detoxification effect of chelating organic compounds in seawater prevented toxic effects of elevated seawater concentrations. But maybe, too, certain species did not survive periods of high mercury concentrations, which at the same time were periods with high concentrations of other heavy metals in general.

8.6 Contamination of the Oceans with Cadmium

A commission of experts of the World Food and Agriculture Organization (FAO) and the World Health Organization (WHO) came to the temporary conclusion in 1972 that the tolerable amount of cadmium for humans is not more than 0.0075 mg/kg body weight or 0.5 mg per person per week. No more than 10 µg/l of cadmium can be permitted in drinking water. These maximum amounts are thus just slightly over those for mercury. Actually, cadmium is a very poisonous element and especially insidious for human beings because it is insufficiently eliminated by the body. Over a period of years, cadmium accumulates in particular in the bones. Fatal cases of cadmium poisoning have occurred in Japan. There is also an apparent connection between cadmium contamination and high blood pressure.

On the average, an inhabitant in the Federal Republic of Germany ingests 0.476 mg of cadmium a week with his food. This is almost as much as the internationally recommended maximum limit (Ernährungsbericht 1976). Agricultural products contaminated by fertilizers which contain up to 30 mg/kg cadmium are under particularly heated discussion. Sea products have nearly no effect on the amount of cadmium a human absorbs because he only receives 0.008 mg of cadmium from the 200 g of sea fish that the average German eats per week. This is one-sixtieth of the total amount of cadmium taken up with food.

In such coastal areas where industrial pollution occurs or where wastes and sludges are dumped into the sea (see Chap. 4.2) in marine seafood other than fish, high cadmium concentrations can show up which might be a human health risk. In the Bristol Channel there is severe pollution with heavy metals from the mining district of Southern Wales (Fig. 74); limpets *(Patella vulgata)* have 119 mg/kg cadmium, on dry weight basis, equal to about 24 mg/kg on wet weight basis (Peden et al. 1973). In Kiel Harbor, mussels *(Mytilus edulis)* have 10–34 mg/kg cadmium (on dry weight basis). Fortunately, mussels from the mussel culture areas in the German Wadden-Sea have concentrations of only 1–3 mg/kg cadmium, on dry weight basis, or 0.1 mg/kg on wet weight basis (Theede et al. 1979a). But in oysters from some oyster culture areas higher concentrations, up to 5 mg/kg on wet weight basis, have been reported; it has to be confirmed whether such local data have to be generalized.

When the cadmium-in-oyster question was discussed among experts in the Federal Republic of Germany, it did not seem reasonable to work out a regulation setting standards for the highest permissible concentration of cadmium in oysters. It seemed a better procedure to inform the press about the risk that oysters, like certain champignons and other mushrooms from unpolluted woodland, due to special physiological capacities, can accumulate cadmium from the low concentrations in the environment to relatively high concentrations in certain tissues of their body. In the Federal Republic of Germany oysters are so expensive that those who like to eat oysters can read the newspapers and decide for themselves whether or not to continue to eat oyster, and if so, to consider them as a delicatesse. Future analyses will better inform about concentrations of cadmium in oysters and other seafood, like cuttlefish, squid, and octopus, or in the brown meat of crabs and shrimps. Fortunately, cadmium concentrations in fin fish, which is the only marine food of great importance, is very low. That is why no regulation in general regarding cadmium is necessary, and why only 0.01–0.2 mg/kg of cadmium, on wet weight basis, is present in livers from seal *(Phoca vitulina)* living in the German Wadden-Sea (Fig. 29).

Astonishingly high concentrations of cadmium accumulate in the liver of molluscs: up to 2000 mg/kg in scallops *(Pecten)*, up to 1100 mg/kg (both data on dry weight basis) in the squid *Symplectotheuthis oualaniensis* which lives in the high seas where industrial pollution cannot play any role (Martin and Flegal 1975; Bryan 1976). Besides cadmium, the liver of these squids contains up to 1900 mg/kg of copper and

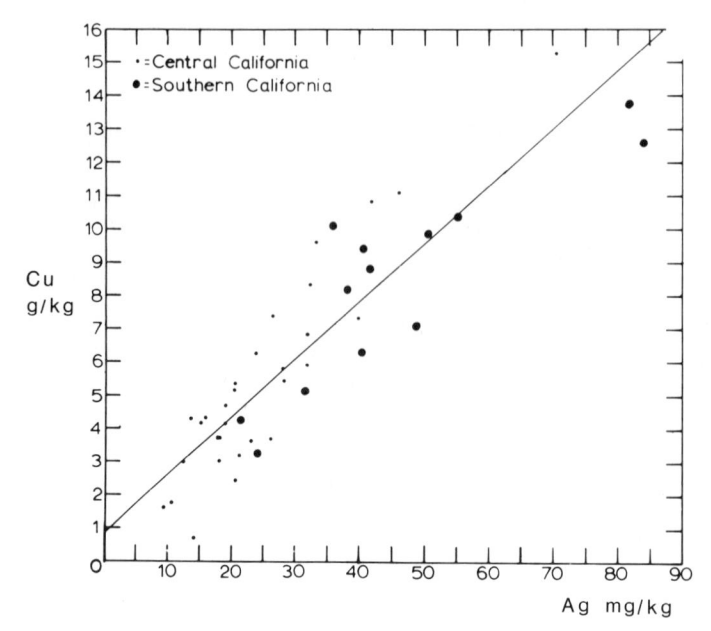

Fig. 83. Squids (Cephalopoda) store up extremely high amounts of cadmium, together with silver and copper, in their liver, There is a strong correlation between concentrations of silver and copper in the liver of the squid *Loligo opalescens* which lives off the coasts of California. Data refer to dry weight (Martin and Flegal 1975)

up to 40 mg/kg of silver (Fig. 83). Copper is a compound of hemocyanine, the characteristic blood pigment of gastropod and cephalopod molluscs. It is possible that cadmium and silver are channeled into the squids by the same mechanism which allows them to take up the copper which is essential for their oxygen transport system.

In Chapter 7.3 it has been explained that high cadmium concentrations occur in the sea-skater *Halobates,* a marine insect from the family of pond-skaters, and in petrels, both animals of the high seas. The explanation of high cadmium concentrations in these animals is probably an enrichment with cadmium of the surface layer of the sea. High cadmium concentrations as in these animals and in oceanic squids have, according to our present knowledge, nothing to do with marine pollution. They are natural phenomena, All the mines of the world together have a cadmium production of only about 15,000 t per year. The amounts of cadmium in the total water masses of the world ocean, at 100 ng/l (Fig. 75), are probably in the range of 140 million t (Table 36). It is highly improbable that man has influenced the cadmium concentration in the oceans. Local introductions into coastal areas, however, must be forbidden, because cadmium may show up in marine seafood.

8.7 Contamination of the Oceans with Lead

From various calculations there is reason to suspect that lead concentrations in all organisms, land, freshwater, and marine, have increased 20-fold during the past few centuries and are now in the range of 0.1 mg/kg wet weight. It seems even that lead concentrations in the average citizen in developed countries are high enough to result in first symptoms of poisoning of the nervous system. However: marine food does not significantly contribute to this chronic lead burden. The average citizen of the Federal Republic of Germany takes up, with food, 3 9 mg of lead per week. But out of this quantity only 0.02 mg are derived from the 200 g of fish the average citizen eats per week (Ernährungsbericht 1976). Lead uptake from other food, from drinking water, from car exhaust and cigarette smoke, from contact with paint and glazes, with paper and metal is much more important. Marine fish contains only small concentrations of lead, and therefore even the fish-eating seal does not accumulate larger concentrations of lead in its liver (Fig. 29).

In Earth crust material lead is much more abundant than mercury and cadmium. Man has been using lead for 4000 years, and there are estimates that even before the year 1850 about 70 million t of lead had been smelted, and 130 million t more from 1850 to 1950. At present about 3.5 million t of lead are mined every year. During roasting and smelting of lead ore in order to fabricate metallic lead, large amounts of lead dust reach the atmosphere, and estimates are that annually about 3000 t of lead from such processes could reach the oceans via the atmosphere. Approximately 10 million t of lead have been converted, during the past 40 years, to tetraethyllead. In 1970, about 400,000 t of lead in tetraethyllead, have been used as an anti-knock agent with car gasoline. After the gasoline has been burned, the lead is left over and reaches the atmosphere as fine dust particles. Estimates are that annually 37,000 t of lead from gasoline reach the oceans (Patterson et al. 1976). More lead comes to the oceans from road runoff, from paints and from weathering of lead surfaces, mostly via the rivers. However, much of the lead mined is used in metallic form and is rather resistant against leaching and weathering; therefore it is not permitted to assume that a large fraction of the mining product will go to the oceans, as is the case with mercury. If modern analytical techniques give reliable results, lead concentration in unpolluted oceanic waters is only about 2 ng/l. The water of all oceans together then contains only 2.8 million t of lead, less than the annual production of the lead mines (Table 36).

Lead, however, is difficult to analyze in seawater. There is a great risk of measuring, instead of the seawater concentration, lead introduced into the measuring system with sampling gear, impurities of chemicals, or from other sources. In 1973, subsamples of the same seawater sample were sent to nine experienced laboratories all over the world. The results of lead analysis were strongly divergent, and some results were ten times higher than the value that had been obtained by isotope dilution methods, i.e., 80 ng/l, in seawater sampled off the polluted coast of California (Anon 1974). Concentrations in unpolluted coastal waters are about 10–40 ng/l but concentrations in water from the open ocean are only 1–2 ng/l (Burnett and Patterson 1980; Schaule and Patterson 1980). The isotopic composition of lead in coastal seawater gives a hint that a large percentage consists of lead from lead mining (Stukas and Wong 1981).

Due to the analytical difficulties, there are no time series of seawater lead concentrations available which could document an increase of lead concentration with time. Analyses should be repeated which have been made with ice from Greenland glaciers and demonstrate that ice which formed before 1750 has only 0.02 μg/kg of lead, and that ice from 1968 contains ten times higher concentrations (Murozumi et al. 1969). Data from anoxic Santa Barbara Basin are not convincing, because an increase in the sedimentation rate of lead from 2–10 mg/m^2 per year to 9–21 mg/m^2 per year can as well be the result of industrial pollution from the nearby city of Los Angeles (Chow 1973; see Chap. 8.5 and Fig. 81 for comparable mercury discussion). Unfortunately, there are, apparently, no time series of lead concentrations in biological material available to demonstrate an increase of lead concentrations during the past century. Therefore one has to rely on general speculations regarding the conclusion that marine organisms, too, at present may have about 20 times higher lead concentrations than some thousand years ago.

Lead is not as toxic for marine organisms as mercury or cadmium. Usually, values of around 0.1 mg/l are indicated as threshold at which adverse effects become apparent; concentrations in seawater are far from that. But lead is an insidious poison, as experiments with mussels (Fig. 84) have shown.

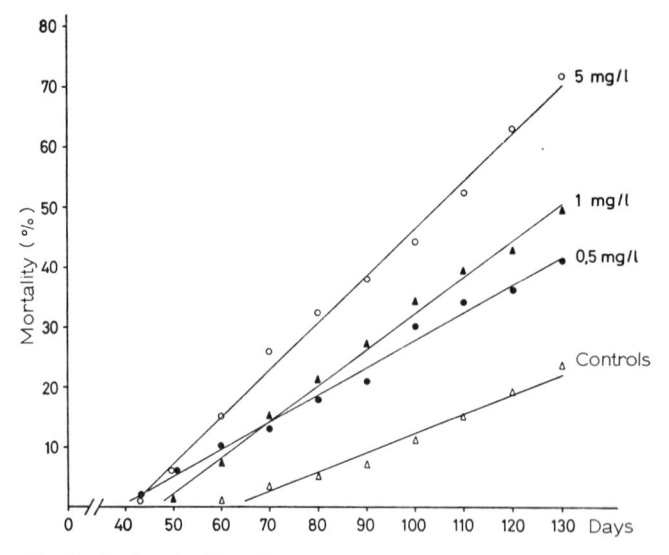

Fig. 84. During the first 40 days, lead is not very poisonous to common mussels *(Mytilus edulis)* in laboratory experiments. Even at maximum concentrations of approximately 5 mg/l of lead as lead nitrate corresponding to saturated conditions in seawater, no increase of mortality resulted in the first 40 days of the experimental period. Only in long-term experiments does mortality occur, which increases at higher concentrations of lead. In the diagram, the cumulative mortality is indicated for experiments with 0.5, 1 and 5 mg/l of lead (Schulz-Baldes 1972)

Adult mussels, or laboratory cultures of plankton algae (Fig. 63) are not the most sensitive test organisms for toxicity experiments. One can suppose that more sensitive organisms under more natural experimental conditions will demonstrate effects of lead poisoning at lower lead concentrations. One should, further, have in mind that probably the highly toxic organic compound tetraethyllead, the same substance which is used as an anti-knock agent in car gasoline, is formed by natural processes in the sediment of tidal flats (Harrison and Laxen 1978), and that more than 25% of the lead which accumulates in fish can be tetraethyllead.

But even if lead, compared to mercury, should be considered as less toxic element, even if it seems that there is no immediate danger that rising lead concentrations in the marine environment could disturb the marine ecosystem or render marine food unfit for human consumption, even then it cannot be countenanced that approximately 8% of the lead produced in mines is distributed so finely and irretrievably by automobile exhaust gases throughout the biosphere that there is no prospect of recovering it. In the Federal Republic of Germany, the maximum allowable concentration of lead in gasoline was lowered in 1976 from 0.4 g/l to 0.15 g/l. Cars are still running. Other countries pretend to be unable to go below the limit of 0.4 g/l which will become standard in the European Community by 1981.

9 Global Pollution of the Oceans with Chlorinated Hydrocarbons

9.1 General

The chemical industry produces, from chlorine, simple volatile chlorinated hydrocarbons with 1–2 carbon atoms, to a certain extent also chlorine-fluorine compounds. The largest amounts are divided among dichlorethane (EDC, $H_3C\text{-}CHCl_2$; around 20 million t produced world-wide in 1973), cancerogenous vinylchloride ($H_2C=CHCl$; 10 million t), solvents used in industry, and byproducts which are not intentionally released, but do, however, turn up in the environment, for example in the water and sediment of the Bay of Liverpool (Pearson and McConnell 1975). Practically the entire production of many compounds reaches the environment, for example the solvent trichlorethylene ($ClHC=CCl_2$; approximately 1 million t produced world-wide in 1973), the degreasing solvent trichlorethane ($CH_3\text{-}CCl_3$; 0.5 million t), the compounds perchlorethylene ($Cl_2C=CCl_2$; 1 million t) and carbontetrachloride (CCl_4; 1 million t) used in the textile dry-cleaning, and chlorofluoro-hydrocarbons of various compounds used as a propellant gas in spray cans and as a coolant (freon, frigene; 1 million t). These substances are almost resistant to microorganisms. In the atmosphere, however, photo-oxydation into hydrochloric acid takes place under the influence of ultraviolet rays. Hydrolysis can be observed in the hydrosphere. The half-life values are in the area of months.

So far, research just starts to reveal increasing concentrations of CCl_3F, CCl_2F_2 and trichlorethane in the air of regions as remote as the South Pole (Rasmussen et al. 1981), but only small accumulation takes place in organisms (Table 37) and an acute threat to life in the sea has not been recognized. But on this question, research is still in the initial stages. Special attention must be devoted to a number of more dangerous byproducts, for example to hexachlorobutadiene which accumulates up to a 1000 times in organisms.

In the process of chlorinated hydrocarbon fabrication quite regularly several compounds are synthesized simultaneously, but only one compound is wanted. This explains the large amounts of unwanted byproducts which, years ago, were released into the environment. Intelligent techniques make it possible to reduce the waste by utilizing waste products; of special importance is the perchloration method by which more chlorine atoms are added to the compounds. Finally, however, there remains a residue, a waste which up to a very few years ago could not be utilized further. This waste is called EDC-tar. It had been dumped into the sea up to the 1960's and it is now incinerated at sea in various sea areas (see Chap. 4.1).

If dichlorethane and vinylchloride are found to be present in a marine area, industrial introduction, marine dumping of wastes or transportation via the atmosphere is

Table 37. The concentration of some halogenated methanes and ethanes are 2–25 times higher in some marine organisms than in the seawater. Analysis of organisms from the area around Port Erin, Isle of Man, Irish Sea, an area not locally polluted. Data in mg/kg dry weight (Dickson and Riley 1976)

Species	$CFCl_3$	CH_3I	$CHCl_3$	CCl_4	$CCl_2=CHCl$	$CCl_2=CCl_2$	CH_3CCl_3
Whelk *(Buccinum)*							
Intestinal gland	0.0003	0.014	0.117	0.008	0.002	0.033	–
Mussel *(Modiolus)*							
Intestinal gland	0.0002	0.010	0.056	0.020	0.056	–	0.004
Muscle	–	0.015	0.200	0.028	0.033	0.016	–
Scallop *(Pecten)*							
Gills	0.0005	0.024	1.040	0.014	+	0.088	–
Muscle	0.004	0.009	0.440	0.006	–	0.024	–
Conger *(Conger vulgaris)*							
Brain	0.0050	0.032	–	0.015	0.062	0.006	0.009
Liver	0.0021	0.042	0.474	0.051	0.043	0.043	–
Muscle	0.0001	0.011	0.219	0.003	0.070	0.001	–
Cod *(Gadus morhua)*							
Muscle	0.0009	0.004	0.168	0.007	0.008	0.002	0.005
Coalfish *(Pollachius virens)*							
Muscle	0.0009	0.004	0.168	0.007	0.008	0.002	0.006
Dog Fish *(Scylliorhinus)*							
Muscle	0.0009	0.031	0.649	0.019	0.041	–	–

usually suspected. With other volatile chlorinated hydrocarbons, particularly those with 1 carbon atom (halogenated methanes), the suspicion is also justified that they could be the product of marine vegetation. In kelp beds, seawater manifests 40 times higher concentrations of methylchloride (CH_3Cl; Lovelock 1975). Numerous halogenated methanes with 1–4 chlorine atoms, 1–3 bromine atoms, and an occasional 1 iodine atom were found in red and brown algae (Su and Goldberg 1976). It is also known that such compounds come into being in some fungi and lichen and as the product of microbial fermentation, and when organic substances are incompletely incinerated. In the marine worm *Lanice conchilega* from the Weser Estuary, brominated phenoles have been analyzed. There are hints that these compounds are synthesized from the worm and have nothing to do with marine pollution by halogenated hydrocarbons (Weber and Ernst 1978). Finally, wherever it is introduced into the environment, chlorine gas reacts to various organic substances, and the chlorination has lead to the formation of chloroform ($CHCl_3$) in the $\mu g/l$ range in the drinking water of many areas in the Federal Republic of Germany. Chlorine gas is also used to keep the cooling water pipes of power plants free of algae and mussel fouling. We do not yet have sufficient knowledge to know whether this application produces a significant amount of chlorinated hydrocarbons in the sea.

PCB's (polychlorinated biphenyles) have been used in industry since 1929 and have been enjoying greater popularity in recent years (Table 38). They are oily substances which barely react with other substances and are also impervious to high temperatures. For this reason, they are used for cooling and heat transfer, for example in electrical transformers and capacitors. But they were also widely used as hydraulic fluid, and as a carrrier of other substances. A few years ago, chloro-rubber paints and PVC contained around 5%–8%, and colorless copy paper 1%–2% PCB's (Jensen 1972). The total amount of PCB's produced up to now by the chemical industry (about 1 million t) is less than the amount of chlorinated hydrocarbon pesticides. Since 1971, PCB production has been drastically reduced (Table 38). In 1973 the Organisation for Economic Co-operation and Development (OECD) made a decision regarding Protection of the Environment by Control of Polychlorinated Biphenyls. Member states reduced production to about 23,000 t, consumption to about 7000 t in 1977 and decreated restrictions of handling, use and disposal. Dissipative uses in paints, plastics and lubricants ceased; unfortunately they continue with small capacitors as used for luminescent tubes. PCB is still used to maintain existing transformators. Diphenylether (Ph_2O) has been proposed to be used in industry as a substitute for PCB's, because some properties are similar. However, this compound seems to accumulate in a similar manner, in the marine environment too; concentrations of 3–130 ng/l have been analyzed in seawater from Halifax, Canada, and concentrations, on fat basis, in the blubber of the grey seal *(Halichoerus grypus)* and in the liver of cod *(Gadus morhua)* are 1–4 mg/kg (Addison 1977).

Table 38. U.S.A. production of PCB's. World production was approximately twice as large. In 1971, 8000 t of PCB's were produced in the Federal Republic of Germany, 7600 t in France, 6800 t in Japan, 5000 t in Great Britain, and around 1500 t apiece in Spain and Italy. Data in 1000 t (Longhurst and Radford 1975; Risebrough et al. 1976b)

1960	1961	1962	1963	1964	1965	1966	1967	1968	1969	1970	1971	1972–1974
17.2	16.6	17.4	20.3	23.1	27.5	29.5	34.2	37.2	34.7	38.6	18.3	18–19

In some countries polychlorinated terphenyls (PCT's) had been produced for similar industrial use as PCB's, as a flame-retardant, and for paints and glues. Two thousand seven hundred t of PCT's had been produced from 1954 to 1972 in Japan (Doduchi 1977), when it became evident that PCT's are toxic, persistent, and accumulating in organisms; production in Japan and in the U.S.A. ceased in 1972. PCT's have been analyzed with 0.5–1 mg/kg in the blubber of Baltic gray seal, along with 48–190 mg/kg of PCB's. It is only in the last few years that modern gas chromatographic techniques have allowed to distinguish between PCB's and PCT's in such samples (Renberg et al. 1978).
DDT (pp'DDT, 2,2-bis(p-chlorophenyl-1,1,1-trichlorethane) has, to be sure, been known for a century as a compound, but it was not put to use as an insecticide until the Second World War. The annual production in the U.S.A. rose from 4400 t in 1944

to 81,300 t in the sixties. The world production for 1970 can be estimated as having been 100,000 t (Goldberg 1975, 1976). There are estimates that a total of 2.8 million t were produced and used up to the year 1974.

In the meantime, the use of DDT has been prohibited or very severely restricted in many countries – in 1972 in the U.S.A., for example.

The DDT problem was never serious in the Federal Republic of Germany because DDT represented only 3% of the products used in insecticides. In the developing countries, however, it is generally believed that the widespread use of DDT cannot be given up because to date no other product can compete with DDT as far as effectiveness, relative safety for humans and animals, as well as manufacturing costs are concerned. For this reason, the DDT problem continues to be acute.

In 1969, the U.S.A. produced a total of 63,400 t of DDT and exported more than half of this figure. Over 60% of this export figure was used not in agriculture, but to combat malaria-transferring mosquitoes. From the amount that was used in agriculture, 70% was sprayed on cotton plantations.

If effective combating of mosquitoes is contemplated in the Third World, approximately 47,000 t of DDT will be needed annually in the coming years, a little more than the 42,000 t that were used for this purpose in 1970 (Goldberg 1975, 1976). If greater amounts of cotton are to be planted, tropical agriculture will need 69,000 t of DDT per annum for this purpose. These estimated amounts, like all projections, may be questionable, but they show that a continued use of DDT has to be counted with in about the same amounts as to date, the difference being that the point of main use is moving to the tropical zones.

Along with DDT, a whole arsenal of additional insecticides of the chlorinated hydrocarbon group has been developed (Fig. 85): aldrin (5900 t produced in the U.S.A. in 1972, forbidden in 1974; used since 1974 in the Federal Republic of Germany only in small quantities, 0.5 t/year against wine pests, 20 t/year against tropical wood-eating insects, now forbidden to be used), dieldrin (HEOD; 270 t produced in the U.S.A. in 1970; occurs in the environment as aldrin, prohibited in 1974); chlordan (9100 t produced in the U.S.A. in 1972; use prohibited in 1976); heptachlor (3600 t produced in the U.S.A. in 1970, use prohibited in 1976); endrin; and methoxychlor (National Academy of Sciences 1975a). Gamma-hexachlorocyclohexane (gamma-HCH, in older papers sometimes called also benzenehexachloride, BHC) is commercially available as insecticide with the name lindan. Alpha- and beta-HCH are isomers that result as waste products. Hexachlorobenzene (HCB) is applied as a fungicide; it is, however, also a waste product in the production of other chlorinated hydrocarbons, is contained in EDC-tar, results from incineration and chlorine fabrication. Toxaphene (polychlorocamphene, PCC) is a mixture of around 200 different polychlorinated camphenes (35,000 t produced in 1972 in the U.S.A.; now forbidden to be used in the Federal Republic of Germany). It is mainly used to combat termites; rather high concentrations have been found in marine fish (Zell and Ballschmitter 1980). Mirex is mainly used in the southern U.S.A. to combat ants. Mirex has been found, as an environmental pollutant, in areas far from those in which it was used (Zitko 1976). In all, the total world production of chlorinated hydrocarbon pesticides for the year 1970 may have been around 200,000 t to 300,000 t.

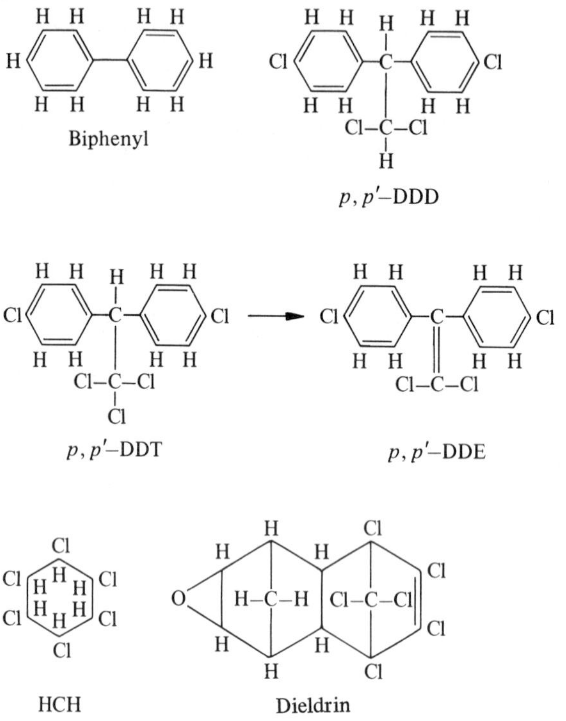

Fig. 85. Structure of some chlorinated hydrocarbons

Herbicides are chemicals to inhibit weeds. They form the bulk of all pesticides produced and are used all over the world. During the Vietnam war at least 1000 km² of mangrove areas were defoliated by U.S.A. troops with 2,4,5-T (2,4,5-trichlorophenoxyacetic acid) which is the herbicide most widely used (Odum and Johannes 1975). Little is known about effects of 2,4,5-T upon other marine life than mangrove trees, which were killed. It is to be hoped that such herbicides disintegrate in the marine environment as they do in terrestrial biotopes. During 2,4,5-T fabrication and by burning it small amounts, up to 0.1 mg/kg of TCDD are produced which later are dispersed into the environment together with 2,4,5-T. TCCD (dioxine, 2,3,7,8-tetrachlorodibenzo-p-dioxine) is a highly poisonous and persistent substance and has a bad name because of the accident at the ICMESA chemical plant which poisoned the area around the Italian village of Seveso in 1976. It was for this reason that the incineration vessel "Vulkan" was denied permission to clean its tanks for chemicals with Emden harbor facilities, in 1978. For years, large quantities of military surplus 2,4,5-T were burned from this vessel, along with other substances. Remnants of dioxine were feared. Dioxine accumulates with a concentration factor of 2000–7000 in aquatic organisms; in principle it is a thread to marine life, but has not been adequately documented. PCB's are regularly contaminated with small impurities of highly poisonous chlorinated dibenzofuranes, which are developed in PCB's especially at 200–300°C and may be the reason of some of the toxicity which PCB's exhibit (Bowes et al. 1975).

Unfortunately, the analysis of chlorinated hydrocarbons is very painstaking work, in spite of gas chromatography and mass spectrometers, especially if not a certain, well-known compound is to be analyzed, but unknown components are to be checked. Only in 1966 was it possible to separate PCB's from DDT and its metabolic products. Many other surprises will result from improved methods of analysis in the future.

Organic bound chlorine concentration in seawater from the Drøbak area, Oslo Fjord, Norway, has been determined by nondestructive neutron activation analysis to be 40 and 195 ng/l, in two samples (Lunde et al. 1975). However, only 1.1 and 1.6 ng/l can be explained by gas chromatographic analysis as PCB's. If such figures can be confirmed, this means that about 99% of the organic bound chlorine is in chlorinated compounds which are at this moment still unknown to science. They may have their origin in pollution from Oslo City, or come from biosynthesis within the natural alga vegetation (Table 37), they may be harmless or toxic. Much has to be learned about organic chemistry of seawater and marine organisms. Other techniques than gas chromatography should be applied (Ernst 1980).

Besides chlorinated hydrocarbons, there are also other organic compounds which are so persistent that they are able to accumulate in the environments of the Earth and in organisms. To these belong the phthalate esters which are mainly used as a softening agent in synthetic fabric production. Approximately 0.5 million t of them are produced annually. In the muscle tissue of various fish taken from Gulf of Mexico waters a good distance from land, scientists found an average wet weight concentration of 0.005 mg/kg di-2-ethylhxylphthalate (DEHP) along with dibuthylphthalate (DBP; Giam et al. 1978). Fortunately, such concentrations in fish can be ignored from the human health risk point of view, because man takes up DEHP, at present, in much larger quantities from plastic food packaging than from contaminated food (Tomiata et al. 1977). But from the environmental health point of view the further tendency of phthalate ester accumulation has to be carefully monitored.

9.2 What Quantity of Chlorinated Hydrocarbons May Be Tolerated in Marine Food for Human Consumption?

The World Health Organization (WHO) has established an acceptable daily intake limit for DDT of 0.3 mg per person, or 0.005 mg/kg body weight [8]. On this basis various countries have different regulations regarding the tolerable concentrations of DDT and other chlorinated hydrocarbon pesticides in fish and other sea-food. The Regulation for Maximum Amounts of DDT and other Pesticides in or on Food of Animal Origin for Human Consumption went into force in the Federal Republic of Germany in 1974. DDT-like substances are not permitted to exceed concentrations of 2 mg/kg in sea fish. An exception is made for eel *(Anguilla anguilla)*, salmon *(Salmo salar)*, and sturgeon *(Acipenser sturio)*. In these fish, 3.5 mg/kg is the limit. For fish liver and fish oil the limit is set at 5 mg/kg.

---

8 Where DDT is mentioned in text and figures, the total amount (Σ DDT) of DDT + DDD + DDE and other metabolic products is meant. If not otherwise indicated, the concentrations are relative to wet weight

An average citizen in the Federal Republic of Germany, who eats 200 g fish per week, would take up, at a concentration of 2 mg/kg, 0.4 mg DDT and similar substances per week or 0.07 mg per day. This is one fourth of the limit set by WHO.

As regards PCB's the World Health Organization concluded (WHO 1976b) that skin effects may occur in man at a daily intake rate of 4.2 mg per person (of 60 kg) or 0.07 mg/kg of body weight. There are no WHO standards about acceptable daily intake of PCB's. In general, intake from normal food is in the range of 0.005 to 0.1 mg per person, which gives about a margin of safety from poisonous skin effects of 1:100. In the Federal Republic of Germany there is at present no regulation regarding maximum permissible concentrations of PCB's in fish. In the U.S.A., the Food and Drug Administration suggested a tolerance level of 5 mg/kg. At this concentration, the average fish eater, consuming 200 g fish per week, would take up 1 mg PCB per week or 0.14 mg per day, much more than with normal food and only about 1:30 away from the risk of skin effects.

The question is what actual risk of poisoning by chlorinated hydrocarbons the fish consumer runs. Fish from locally polluted regions may have concentrations of chlorinated hydrocarbons which make them unfit for human consumption (see Chap. 3.3). Fortunately concentrations in fish from the main fishery regions are much less.

Before 1970, fish landed in British fishery ports had the following DDT concentrations: cod *(Gadus morhua)* caught near British coasts up to 0.68 mg/kg in muscle tissue, herring *(Clupea harengus)* up to 1.4 mg/kg; cod from open Atlantic regions far away from the coast had a maximum of only 0.03 mg/kg of DDT (Agricultural Research Council 1970). From data for fish landed in German fishery ports (Huschenbeth 1973; Huschenbeth 1977), it seems that DDT concentrations have a decreasing tendency. In 1970–1972 herring from the North Sea had 0.16 mg/kg, in samples from 1973–1976 concentrations did not surpass 0.07 mg/kg. Cod and coalfish *(Pollachius virens)* had only 0.013 mg/kg; the highest concentrations were found with 0.13 mg/kg in mackerel *(Scomber scombrus)*. Fish from the Baltic Sea had, in 1975, somewhat higher DDT concentrations, 0.4 mg/kg in herring, 0.3 mg/kg in flatfish (Pleuronectidae), below 0.1 mg/kg in cod. In some other areas concentrations have been higher; for example, sardines from the Black Sea, in 1975, had 1.15 mg/kg (Huschenbeth 1977). Critical values, compared with the German regulation on maximum permissable concentrations, are only in fish liver. Cod liver from the North Sea with 0.9–3 mg/kg DDT concentration is legal, but the liver of cod caught in the Baltic (Gotland and Bornholm Deeps) contains in general concentrations of above 5 mg/kg, with maximum figures in 1973 of 38 and 77 mg/kg.

What do these concentrations mean for the human consumer of sea fish? It seems reasonable to conclude that sea fish landed in North Sea fishery ports has an average DDT concentration of about 0.01 mg/kg. The citizen who eats an average 200 g of fish per week takes up, with the fish, not more than 0.002 mg DDT per week or 0.0003 mg DDT per day. This is one thousandth of the daily intake accepted by WHO. Concentrations in fish are only one tenth of concentrations in meat, sausages, and other food of domestic animal origin (Ernährungsbericht 1976).

As regards fish liver and fish oil, there is at present no evaluation available about average consumption and about the risk to human health by direct intake. One should, however, keep in mind that fish oil and fish meal are largely used to produce animal food, so that DDT and other chlorinated hydrocarbons contained in fish will reach, via chicken, pork, veal, and trout, the human consumer.

More data on DDT and PCB concentrations are given in Tables 39–41. PCB concentrations in sea fish from the Atlantic and the North Sea are, as a rule, 2–5 times higher than DDT concentrations in the same fish: cod up to 0.8 mg/kg, plaice *(Pleuronectes platessa)* up to 0.6 mg/kg, herring up to 0.4 mg/kg. Sole *(Solea solea)* from the Dutch coast had up to 1 mg/kg, and this figure did not change significantly between 1972 and 1976 (Hagel and Tuinstra 1978). It is a general phenomenon that, contrary to the decline in DDT concentrations, PCB concentrations seem to remain at about the same level in spite of the fact that use of PCB's has been reduced for the last few years.

Summarizing results from Tables 39–41 it can be stated that PCB concentrations in the average fish that reaches the consumer from North Sea fishery ports are below 0.1 mg/kg. This means that the citizen with average fish consumption takes up, from fish, less than 0.003 mg PCB's per day. This is only a fraction of the PCB intake from other sources, and PCB concentrations in fish are less than in meat (about 0.3 mg/kg; Ernährungsbericht 1976).

Table 39. Concentrations of DDT and PCB's in marine food from St. George's Bank, east coast of the U.S.A. and possibly heavier polluted via the west wind regime than marine food from the Denmark Strait between Iceland and Greenland. It is usual to find high concentrations of chlorinated hydrocarbons in redfish; the reason is that they live in deep cold water, grow slowly and are, therefore, when caught, relatively old. In addition, their meat contains a high percentage of fat. Data from 1971, in mg/kg wet weight (Harvey et al. 1974)

		George's Bank		Denmark Strait	
		PCB's	DDT	PCB's	DDT
1.	Deep sea Prawn *(Pandalus borealis)*	0.36	0.007	0.018	0.001
2.	Muscular system				
	Cod *(Gadus morhua)*	0.038	0.011	0.002	0.003
	Coalfish *(Pollachius virens)*	0.037	0.003	–	–
	Haddock *(Melanogrammus aeglefinus)*	0.030	0.002	–	0.003
	Black halibut *(Reinhardtius hippoglossoides)*	–	–	0.068	0.021
	Redfish *(Sebastes marinus)*	0.190	0.073	0.360	0.032
3.	Liver				
	Cod	22.0	2.7	0.73	0.17
	Coalfish	2.8	1.1	–	–
		45.0	3.0	–	–
		1.5	1.0	–	–
	Haddock	8.8	0.4	0.48	0.26
		3.9	1.6	–	–
		2.2	1.1	–	–
	Black halibut	–	–	0.10	0.33
	Redfish	1.5	1.3	–	–

Table 40. Concentrations of DDT and PCB's in marine organisms from European waters. Data in mg/kg wet weight [Ernst et al. 1976 (English Channel); Schaefer et al. 1976 (Central North Sea); Eder et al. 1976 (Skagerrak)]

	Liver		Muscle	
	PCB's	DDT	PCB's	DDT
1. English Channel 50°N; 0°–4°W, 60–70 m, 1971				
Grey gurnard *(Trigla lucerna)*	0.2–1.7	0.1–0.5	0.03–0.12	0.01–0.03
Plaice *(Pleuronectes platessa)*	0.4–2.2	0.12–0.5	0.01–0.04	0.003–0.007
Brill *(Scophthalmus rhombus)*	2.8	0.9	0.03	0.005
Scallop *(Chlamys opercularis)*			0.02–0.05	0.003–0.007
Squid *(Loligo forbesi)*			0.08–0.18	0.01–0.06
2. Central North Sea 55°N; 6°E, 40–50 m, 1972				
Herring *(Clupea harengus)*	0.13–0.33	0.04–0.08	0.03–0.11	0.01–0.06
Long rough dab *(Hippoglossoides platessoides)*	0.4–0.6	0.05–0.1	0.03–0.05	0.004–0.005
Dab *(Limanda limanda)*	0.4–0.7	0.05–0.2	0.03–0.06	0.003–0.01
Cod *(Gadus morhua)*	0.3–8.1	0.03–1.8	0.03–0.07	0.006–0.01
Cockle *(Acanthocardia tuberculata)*			0.01–0.04	0.001–0.003
3. Skagerrak 58°N; 7°E, 400 m, 1972				
Pole-dab *(Glyptocephalus cyneglossus)*	0.4–0.6	0.16	0.03–0.05	0.008
Octopus *(Benthoctopus piscatorum)*			0.01–0.03	0.001–0.003
Deepsea Prawn *(Pandalus borealis)*			0.01–0.02	0.001–0.003
Crab *(Munida tenuimana)*, Eggs	0.13–0.24	0.09–0.13	0.01–0.03	0.004–0.008

It is beyond the scope of this chapter to discuss the actual position of science as regards adverse effects through regular intake, over years, of small amounts of DDT, PCB, and other chlorinated hydrocarbons, or through the concentrations which accumulate in human tissues. Therefore in this chapter standards of "accepted daily intake" (ADI) are not further questioned. From the point of view of marine food resources one could be satisfied with the knowledge that concentrations of chlorinated hydrocarbons in fish are one order of magnitude less than concentrations in the meat of domestic animals. However, it remains as a remarkable fact that out of very small concentrations in food, man and animals accumulate rather large amounts of chlorinated hydrocarbons which then are stored in tissues.

Mother's milk in the Federal Republic of Germany has average concentrations, on fat basis, of 1–2 mg/kg PCB's, 1 mg/kg DDT, 0.5 mg/kg of HCB, 0.2 mg/kg of beta-HCH. Maximum figures were 5 mg/kg of DDT, 5 mg/kg of HCB, 1 mg/kg of alpha plus gamma-HCH, 0.3 mg/kg of beta-HCH. If a baby of 5 kg body weight drinks, over a period of 100 days, 0.5 kg of mother's milk per day, and if the milk has a fat content of 4%, then 0.5 kg of fat are eaten in 100 days, with a content of 0.5 mg of DDT, 0.25 mg of HCB and 0.1 mg of beta-HCH. The daily intake of DDT, during

Table 41. Chlorinated hydrocarbon concentrations in organisms off the coast of the Netherlands and from the southwestern North Sea. Phytoplankton and zooplankton collected in the Delta area in 1973. *Crangon crangon* shrimp and *Mytilus edulis* mussels collected in the Wadden-Sea in 1972. Young herring and plaice caught near Den Helder in 1973. Three-year-old herring, plaice, and cod caught in the southwestern North Sea off Den Helder in 1972. Data in µg/kg wet weight (Ten Berge and Hillebrand 1974)

	Penta-chloro-benzene	HCB	Alpha-HCH	Gamma-HCH (lindan)	Dieldrin	Endrin	Total DDT	PCB's	Fat content
Phytoplankton	0.05	0.06	0.09	0.21	0.30	0.12	0.23	3.5	0.04%
Zooplankton	0.12	0.21	0.50	0.72	1.48	0.60	1.31	20.0	0.23%
Shrimp	1.8	0.70	4.5	1.2	2.4	–	2.91	83.0	1.2 %
Mussels	0.40	0.53	4.7	3.2	8.9	8.6	9.0	237	1.6 %
Young herring	2.6	9,5	17.1	11.3	45.2	10.6	45.7	765	3.5 %
Young plaice	1.13	2.2	5.0	3.3	10.5	3,1	11.0	290	1.7 %
3-yr-old herring	–	8.3	11.3	6.4	33.7	–	76.4	413	6.5 %
3-yr-old plaice	–	3.0	2.5	2.0	10.6	–	43.4	331	2.2 %
3-yr-old cod	0.28	1.0	0.27	0.38	1.4	–	5.3	51.7	0.26%

this period, is 0.001 mg/kg of body weight, one fifth of the ADI established for adult man. An expert commission in the Federal Republic of Germany realized that this might represent a health risk, which according to present knowledge, however, is so small that it does not compensate the physiological, immunological, and psychological advantages a baby has from being nursed (Hapke 1980). Concentrations of chlorinated hydrocarbons in mother's milk quite often exceed the figure 1 mg/kg, on fat basis, set for cow's milk by the Regulation for Maximum Amounts of DDT and other Pesticides in or on Food of Animal Origin for Human Consumption, existing since 1974 in the Federal Republic of Germany. Meat, sausages, and animal fat must not have more than 3 mg/kg, on fat basis, and the DDT concentration in hen's eggs is restricted to about 5 mg/kg on fat basis, or 0.5 mg/kg on wet weight basis. Quite regularly higher concentrations are found in marine mammals and birds even from ocean regions without any hint as to a local pollution source. From very small daily intake figures over the years, fish-feeding warm-blooded animals, and terrestrial predators accumulate during their life-span astonishing concentrations. In the eggs of Norwegian herring gulls *(Larus argentatus)*, 1.2 mg/kg of DDT and 5.4 mg/kg of PCB's were found, 2.1 mg/kg of DDT, and 7.7 mg/kg of PCB's in the eggs of the gannet *(Sula bassana)*, on wet weight basis. The figures for the colonies in the north of Norway are no smaller than those for the south (Fimreite et al. 1977) or in the German North Sea (Table 42). The especially high concentration in Arctic seagulls is astounding and for the moment inexplicable. Definite symptoms of disease were noted in an Iceland gull *(Larus hyperboreus)* from Bear Island containing 67 mg/kg DDT and 555 mg/kg PCB's relevant to extractable body fat (Bourne and Bogan 1972). Polar bears *(Ursus maritimus)* in the Canadian Arctic also have between 0.2 and 80 mg/kg PCB's in their fat, ten times as much as DDT (Bowes and Jonkel 1975). This is curious, because seals, from which polar bears live, have somewhat less PCB in their fat than DDT. Obviously, polar bears have a particularly selective metabolism

Table 42. In general, chlorinated hydrocarbons in sea birds from the German coasts are at the same levels as in other North Atlantic and North Sea populations. For example, eggs of German herring gulls contained about 7 mg/kg PCB's, 0.4 mg/kg DDT, 0.2 mg/kg dieldrin, 0.05 mg/kg total HCH and 0.01 mg/kg HCB, on wet weight basis (Hoerschelmann et al. 1979). High figures can be found when chlorinated hydrocarbons analyzed in muscle tissue are calculated relative to extractable fat. Data are from Helgoland (North Sea) birds 1974–1975. Kittiwakes, guillemots, and herring gulls breed on the island. The other species are visitors. Data in mg/kg relative to extractable fat (Vauk and Lohse 1978)

	n	total HCH	HCB	DDT	PCB's	Heptachloro-epoxide
Kittiwake *(Rissa tridactyla)*	8	0.5	4.6	32	203	2.5
Guillemot *(Uria aalge)*	16	1.2	11.8	90	354	7.1
Herring gull *(Larus argentus)*	24	0.8	2.6	58	225	1.5
Razorbill *(Alca torda)*	1	3.3	21.3	75	47	6.2
Red-throated diver *(Gavia stellata)*	1	4.5	19.4	82	65	8.3
Great black-backed gull *(Larus marinus)*	4	0.7	8.0	80	1250	2.3
Gannet *(Sula bassana)*	1	4.4	3.7	185	330	14.3
Fulmar *(Fulmarus glacialis)*	1	–	4.9	202	70	–

for chlorinated hydrocarbons. 5–6 mg/kg DDT and equal amounts of PCB's were found in all adult fur seals *(Callorhinus ursinus)* analyzed from the Pribiloff Islands, 63 mg/kg DDT and 33 mg/kg PCB's in two-month-old pups. The higher amounts in the pups can be explained by the fact that they are nursed and that their mothers' milk contains a much higher concentration of chlorinated hydrocarbons than the fish they feed on later (Kurtz and Kim 1976). The fat of whales off the east coast of the U.S.A. was found to contain between 1 and 268 mg/kg DDT and between 0.7 and 114 mg/kg PCB's (Taruski et al. 1975). In the blubber of seals *(Phoca vitulina)* from the North Sea coast of Germany, the concentrations vary between 3 and 30 mg/kg DDT and 30 and 300 mg/kg PCB's on wet weight basis (Fig. 29). Finally, Baltic gray seals are known to have very high concentrations of DDT, PCB's and other chlorinated hydrocarbons (Table 43).

Even when concentrations in many cases exceed the tolerance set for human food, the public health risk for citizens in the Federal Republic of Germany is very small, because marine mammals and birds, or bird eggs are only exceptionally eaten, and, if at all, in small quantities. However, it would be interesting to have a better knowledge on intake of chlorinated hydrocarbons and on concentrations in tissues and mother's milk of eskimos, because some eskimo still depend largely upon a diet of seafood, including seal, whale, and bird.

9.3 Ways of Transport, Transformations and Concentrations

Except in the antifouling paint for ships and in hydraulic oil, PCB's were not used on sea. In the shipping industry pesticides are used for antifouling and wood protection, but the major part of pesticides is being used and spread on land. Nevertheless, what in 1966 came to be known as a sensation has now been confirmed: even in the

Fig. 86. Between 1970 and 1972 probes of zooplankton were sampled by various American research ships in order to find out the contents of PCB's. An average of 0.2 mg/kg wet weight was found, generally 30 times more PCB than DDT (Harvey et al. 1974)

● > 1 mg / kg
● 0.5 - 1
● 0.1 - 0.5
· < 0.1

Antarctic DDT compounds are to be found: the eggs of penguins and other seabirds contain 0.1–1 mg/kg (in fat); in snow some 100 km from the edge of the ice up to 2 ng/kg PCB's are found. The PCB concentration is 2–4 times higher than that of DDT in penguins, six times higher in snow (Risebrough 1977). In the water and in other organisms of the open northern Atlantic Ocean, too, the concentration of PCB's is usually higher than that of DDT (see Fig. 86). Up to now, the claim that with the help of the latest analytic methods it is possible to detect chlorinated hydrocarbons in all organisms of the ocean, even in the deep sea, has not been contradicted. How did it come to this world-wide distribution?

Even at a very high temperature PCB's are resistant, only very much above 800°C do they oxidize into carbondioxide and hydrochloric acid. At lower combustion temperatures the PCB's enter the atmosphere unchanged. If DDT is sprayed to destroy parasites only part of it reaches the plants and the soil, the rest remains finely distributed in the atmosphere and is then spread further by air currents. Even though the vapor pressure may be low the DDT still passes into the atmosphere, under the influence of a number of complicated mechanisms, like, for example, transport together with water evaporation. This is, however, a very slow process and if one traces the concentration of DDT in the soil, it is only after 4–30 years, or an average of 10 years, that one can detect that the DDT has been reduced to 5% of the original amount (Edwards 1966). Of course the fact that the DDT has broken down into another substance could also be the cause of its disappearance from the soil.

In experiments it has sometimes been observed that the DDT decomposes to a certain extent under microbial influence, these results are, however, still contradictory (Fries 1972). The breaking down of dieldrin, lindan and endrin is more effective under certain conditions, with the participation of bacteria and soil fungi. In the ocean, too, microorganisms to a large extent break down DDT into DDD [also known as TDE; 1,2-dichlor-2,2-bis(p-chlorophenyl)ethane]. In animals and other organisms, as well as under the influence of sunlight, DDT is broken down to DDE [1,1-dichloro-2,2-bis(p-chlorophenyl)ethylene]. DDD and DDE, however, differ only slightly from DDT in their toxicity and are largely persistent, except to further decomposition into similar chemical compounds. After a careful analysis one also discovers polar components that signalize a further decomposition of DDT compounds (Ernst and Goerke 1974). Such components were found in experiments with sole *(Solea solea)* but there are indications that these components are not produced by the sole themselves but rather by the bacterial flora in the sole's gut. In the feces of guillemots *(Uria aalge)* and grey seals *(Halichoerus grypus)* of the Baltic Sea phenolic metabolic products of DDT and PCB's were found that indicate a special metabolism of these warm-blooded animals. The PCB compound here also differs from that of their prey, the herring *(Clupea harengus)* (Jansson et al. 1975). If one feeds polychaetes *(Nereis virens)* with PCB's, most of them are excreted unaltered with the feces, but polar break-down products can also be found in the feces, among others trichlorobiphenylol (Ernst et al. 1977; Ernst 1980).

Under anaerobic conditions mixed cultures of marine bacteria found near the coast metabolize PCB's into an acid lactone metabolite (Carey and Harvey 1978). In general, bacteria that decompose PCB's can be found in estuaries, especially near industrial plants (Sayler et al. 1978).

It is also known that under anaerobic conditions, such as can be found in sewage sludge in sewage treatment plants as well as in the deeper layers of marine sediments, bis-(p-chlorophenyl)-acetonil develops from DDT (Jensen et al. 1972). These are, however, conversions that only affect a minor part of the DDT, and the question is what significance these conversions observed in the laboratory have in natural surroundings (Addison 1976). For the moment the allegation that a number of chlorinated hydrocarbons, among others DDT and PCB's, are to a large extent persistent, remains valid: they are broken down to harmless components neither through hydrolysis in water, nor through microbes nor in the metabolism of other organisms, and the ultraviolet rays of the atmosphere have only a limited effect.

Various speculations have been made about the amount of DDT that may have entered the world oceans over the last decades. Supposing 50% of the DDT used enters the atmosphere (Butler et al. 1972), then 700,000 t of the 2.8 million t of DDT produced until 1974 still ought to be in the oceans. With an ocean area surface of $3,6 \times 10^{14}$ m^2 this amounts to approximately 2000 μg/m^2 or with a median water depth of 3700 m this means 0.5 μg DDT/m^3 water, resulting in a concentration of 0.5 ng/l. In 1970 approximately 100,000 t of DDT were used; if 25% of this, or 25,000 t, entered the ocean it would imply a fall-out of 70 μg/m^2. In the biologically important, 100-m-deep surface layer of the ocean, this would imply an input of 0.7 ng of DDT per liter of seawater. These figures correspond with those found by measurements of DDT contents in rainwater. The annual precipitation into the ocean amounts to 3×10^{14} m^3, nearly 1 m^3/m^2 of the ocean surface area, In the western part of Sweden lower figures were registered, the fall-out of PCB's from the atmosphere amounted to 6–120 μg/m^2 per annum, but only 1.2–2.4 μg/m^2 DDT per annum (Södergren 1972). Similar figures were found for PCB's which accumulate in sedimentation traps together with particles suspended in the seawater. In the Kiel Bight the PCB figures are 24–112 μg/m^2 per year (Osterroth and Smetacek 1980).

Of course, global figures are still very uncertain and do not represent well-grounded scientific results. There is also insufficient knowledge about the extent to which DDT by way of sedimentation is again eliminated from the biosphere, that is, to what extent it can be found in such ocean sediments as are excluded from the biological cycle. It is, however, alarming that the figures of different calculations coincide with each other so that there is an increasing possibility that they might be correct.

An analysis of chlorinated hydrocarbons in seawater is problematic for various reasons. On the one hand the salty ocean water is a difficult medium for chemists, and on the other it is not clear in which form DDT, for instance, is found in the seawater. It is certain that only a minor part is really dissolved, if at all, and that the larger part is either found in the form of aggregates that are fine enough to pass through filters, or is attached to particles with a diameter of less than 1 μm. It is also known that chlorinated hydrocarbons attach themselves intensively to organic matter and especially dissolve in fat. This explains the enormous accumulation of chlorinated hydrocarbons in organisms. In not specially polluted ocean water concentrations of approximately 0.1–1 ng/l DDT and about 0.5–2 ng/l PCB's have been analyzed. Higher figures have also been given, but it is questionable whether the analytic methods were appropriate, and there are also experts who doubt whether all PCB analyses that state more than 1 ng/l are reliable, simply because from the relatively low amounts of PCB's produced no higher concentrations can possibly develop in large areas in the Atlantic and Pacific Ocean (Risebrough et al. 1976a). The present knowledge on chlorinated hydrocarbon concentrations in seawater is summarized in Fig. 87.

Instead of analyzing chlorinated hydrocarbons from the very low concentrations in seawater, it is normally much easier to identify DDT, PCB's, and other compounds from the higher concentrations which organisms have built up within their tissues. Concentration factors, however, vary from substance to substance, and from organism species to species; comparisons are therefore difficult to make. Some generalizations are presented in Fig. 64. If one considers results from mussels in the laboratory, concentration factors are about 10,000 for DDT, about 50,000 for PCB's. However, if one takes into account analyses from the natural environment, it seems that concentration factors are much higher (Table 28). It may be important that chlorinated hydrocarbons, aside from seawater, are accumulated from food and from contact with contaminated suspended particles. Such conditions are difficult to simulate in the laboratory.

Even if, on the whole, concentrations of chlorinated hydrocarbons in sea birds and seals seem to be rather uniform throughout the Atlantic and may be even higher in Arctic latitudes than closer to the European coast (see Chap. 9.2), some regional differences have been observed, apart from high concentrations in organisms from locally polluted areas (see Chap. 3.3). Seals from the Baltic Sea seem to be specially

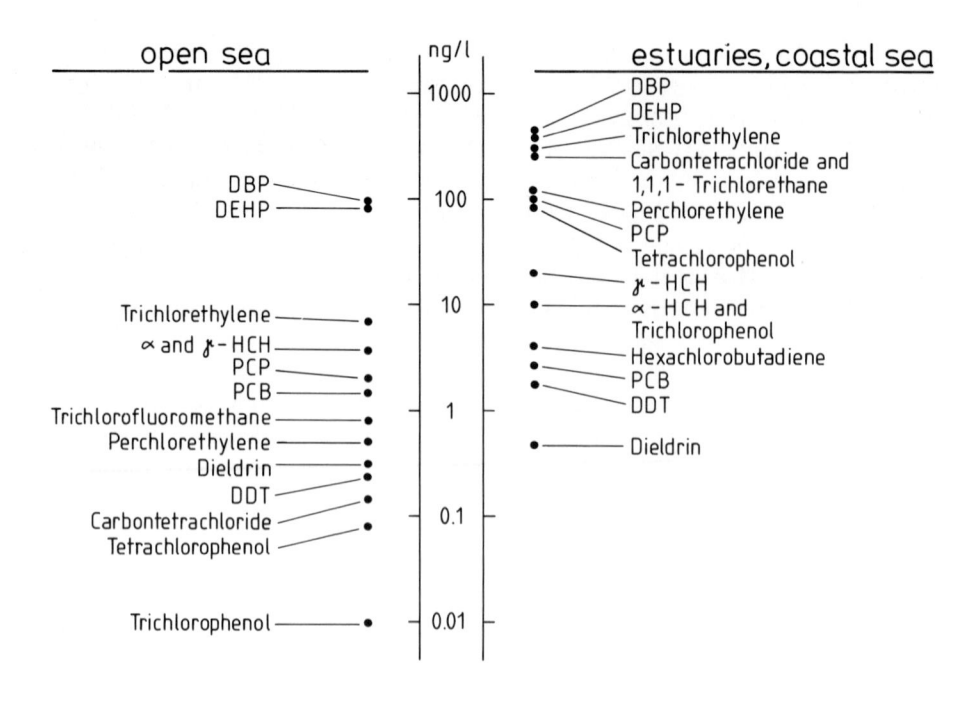

Fig. 87. Concentration of various organic pollutants in seawater of the open sea and from estuaries and coastal areas, including data of many authors (Ernst 1980)

Table 43. The concentrations of DDT and PCB's, found in the blubber of grey seals *(Halichoerus grypus)* in the Baltic Sea are just as high as those found in Californian sea-lions, that suffer from premature and stillbirths. There have been first indications that the baby grey seals in the Baltic Sea have been born dead two months too early. A grey seal weighing 150 kg eats roughly 7 kg of fish a day. With a concentration in the fish of 1.34 mg/kg DDT and 0.63 mg/kg PCB's this means a daily intake by the seal of 0.06 mg of DDT and 0.03 mg PCB's per kg of body weight (Olsson et al. 1975)

Region	DDT	PCB's
	(mg/kg of extractable fat)	
Northern Baltic Sea	470 (230–800)	150 (72–250)
	400	260
	820 (670–970)	260 (230–290)
Åland Sea	230 (68–490)	97 (20–170)
	320 (70–850)	130 (21–320)
	330	120
Gulf of Bothnia	210	80
	180 (139–230)	120 (82–180)
	280 (110–560)	170 (73–330)

contaminated (Table 43). The concentration of DDT in the muscle tissue of cod *(Gadus morhua)* from various regions of the Baltic Sea is 2–10 times higher than that in those from the Kattegat-Skagerak area, which is less influenced by pesticides used in forestry. The figures of PCB concentration are 20–30 times higher in cod from the Baltic Sea than those from the Kattegat-Skagerak area, where 8–21 µg/kg are found (Dybern and Jensen 1978). The southern hemisphere, at least a few years ago, was less polluted by DDT than the North Atlantic. Terns which returned in spring from their southern wintering areas had much less DDT in their tissues than a few weeks later when they had fed upon northern food (Fig. 88).

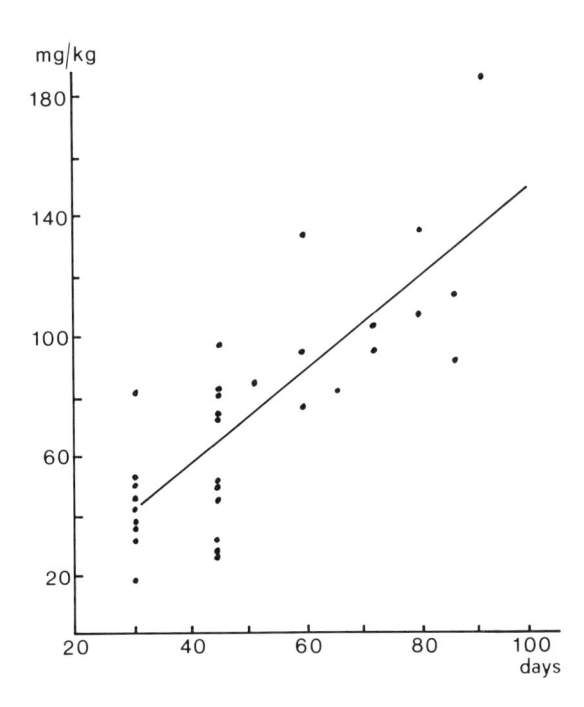

Fig. 88. Terns *(Sterna hirundo)* migrate to the southern hemisphere during winter where they obviously have little contact with DDT. The first eggs they lay in the colony near Hamilton Harbor in Canada show low concentrations of DDT. During the following weeks the content of DDT in freshly laid eggs increases. This is the consequence of the fact that the Canadian fish which the tern live on are contaminated with DDT. Figures in mg/kg egg dry weight (Gilbertson 1974)

9.4 Effects of Chlorinated Hydrocarbons

Apparently certain species of sea birds are specially vulnerable to DDT intoxication. Pelicans react noticeably on increasing concentrations of DDT in their fish prey. A classical example are the pelicans breeding on the isles in the Bay of California, U.S.A. (see Chap. 3.3). Only after DDT concentrations in sardines dropped below 0.1–1 mg/kg wet weight did pelican reproduction start to become positive again.

It is not so easy to explain other observations with a direct correlation between DDT concentrations and breeding success, because along with DDT a number of other toxic compounds are found in the pelican's environment and in its tissues.

Thus the rate of successful breeding of the brown pelican *(Pelecanus occidentalis)* on Marsh Island (South Carolina, U.S.A.) dropped strongly between 1968 and 1972.

In 1971 and 1972 only those pelicans bred successfully, where, referring to wet weight, the eggs contained less than 2.5 mg/kg of DDT and less than 0.54 mg/kg of dieldrin. During this period, however, the average concentration of harmful substances in pelican eggs decreased from 7.3 in 1969 to 2.5 mg/kg of DDT in 1973 and from 1.16 in 1969 to 0.45 mg/kg of dieldrin in 1973, on the average. In 1973 the rate of successful breeding of the pelicans was for the first time excellent, 2726 young pelicans fledged, that is 1.66 chicken per nest, after the number of fledglings had been between 0.7 and 0.9 per nest from 1969 to 1972. In 1973, however, the food supply and weather conditions had also been favorable (Blus et al. 1974, 1977). PCB concentrations in pelican's eggs remained more or less the same over the entire period. Therefore PCB's seem not to be responsible for the catastrophic failure in breeding. On the other hand it is difficult to conclude with significance whether DDT, dieldrin, the weather, or the food supply really was the key factor.

In the osprey *(Pandion haliaëtus)* from the eastern coast of the U.S.A. (Connecticut and Long Island) breeding was very poor during the 1960's. DDT concentrations in the eggs decreased, however, between 1969 and 1976 to one fifth, and since they were below 12 mg/kg (wet weight, or 60 mg/kg dry weight) breeding success improved. In 1976 there were 1.2 fledglings per nest, equal to conditions in the 1950's (Spitzer et al. 1978). However, in this case too, there was a parallel decrease of dieldrin, so that it cannot be said with significance that breeding success is only correlated with DDT. There is a chance that DDT and dieldrin act in a similar way.

As in pelicans, the effect of DDT (or of DDT and dieldrin) on ospreys in the 1960's was egg shells which were 15%–20% thinner than normal. In the double-crested cormorant *(Phalacrocorax auritus)* from the Canadian lakes, an egg-shell thinning of 20% was critical (Vermeer and Peakall 1977). During the 1960's, cormorant populations *(Phalacrocorax carbo)* in the Netherlands declined, possibly as a consequence of egg-shell thinning. Eggs collected in 1971, which had shells 0.45 mm thick, had concentrations of 2 mg/kg (wet weight) DDT and 0.3–0.5 mg/kg dieldrin, but in eggs with thinner (0.35 mm) egg shells concentrations were higher: 12 mg/kg DDT and 5 mg/kg dieldrin (Koemann et al. 1973b).

In the meantime, there is experimental proof that birds fed with DDT lay eggs with thinner egg shells (Fig. 89). Chickens may hatch from such thin-shelled eggs, when the eggs are carefully treated in an incubator. But under natural conditions, in the nest, even eggs with normal egg shells sometimes break. The chance that eggs with 20% thinner egg shells do not break is very poor. The conclusion is that relatively small amounts of DDT which apparently have no other toxic effects upon the bird disturb the enzymatic activity necessary for the process of egg shell formation.

It has not been completely proved that high concentrations of PCB's in blubber and other body tissues of seal lead to a decrease of reproductive success. After the emission of DDT into the California coastal area had been stopped, stillbirths in the sealion colonies continued. This could be due to PCB effects, however, DDT is still present in the sediment (see p. 49), and concerned females have high, unbalanced mercury and selenium concentrations, too (see p. 135). There is no proof that PCB's have a harmful effect on the harbor seal *(Phoca vitulina)* of the German North Sea coast. In comparing the concentrations of harmful substances in dead or sick animals and such that had been shot dead, no significant differences could be found (Drescher

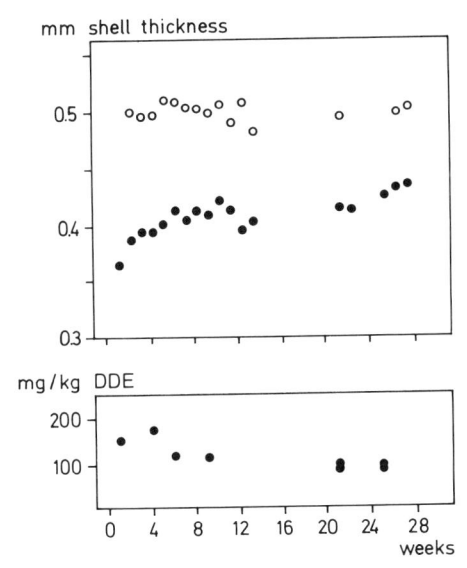

Fig. 89. Even just a short-term feeding with DDT affects the thickness of the egg-shells of birds for a long period. In the experiment domestic white Peking ducks were fed for 10 days in such a way that each duck took in 0.5 g of DDT. When, 2 months later, the ducks started laying eggs, these eggs had considerably thinner shells *(closed circles)* compared to those of ducks that had not been fed DDT *(open circles)*. This difference was still significant 27 weeks after the first eggs had been laid. The *lower diagram* shows the DDE contents of the egg-yolk (mg/kg referring to dry weight; referring to wet weight figures have to be multiplied by 0.32; referring to fat weight by 1.7). The concentration of DDE in the egg-yolk decreases roughly by the same extent as DDT is eliminated from the duck's organism through egg laying. (After Peakall et al. 1975)

et al. 1977). But seals from the Dutch coast have higher concentrations of PCB's than seals from Schleswig-Holstein, as a consequence of pollution by River Rhine, and the decline of their population could well be a consequence of PCB contamination.

Based on observations made in the natural environment it is difficult to prove whether there are harmful effects that cause just a lower rate of offspring. Since these effects do not immediately lead to the death of older birds or seals, it is only after several years, when these have been decimated through natural causes and natural mortality, that it becomes apparent that there are not enough young birds or seals to permanently secure the population's existence. Especially sea birds and seals almost yearly change the site of their breeding colonies. so that it is hardly possible to count the population accurately. Because of this, it is not clear whether the reduced number of seal *(Phoca vitulina)* in the Dutch Wadden-Sea is due to a higher death rate, or because the animals have migrated to the German coast. It would be necessary to have information about the total population development in the North Sea area, but this requires a well-organized observation network. It is no pleasant thought, that, possibly right under our eyes, the pollution of the sea is causing damage to the sea bird and seal populations to such an extent, that every year they are having less offspring and we do not even realize this at first.

All Baltic seals have high concentrations of DDT and PCB's in their tissue (Table 43). The population of grey seals *(Halichoerus grypus)* in the skerries of southeastern Sweden declined from 20,000 individuals in 1940 to a few thousand. Only 27% of Bothnian Bay ringed seal *(Phusa hispida)* females of reproductive age are pregnant (Anon 1980b). Half of the non-pregnant females have sealed uterus horns, proof that implanted eggs have been aborted or resorbed. In these females the blubber contains 130 mg/kg DDT and 110 mg/kg PCB's, on fat basis, compared with 88 mg/kg DDT and 73 mg/kg PCB's in pregnant females (Helle et al. 1976; Leppäkoski 1980). Experiments have not been done with seals, but in another fisheating mammal, the mink

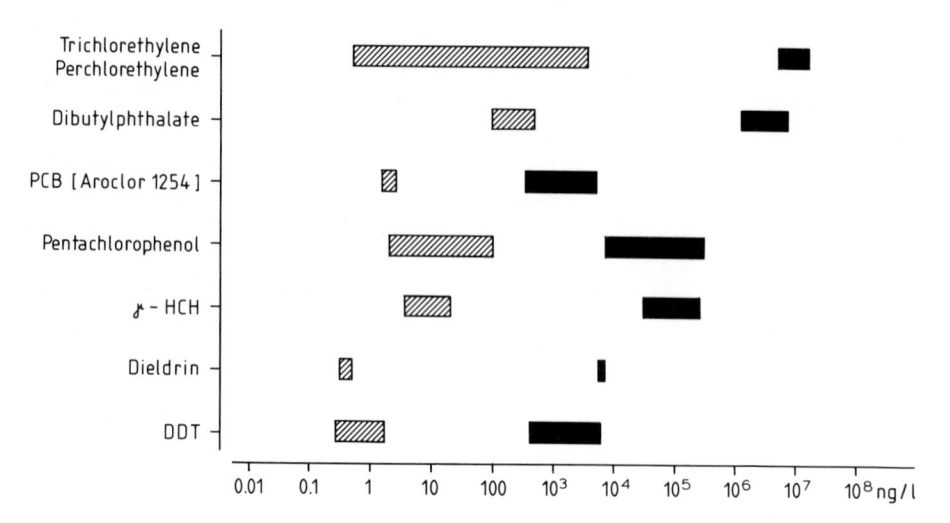

Fig. 90. Range of concentration of organic pollutants in seawater *(hatched areas)* in comparison to range of concentration which exhibits acute toxicity effects in experiments with fish and shrimps *(black bars)* (Ernst 1980)

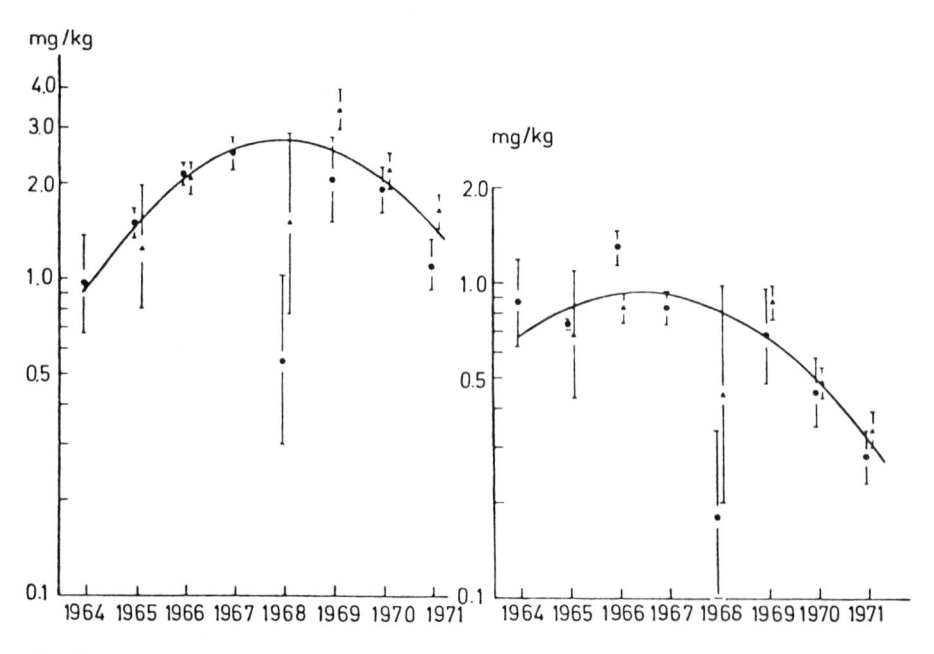

Fig. 91. From 1964 to 1971 the amount of pesticides in the eggs of the shag *(Phalacrocorax aristotelis)* was being regularly examined in two British bird colonies: Farne Island (Northumberland, *points*) and Isle of May (Scotland, *triangles*). Apparently the reduced application of pesticides had already led to a reduction of pesticide concentrations in sea birds living on fish: shown here are, on the *left*, DDT and on the *right* dieldrin. The analyses of 1968 are irrelevant, since after wide-spread deaths of adult shags, only a few eggs could be collected in that year. (After Coulson et al. 1972)

(Mustela vision) it could be shown that PCB's, not DDT in the diet are responsible for an increase of the percentage of non-pregnant females and for a reduction of the whelp number (Jensen et al. 1977).

Recently it could be demonstrated that DDT and PCB concentrations in flounder *(Platichthys flesus)* caught in the Western Baltic vary largely from fish to fish (0.005– 0.73 mg/kg in liver, 0.005–0.317 mg/kg in ovaries, on wet weight basis). 100% streight and healthy appearing larvae could only be obtained from females which had PCB concentrations less then 0.12 mg/kg. When eggs from females with higher concentrations of PCB's were incubated, there was a significantly reduced survival of hatched larvae (Westernhagen et al. 1981). These results mean that the presently existing PCB concentrations in the water of the Baltic are such that females which by some circumstance of their life history or by their physiological conditions have accumulated higher PCB concentrations in their tissues than other females, exhibit toxic effects of PCB in their offspring. Unfortunately there is evidence that this toxic effect may not be restricted to the Baltic. For example cod from the North Atlantic and the North Sea have PCB concentrations in tissues which are only slightly less than in Baltic cod (Ernst 1981).

It is also not possible to exclude that plankton life in the oceans is adversely affected through chlorinated hydrocarbons. In experiments, concentrations of 1 μg/l of DDT have a toxic effect on the *Cyclotella* diatom from the Sargossa Sea (Menzel et al. 1970), and just 0.1 μg/l of PCB's apparently has a toxic effect on the *Thalassiosira* diatom (Fisher et al. 1974). Existing data shows that 1–10 μg/l of PCB's have a negative effect on the biomass and cell size of phytoplankton cultures of coastal areas (O'Connors et al. 1978), and that the frequency of cell division, as well as photosynthesis efficiency, are therefore impaired (Harding and Phillips 1978). Such concentrations are only 100 times higher than in the ocean (Fig. 90).

Fortunately there are indications that the concentrations of DDT and other pesticides in marine organisms have been decreasing in the last few years. This could be demonstrated with sea fish (see Chap. 9.2), with oysters on the coasts of the U.S.A. (Butler 1973), with the shag of British bird colonies (Fig. 91), and with the double-crested cormorant *(Phalacrocorax auritus)* in the Bay of Fundy, Canada (decrease from 10 mg/kg to 2 mg/kg DDT concentration in eggs, on dry weight basis; Zitko 1976). Examples from other cormorants and pelicans have been given in this chapter.

Unfortunately however, in general, PCB concentrations in fish from the North Atlantic, the North Sea and the Baltic up to now did not show significantly decreasing trends. World wide a reduction of PCB production from some 80,000 t in 1970 to some 20,000 t in 1977 has been achieved, but factories in France, in the Federal Republic of Germany, in Italy and Spain are still producing. The development of trends of PCB concentration should be carefully watched, and stricter measures should be taken.

One should, too, carefully watch trends of pesticide concentrations in organisms and sea water from shores in tropical areas where persistent chlorinated hydrocarbons are all the time used in large quantities. In air collected at Enewetak Atoll far away from any pollution source in the Pacific, concentrations of organic pollutants were, in 1979: DEHP 1.4 ng/m^3, DBP 0.87 ng/m^3, PCB 0.54 ng/m^3, HCH 0.25 ng/m^3, HCB 0.10 ng/m^3, chlordan 0.013 ng/m^3, dieldrin 0.010 ng/m^3, DDE 0.003 ng/m^3 (Atlas and Giam 1981).

10 Laws Against the Pollution of the Oceans

It is legitimate that everyone expresses his opinion about the problems connected with the pollution of the oceans according to his own interests. A conservationist who is concerned with preservation and tending of the variety of species will have a different point of view than an environmentalist who wants to maintain the quality of living for mankind in the long run; a farmer has a different opinion about the problems resulting from residues than a fisherman. The head of a chemical plant has to be concerned with the productivity of the plant, the local politician with the maintenance of tax revenues and jobs. Within a government, too, the positions of the different ministries are not identical.

A scientist ought to be free from such preoccupations, but this is only seldom the case: instructed by authorities, through economic stimulus to cooperate in coordinated research programs, or through personal commitment to environmental protection, he chooses from among the variety of possible research projects the one which suits him best. When, from the obtained data, he draws such conclusions which fit into his concept, this is legitimate as long as the scientific truth is not disturbed by it.

Every law against marine pollution develops from the preparatory work of science, from individual activities, and from the work of civil activist groups, but before it is passed, the confrontation of contracting opinions must be ended and the formation of democratic will completed.

The problem here is that the assessment of problems associated with marine pollution and of possible countermeasures depends on the actual progress of science and the formation of scientific opinion. By the time a law has finally been passed, science has usually advanced. This is why laws against marine pollution will also be changed again in the future, partly because restrictions will be tightened and new dangers must be taken into account, and partly as there will also be relief when dangers have initially been overestimated.

In the regional and national field every state is responsible for its own coastal waters, meaning, in the past, that it is responsible for a 3-nautical mile zone of sovereign power and a 12-mile fishing zone. At present there is a tendency to enlarge the zone of sovereign power to 12 nautical miles, and the zone of commercial interest to either 200 nautical miles, or the entire shelf off-shore.

On August 15, 1967, the Third Law for the Modification of the Water Regulation Law was passed in the Federal Republic of Germany; from that time on this law has applied to coastal as well as inland waters. Waste water treatment plants have to be built if disadvantages for the public are to be expected from the discharge of untreated or poisonous waste water. The four coastal states Niedersachsen, Bremen, Hamburg, and Schleswig-Holstein are responsible for the enforcement of this law; every industrial plant in the coastal area has to be submitted to experts and sanctioned with regard to possible problems arising from the release of wastes.

At the national level, however, laws of this kind are double-edged: they might lead to the fact that an industrial plant settles down in the neighboring county if the requirements for the quality of waste water can be more easily met there. Moreover, the harmful substances are carried from one coast to the other by ocean currents. For this very reason the Paris Convention has been worked out in the framework of the European Community which is intended to uniformly regulate the "prevention of marine pollution from land-based sources". This convention of 1974 has also been signed by the Federal Republic of Germany, but it has not yet entered into force, because it took a long time to reach an agreement between the member states of the European Community on the question whether emission standards or water quality standards should serve as a basis for judgement. Special guidelines were worked out by the European Community for the judgement of pollution of sea- and freshwaters for bathing, for quality requirements of seawater in shellfish-cultivating areas and for wastes resulting from titanium dioxide production.

The Paris Convention will oblige the member states of the European Community to protect the Atlantic Ocean and North Sea coasts from harmful substances which enter the oceans through water courses, pipelines, water works in sovereign territory, or otherwise from the coast.

There will be a "black list" of substances which because their persistence, toxicity, or capacity to accumulate are so dangerous that their introduction is not permissible and immediate action must be taken against existing contamination. Among these substances are chlorinated hydrocarbons, other halogenated organic compounds, and substances which can form organohalogen compounds in the marine environment, unless they are decomposable or biologically innocuous substances. Included are: mercury and cadmium compounds, plastic driving or floating in the seawater and consistent oils and hydrocarbons derived from petroleum.

A "gray list" includes substances whose introduction into the oceans must be strictly limited. Only with special permission may these substances be released; among them organophosphorus, organosilicium and organo-tin compounds, pure phosphorus, degradable oils and hydrocarbons derived from petroleum, compounds of arsenic, chromium, copper, lead, nickel, and zinc, as well as substances which impair the taste of fish-meat and other seafood.

Moreover, the Paris Convention stipulates that the member states gradually establish a control system in their coastal waters in order to realize the degree of marine pollution and to reexamine the efficiency of measures taken so far. The Paris Convention does not apply to the Baltic Sea and the Mediterranean.

In 1974 the Helsinki Convention was elaborated between Denmark, the German Democratic Republic, the Federal Republic of Germany, the People's Republic of Poland, Finland, Sweden, and the Soviet Union. The Convention on the Protection of the Marine Environment of the Baltic Sea Area was enforced on May 3, 1980.

This convention includes all types of marine pollution, not only originating from the coasts and from navigation, but also from the atmosphere. Thus it is laid down in Annex I that it is not only prohibited to dump DDT and PCB's into the Baltic Sea, but that prohibitions and regulations will also be necessary for trade, and for the application and deposition of these substances in the countries which are subject to this convention, because the Baltic Sea can only be protected against these substances when the atmospheric transport is stopped. Annex II comprises a variety of substances which may not be released into the Baltic Sea without special permission, and since mercury, cadmium, and persistent chlorinated hydrocarbons are listed in this category, one can presume that in many cases permission will not be given. In the case of domestic sewage it is obligatory that it must be treated, so that neither the oxygen content of the Baltic Sea is reduced nor eutrophication of the Baltic Sea takes place. There are special regulations for oil and dangerous shiploads; and ships above a certain size which are closer than 12 nautical miles to the land are not permitted to let unpurified sewage overboard. If they are not equipped with small sewage treatment plants of their own, they have to discharge sewage into corresponding plants in the harbors. In addition garbage may not be thrown overboard in the Baltic Sea, apart from kitchen leftovers, and then only when the ship is more than 12 nautical miles away from land. This regulation holds even for small craft.

For the Mediterranean the neighboring states worked out the Barcelona Convention or the Convention for the Protection of the Mediterranean Sea Against Pollution, on February 16, 1976, which principally refers to all types of marine pollution, but which is primarily applicable to the dumping of harmful substances by ships, as well as to cooperation in case of dangerous oil-spills, and in case of shipwrecks with dangerous cargoes.

The Oslo Convention which is applicable to the area of the Atlantic Ocean between 42°W and 51°E, and north of 36°N was already signed by 13 contracting states on February 15, 1972. This Convention for the Prevention of Marine Pollution by Dumping from Ships and Aircraft was internationally enforced on April 7, 1974. In the meantime a similar agreement has also been worked out on a world-wide scope, the Convention for the Prevention of Marine Pollution by Dumping of Wastes and Other Matter from December 29, 1972 (London Convention) which was internationally enforced on August 30, 1975. Ninety one states are participating in this Convention.

The "black lists" in the annex to both conventions include the substances that may not be dumped into the ocean, because they are at the same time poisonous, persistent, and bioaccumulating, or remain floating in the water, such as organohalogen compounds, unless they are decomposable of innocuous (the Oslo Convention also mentions organosilicium compounds), mercury and cadmium compounds, persistent plastic materials, cancerogenic substances (Oslo Convention only); oil and corresponding hydrocarbons, highly radioactive substances, biological and chemical weapons (London Convention only). However, if such forbidden substances are only contained in trace concentrations, as, for example, in sewage sludge, they do not belong to the "black list" category.

The "gray lists" comprise substances which are either poisonous, persistent, or bioaccumulating, but which show such properties that they may be dumped into the sea after thorough examination and with special permission; these are: arsenic, lead, copper and zinc compounds, cyanides and fluorides, furthermore radioactive substances, pesticides, and organosilicon compounds unless already included in the "black list";

containers, scrap, and tarry substances which impair fishing on the sea bed; and finally substances which may not be very toxic, but which may negatively affect the environment because of the amount dumped. According to the London Convention it has to be examined what amounts of toxic trace elements like arsenic, lead, copper, zinc, beryllium, chromium, nickel, and vanadium are being dumped into the sea, even if the concentrations themselves are very low.

The Oslo Convention is administered by OSCOM, the commission of the contracting states which meets once a year. The decisions of OSCOM are prepared by SACSA, the Standing Advisory Committee for Scientific Advice, which has established working groups which do not only deal with issues of the Oslo Convention, but also with issues of the London Convention and the Paris Convention. The elaboration of a guide-line Regulations for the Control of Incineration of Wastes and other Matter at Sea 1978 was of particular importance.

The conventions restrict the traditional freedom of the high seas, for they oblige the contracting states to control all ships and airplanes running under their flag, loaded in their harbors, and passing through their territory. Violations of the agreements by foreign ships or airplanes on the high seas are to be reported by the contracting states when they receive such information.

In the Federal Republic of Germany the High Sea Dumping Law was passed in 1977 and subsequently the Oslo Convention and the London Convention were ratified. Responsible for the enforcement of the law in the Federal Republic of Germany is Deutsches Hydrographisches Institut in Hamburg which proceeds according to the High Sea Dumping Administrative Regulations from December 22, 1977. If an application has been filed for the dumping of harmful substances into the sea, it will first be checked whether the project conflicts with an international law; later the Umweltbundesamt in Berlin checks whether the wastes might not be deposited or destroyed on land as well without causing any damage to the environment, or how they could best be put to further use; then 13 authorities of the Federal Ministries and of the coastal states are asked for their opinion, and research institutes are consulted as well (Offhaus 1980).

The fact that lawyers with their necessarily formal methods of thinking cooperate with scientists in the making of laws against pollution of the oceans sometimes results in grotesque situations. Out of financial and emotional considerations more and more people decree by will that they want to be buried in the sea after their death. But is there not good reason to define a cinerary urn with human ash as "noxious waste which has to be removed" and does it not impair "the amenities of the marine environment"? In 1978 the German officials had to deal with this question, but were clever enough then not to issue any regulation concerning this matter, but agreed with the sea-burial establishments that the urns have to be transitory, not made out of bronze, but out of clay or rock salt. The heavy metal content of human ash is tolerated.

In the Federal Republic of Germany it has to be checked principally whether harmful substances might not be deposited on land or be neutralized. If expertises are positive, dumping into the sea is out of the question, and permission may not be given. This point of view differs from that of Great Britain, where one does not seek to avoid, by all possible means, negative effects on the marine environment, and thus to a reasonable extent agrees to damages, when the disposal of waste on land seems to be unfavorable in an economic or ecological sense. Thus in the framework of the European Community and within the committees which elaborate the different conventions on the protection of the oceans there are vehement disputes as to what is the right philosophy. Further, in the Federal Republic of Germany instead of well-established proof, concern that harm may originate from marine waste disposal is sufficient to deny a permit for dumping.

The oldest international agreement on the prevention of marine pollution was established in 1954 by the Intergovernmental Maritime Consultative Organization (IMCO): it is the Convention for the Prevention of Pollution of the Seas by Oil which since 1958 has been effective under international law, and in which 23 states participated. A modified version was elaborated in 1962 and has been effective since 1967, a further version from 1969 became effective in 1978; new versions with different amendments and modifications were elaborated in 1971, the main goal being the limitation of the size of oil tanks. Initially restricted areas were established in different oceans by these conventions, where no tanker of more than 150 gross tons and no other ship of more than 500 gross tons was allowed to drain off oil. But drifting oil knows no boundaries. It moves with approximately 4% of the wind's speed until it is decomposed or otherwise disappears from the water's surface. Therefore, in 1968 the plan of the restricted areas was cancelled and within a coastal zone of 50 nautical miles discharging of oil is prohibited. Outside this zone only 60 l per nautical mile may be discharged. Ships built after 1967 with a deadweight tonnage of more than 20,000 t, among them all modern tankers, have to be constructed in such a way that the empty oil tanks are cleaned on the return trip, whereby the oily washings are stored in a tank (slop-tank) and not let overboard (load-on-top system).

But as long as waste-oil depots in the harbors impose fees which are just about as high as the price of oil, as long as a shipowner has to pay fees for waste oil just as for freight to the Suez Canal Administration, and as long as no efficient control and no deterrent punishments are practicable, captains of tankers will very easily be tempted to secretly empty the slop-tanks into the ocean. Nobody talks about it, for shipowner, captain, and crew all have the same interest in this inexpensive illegal measure.

Inexpensive waste-oil depots are necessary not only in oil harbors, but also in all other harbors, so that ships can dispose of their waste oil after an oil change and will not, as it is still done widely, leave it in the sea. In addition, the installation of separators on ships which free the bilge water from leakage oil is required so that only relatively clean bilge water with less than 100 mg oil per liter is pumped overboard, and not more than 50 liters of oil per nautical mile. An oil log has to be kept so that a control of regulations is possible. In the meantime more regulations have been worked out and it is stipulated that oil tanks are no longer washed with water, but with oil, or that separate ballast-water tanks are available.

In 1973 IMCO, with the participation of 79 states, elaborated a new Convention for the Prevention of Pollution from Ships which also includes the hitherto existing agreements against pollution of the ocean caused by oil. It is, however, more comprehensive, and also includes sources of marine pollution other than oil. The convention regulates the transport of chemicals and other dangerous cargoes as well as the measures taken for the cleaning of tanks containing oil. Therefore it was necessary to categorize chemicals according to how dangerous they are. Tanks containing dangerous chemical substances of category A have to be treated in such a way that all residues can be deposited on land. For less dangerous substances dilutions are laid down which are permissible in the wake of the ship. Sections of the convention which deal with garbage and sewage produced by the ship's crew are only regarded as recommendations. So far only few states have ratified this convention (Portmann 1977), the Federal Republic of Germany passed a relevant law on June 26, 1981.

The IMCO conventions also deal in detail with the security measures intended to reduce the number and extent of oil-spills (see Chap. 5.3). But oil-spills will still happen as long as oil is being transported across the oceans. It is, therefore, necessary to supplement the conventions with further agreements concerning compensation. The 1969 Convention on Civil Liability for Oil Pollution Damage was signed by 49 states and became effective in 1975. The Convention on the Establishment of an International Fund for Compensation for Oil Pollution Damage was elaborated in 1971, and in 1974 the corresponding agreement on drilling platforms was worked out. The efficiency of these agreements is, however, determined by the number of states that have ratified them. Therefore, voluntary private agreements which in the meantime apply to most tankers and to more than 90% of the oil transported in world trade are of special importance. According to the TOVALOP Agreement (Tankers Owners Voluntary Agreement concerning Liability for Oil Pollution) compensation payments of up to 16.8 million US $ can be paid in case of an oil-spill. According to the CRISTAL Agreement (Contract Regarding an Interim Supplement to Tanker Liability for Oil Pollution) signed by mineral-oil companies compensation sums might amount to an additional sum of 36 million US $, and in the future probably up to 72 million US $. The OPOL Agreement (Offshore Pollution Liability Agreement) provides 25 million US $ for damages caused by the exploration of oil in the North Sea.

The Convention Relating to Intervention on the High Seas in Case of Oil Pollution Casualities (Brussels Convention) was elaborated in 1969; in case of oil-spills this agreement, which became effective in 1975, entitles the states to take all required measures in order to diminish the dangers, and if necessary also to sink ships outside of their territorial waters. This agreement was extended in 1974, and now also applies to accidents of ships with other dangerous cargoes. On a regional level concerning the North Sea eight states agreed in 1969 on a closer cooperation concerning oil-spills by means of an Agreement on Cooperation in Dealing with Pollution of the North Sea by Oil (Bonn Convention); the agreement became effective in 1970.

On the national level similar agreements are necessary as well, since different authorities have to cooperate in case of oil-spills. In the Federal Republic of Germany the Agency for Waters and Navigation of the Federal Minister of Transport is responsible for navigation and thus also in the case of oil-spills. If the oil is, however, driven toward the coast, coastal states are responsible. For this reason the Oil-Spills Committee Sea/Coast was created which elaborated technical recommendations and organizational proposals for the control of oil-spills in the German coastal region. It also elaborated an administrative agreement between the Federal Government and the coastal states in order to have an organization for the control of oil-spills on the high seas, in the coastal waters, in the Kiel Canal, and in the estuaries as far as up to

Emden, Bremen, Hamburg, and Lübeck. Up to 500,000 DM can be used immediately for the prevention and control of oil-spills, in case of higher expenses the contracting parties have to agree (Boe 1975).

There are at least 36 international and regional agreements which include questions of marine pollution: of these 2 deal with the dumping of garbage originating from ships, 10 with problems of contamination caused by oil, and 12 with the ocean's exposure to radioactivity. The consequences of the exploitation of offshore natural resources have not been regulated by international agreements so far.

Regulations concerning marine pollution from land are inadequate, at least on the supranational level. They are difficult to enforce because important national economic interests are concerned (Waldichuk 1978). Absolutely no regulations, apart from for incineration at sea and for regional Baltic problems, exist concerning harmful substances which enter the world oceans from the atmosphere, in spite of the fact that world-wide pollution of the oceans with chlorinated hydrocarbons has its source in atmospheric contamination. But if one wants to reduce the concentrations of pollutants in the marine atmosphere, one has to regulate the fabrication and the use of such harmful substances inland. Effects upon industry, agriculture, and health standards would be manifold, and the argument from marine biology is just one among many conflicting arguments. However, the health of the biosphere, of all life on the globe, is well-reflected in the oceans and in marine life. Therefore, marine science has an important task at least to monitor trends in the future, and to give warnings if pollution should increase or should not decrease with time.

11 Diagnosis and Therapy

Chapter	Diagnosis	Therapy
2.1	When biodegradable organic substances from untreated domestic sewage or from other sources are introduced into estuarine or seawater, they reduce the oxygen content and create adverse conditions in such bodies of water which suffer from insufficient oxygen renewal.	Waste water treatment plants reduce the content of biodegradable organic substances in waste water.
2.2	With domestic sewage, fecal bacteria and pathogens are introduced into estuaries and coastal areas. Even if the bactericidal effect of seawater reduces their number and vitality, there is a risk of infection when swimming or when eating seafood.	Waste water treatment plants reduce the content of pathogens in the effluent. More research on the fate of dangerous pathogens, bacteria and viruses, and their fate in the marine environment is necessary.
2.3	Plant nutrients are introduced into the sea with sewage, from industry and from agriculture. Such fertilization changes the composition of phytoplankton and larger algae. Increased plant production results in more biodegradable organic matter and in stressed oxygen conditions.	Replacing phosphates in washing powders by zeolites will result in a remarkable reduction of plant nutrients in waste water. If necessary, more nutrients could be removed from sewage by chemical treatment in waste water treatment plants. More research is necessary to clarify at what point costly measures against eutrophication are really necessary. It is essential to get a better understanding of the possible correlation between eutrophication and dinoflaggelate plankton blooms (red tides).

Chapter	Diagnosis	Therapy
2.4	There is a trend in the deep water of the Baltic Sea toward anoxic conditions.	If this trend is mainly caused by natural clima and sea level changes, nothing can be done. If it is caused to a reasonable extent by man-made eutrophication, this could be stopped.
2.5	The surface waters of the warm oceans are mostly poor in plant nutrients, i.e., phosphorus and nitrogen.	In principle, every amount of domestic sewage could be introduced into the open ocean without adverse effects. Whether this is practicable, economic considerations could tell.
2.6	When either biologically treated or untreated sewage is introduced into the water of an open ocean coast with good oxygen renewal, the effect upon the system is about equally unproblematic.	Further scientific research is necessary to affirm this statement.
2.7	Detergents are toxic.	Detergents with less toxicity can be developed.
2.8	While in temperate zones a 2°C rise of temperature in estuarine and coastal waters seems tolerable, organisms in tropical zones live very close to their temperature limit; they cannot stand a rise in water temperature.	Before well-founded conclusions can be drawn, more case studies on thermal pollution should be published.
2.9	In the vicinity of coastal cities, in estuaries and around pollution sources, the natural flora and fauna living at the shore undergo changes, sensitive species are replaced by resistant ones.	More research is necessary to clarify whether this is the effect of one dominating pollution factor, or is the result of several stress factors acting together. Marine conservation parks should be established, before the rapidly proceeding utilization of coastal resources all over the world has destroyed suitable areas.
3	Elevated concentrations of mercury in fish and other seafood are deadly for human consumption.	Mercury effluents from industry are forbidden. More research is necessary to evaluate the effect of small traces of mercury in sewage upon the marine system.

Chapter	Diagnosis	Therapy
	Elevated concentrations of DDT make fish unfit for human consumption and endanger the survival of sea birds and mammals.	Chlorinated hydrocarbons and other toxic organics in effluents from industry have to be forbidden. More research is necessary to evaluate the effect of small traces of such substances in sewage introduced to the marine system.
4.1	Before the 1960's, large amounts of dangerous materials were dumped into the sea.	Dumping of dangerous materials is forbidden, dumping of other materials should be controlled.
4.2	In several parts of the world large quantities of wastes from the titanium dioxide pigment industry are dumped in the sea.	More research on the effects of these wastes upon marine life is necessary.
	It is difficult to draw conclusions from population studies of organisms living in regions with pollution stress, because of natural fluctuations.	As it is risky to conclude from laboratory experiments as to effects in nature, monitoring of natural populations of organisms has to be continued. Long-term observations and adequate research strategies will hopefully give better tools to distinguish between natural and pollution-induced fluctuations.
4.3	Dumping of sludge from waste water treatment plants into the sea results in oxygen depletion and in higher concentrations of toxic heavy metals and organic compounds.	Sewage sludge could be introduced into the high seas without adverse effects. In restricted coastal situations dumping should be forbidden.
4.4	Antifouling paints add to the heavy metal pollution and may have adverse effects in restricted bays and estuaries.	The invention of nonpersistent toxic substances for antifouling paints should be encouraged.
4.5	Garbage from ships is a nuisance.	It should be forbidden that garbage from ships, except kitchen leftovers, goes into the sea in coastal areas.
5	Oil is a complex material, and each component has a special fate when spilled on the sea's surface. Many oil components are either toxic or may possibly change the pattern of	Scientific research on oil in the sea is hampered by the different types of oil and the different properties oil acquires in the course of weathering. A standard oil quality for

Chapter	Diagnosis	Therapy
	chemical signalling among marine organisms. Largest damage is to marine birds, and to coasts. Wave-protected fine grade sediments at the high water mark suffer most. Oil is degraded by a number of microbes. There is a discrepancy of opinions regarding chemical dispersants which bring an oil slick into the water column, and on different strategies of shore clean-up after an oil-spill.	experiments would render results better comparable. When not buried in anoxic sediment layers, oil will disappear from the marine environment after some time. Careful considerations are necessary whether or not to fight oil-spills. When oil-spill fighting is necessary, efficient equipment must be available. More pilot projects should be done in mechanical clean-up devices and efficient oil booms.
6	Concentration of man-made radioactivity in surface ocean waters is about 15% of natural radioactivity. Waste radioactivity released from reprocessing plants is of regional importance.	Local sources of radioactivity pollution should be stopped. Nuclear weapons must not explode. More scientific evidence is necessary whether present concentrations of fall-out radioactivity present a hazard for small marine organisms and for man.
7	Toxicity experiments with marine organisms come to different results depending on experimental conditions. The same holds true with accumulation experiments. Elimination of pollutants from the sea by sedimentation is counteracted by burrowing fauna. Complicated interactions occur at the air-sea and at the water-sediment interfaces, and in the metabolism of organisms.	Much more basic research on pollution-related topics is necessary before a reliable understanding of processes can be achieved, which are similar with natural substances as with man-made pollutants. If the trend of the past decade continues there will be a lot of surprises from future research.
8	Mercury is concentrated, by natural processes, in large fish, seal, whale, and marine birds. Concentrations are such that unlimited consumption by humans is risky. Concentrations in normal sea fish are small. There is a lot of evidence for the thesis that mercury concentrations in the water of the oceans distant from point sources of pollution are natural and have not been influenced by man.	Man should eat high mercury level seafood as a delicatesse, or leave it. Nothing could be done against these mercury concentrations.

Chapter	Diagnosis	Therapy
8.6	Likewise, distant from local points man has not altered cadmium concentrations in the marine environment. There is a possible health risk only from molluscs, but not from fish.	Nothing to be done.
8.7	The quantity of lead mined in the lead mines of the world per year is about equal to the total lead dissolved in ocean water. In principle man could have changed the lead concentrations in the water of the world's oceans.	More scientific results on long-term trends of marine lead contamination are necessary. Lead in car gasoline should be forbidden.
9	Mostly via the atmosphere, many chlorinated hydrocarbons, among them toxic pesticides, have been introduced into the oceans and can be analyzed in marine organisms including seafish and other seafood. Fortunately concentrations are lower than in meat from cattle. High concentrations are present in marine birds and mammals. There is evidence that present day PCB concentration in Baltic seawater causes reduced reproductive success in fish and seal.	It seems that due to prohibitive measures concentrations are already decreasing. Better scientific monitoring of trends on a world-wide scale is necessary, and more research on the fate and effects of such pollutants. It ought to be confirmed that with analytical techniques presently used only a small fraction of the total variety of chlorinated organics has been identified. Atmospheric pollution by toxic and persistent organics must be reduced.
10	An impressive series of environmental laws on international and national level has been issued.	Obviously, through this legislation, impact of pollution upon the marine environment has decreased, and this trend should continue.

References

Aarkrog A (1977) Environmental behaviour of plutonium accidentally released at Thule, Greenland. Health Phys 32:271–284
Ackefors H, Löfroth G, Rosen CG (1970) A survey of the mercury pollution problem in Sweden with special reference to fish. Oceanogr Mar Biol Annu Rev 8:203–224
Addison RF (1976) Organochlorine compounds in aquatic organisms: their distribution, transport and physiological significance. In: Lockwood APM (ed) Effects of pollutants on aquatic organisms. Cambridge Univ. Press, Cambridge, pp 127–143
Addison RF (1977) Diphenylether – another marine environmental contaminant. Mar Pollut Bull 8:237–240
Agricultural Research Council (1970) Third report of the research committee on toxic chemicals. London, pp 1–69
Albers PH, Szaro RC (1978) Effects of no. 2 fuel oil on common eider ducks. Mar Pollut Bull 9:138–139
Anas RE (1974) Heavy metals in the Northern Fur Seal, *Callorhinus ursinus*, and Harbour Seal, *Phoca vitulina richardi*. Fish Bull 72:133–137
Andelman JB, Snodgrass JE (1974) Incidence and significance of polynuclear aromatic hydrocarbons in the water environment. Crit Rev Environ Control 4:69–83
Andersin A-B, Lassig J, Parkkonen L, Sandler H (1978) The decline of macrofauna of the deeper parts of the Baltic proper and the Gulf of Finland, Proc 5th Symposium of the Baltic Marine Biologists. Kieler Meeresforsch Suppl 4:23–52
Anderson DW, Jurek RM (1977) The status of Brown Pelicans at Anacapa Island in 1975. Calif Fish Game 63:4–10
Anderson DW, Jehl JR Jr, Risebrough RW, Woods LA Jr, Deweese LR, Edgecomb WG (1975) Brown pelicans: improved reproduction off the Southern Californian coast. Science 190:806–808
Anger K (1975) On the influence of sewage pollution on inshore benthic communities in the south of Kiel Bay. 2. Quantitative studies on community structure. Helgol Wiss Meeresunters 27:408–438
Anon (1971) Dumpers foiled. Mar Pollut Bull 2:114
Anon (1974) Meeting report interlaboratory lead analyses of standardized samples of sea water. Mar Chem 2:69–74
Anon (1975) Deep sea dumping furore. Mar Pollut Bull 6:68
Anon (1978) Another oil supertanker disaster narrowly averted. Mar Pollut Bull 9:199
Anon (1980a) (Newspaper information). Gebremster Schaum. Kapital 12/80:93–96
Anon (1980b) The case of the Baltic seals. Ambio 9:182
Anon (1981) Titandioxid – Dünnsäure – Problematik. Umwelt 82:17
Appelquist H, Jensen KO, Sevel T, Hammer C (1978) Mercury in the Greenland ice sheet. Nature (London) 273:657–659
Atlas E, Giam CS (1981) Global transport of organic pollutants: ambient concentrations in the remote marine atmosphere. Science 211:163–165
Atlas RM, Bartha R (1973a) Abundance, distribution and oil biodegradation potential of microorganisms in Raritan Bay. Environ Pollut 4:291–300
Atlas RM, Bartha R (1973b) Effects of some commercial oil herders, dispersants and bacterial inocula on biodegradation of oil in seawater. In: Ahearn DG, Meyers SP (eds) Microbial degradation of oil pollutants. Louisiana State Univ, Baton Rouge, pp 283–289

Atlas RM, Boehm PD, Calder JA (1981) Chemical and biological weathering of oil, from the Amoco Cadiz spillage, within the littoral zone. Estuar Coast Shelf Sci 12:589–608

Audonson T (1978) The Bravo blow-out. Field observations, results of analyses and calculations regarding oil on the surface. IKU Report 90 (Institutt for Kontinentalsokkelundersökelser, Trondheim), 1977, 1–287

Avcin A, Meith-Avcin N, Vukovic A, Vriser B (1974) A comparison of benthic communities of Strunjan and Kopper Bays with record to their differing exposure to pollution stress. Biol Vestn 22:171–208

Baker CW (1977) Mercury in surface waters of seas around the United Kingdom. Nature (London) 270:230–232

Barber RT, Vijayakumar A, Cross FA (1972) Mercury concentrations in recent and ninety-year-old benthopelagic fish. Science 178:636–639

Bargmann RD (1975) Problems of meeting changing water quality requirements for large waste disposal systems. In: Pearson EA, Frangipane E de Fraia (eds) Marine pollution and marine waste disposal. Pergamon Press, Oxford, pp 155–161

Bellan G (1976) La pollution par les tensio-activs. In: Perez JM (ed) La pollution des eaux marines. Gauthiers-Villars, Paris, pp 31–50

Ben-Bassat D, Mayer AM (1975) Volatization of mercury by algae. Physiol Plant 23:128–132

Benjamin CYL, Polak J (1973) A study of the solubility of oil in water, Rep EPS-4-EC-76-1 (Environmental Protection Service, Ottawa), pp 1–25

Berge G, Ljøen R, Palmork KH (1972) The disposal of containers with industrial waste into the North Sea: a problem to fisheries. In: Ruivo M (ed) Marine pollution and sea life. Fishing News, London, pp 474–475

Bewers JM, Yeats PA (1977) Oceanic residence times of trace metals. Nature (London) 268: 595–598

Beyer AH, Painter LJ (1977) Estimating the potential for future oil spills from tankers, offshore development, and onshore pipelines. Proc 1977 Oil Spill Conf, March 8–10, 1977. Am Petr Inst, New Orleans Louisiana Washington, pp 21–30

Billings CE, Matson WR (1972) Mercury emission from coal combustion. Science 176:1232–1233

Blaikley DR, Dietzel GFL, Glass AW, van Kleef PJ (1977) "Sliktrak", a computer simulation of offshore oil spills, cleanup, effects and associated costs. Proc 1977 Oil Spill Conf, March 8–10, 1977. Am Petr Inst, New Orleans Louisiana Washington, p 45

Blundo R (1978) The toxic effects of the water soluble fractions of no. 2 fuel oil and of three aromatic hydrocarbons on the behavior and survival of barnacle larvae. Contrib Mar Sci 21:25–37

Blus LJ, Neely BS Jr, Belisle AA, Prouty RM (1974) Organochlorine residues in Brown Pelican eggs: relation to reproductive success. Environ Pollut 7:81–91

Blus LJ, Neely BS, Lamont RG, Mulhern B (1977) Residues of organochlorines and heavy metals in tissues and eggs of Brown Pelicans, 1969–73. Pestic Monit J 11:40–53

Board PA (1973) The fate of rubbish in the Thames. Mar Pollut Bull 4:165–166

Boe C (1975) Bund-Länder-Vereinbarung zur Bekämpfung von Ölunfällen an der Küste und auf der Hohen See. Dtsch Gewässerkdl Mitt. Sonderh 127–128

Boesch DF, Hershner CH, Milgram JH (1974) Oil spills and the marine environment. Ballinger, Cambridge Mass, 114 pp

Boetius J (1968) Toxicity of waste from a parathion industry at the Danish North Sea coast. Helgol Wiss Meeresunters 17:182–187

Bonka H (1980) Nordseestrand mit erhöhter Radioaktivität. Umschau 80:305–306

Bourne WRP (1976) Seabirds and pollution. In: Johnston R (ed) Marine pollution. Academic Press, London New York San Francisco, pp 403–502

Bourne WRP (1977) Ekofiasco. Mar Pollut Bull 8:121–122

Bourne WRP, Bibby CJ (1975) Temperature and the seasonal and geographical occurrence of oiled birds on West European beaches. Mar Pollut Bull 6:77–80

Bourne WRP, Bogan JA (1972) Polychlorinated biphenyls in North Atlantic sea birds. Mar Pollut Bull 3:171–175

Boutron C, Lorius C (1979) Trace metals in Antarctic snows since 1914. Nature (London) 277: 551–554

Bowes GW, Jonkel CJ (1975) Presence and distribution of polychlorinated biphenyls (PCB) in Arctic and Subarctic marine food chains. J Fish Res Board Can 32:2111–2123

Bowes GW, Mulvihill MJ, Simoneit BRT, Burlingame AL, Risebrough RW (1975) Identification of chlorinated dibenzofurans in American polychlorinated biphenyls. Nature (London) 256: 305–307

Boyd BD, Bates CC, Harrald JR (1977) The statistical picture regarding discharges of petroleum hydrocarbons in and around United States waters. In: Proc Symp Sources, Effects and Sinks of Hydrocarbons in the Aquatic Environment. Washington DC, 9–11 August 1976. Am Inst Biol Sci, pp 37–53

Breen PA, Mann KH (1976) Changing lobster abundance and the destruction of kelp beds by sea urchins. Mar Biol 34:137–142

Brewer PG (1975) Minor elements in sea water. In: Riley JP, Skirrow G (eds) Chemical oceanography, 2nd ed, vol I. Academic Press, London New York San Francisco, pp 415–496

Brown RA, Searl TD (1977) Nonvolatile hydrocarbons in the Pacific Ocean. In: Proc Symp Sources, Effects and Sinks of Hydrocarbons in the Aquatic Environment. Washington DC, 9–11 August 1976. Am Inst Biol Sci, pp 239–255

Bruland KW, Knauer GA, Martin JH (1978) Cadmium in northeast Pacific waters. Limnol Oceanogr 23:618–625

Bryan GW (1974) Adaptation of an estuarine polychaete to sediments containing high concentrations of heavy metals. In: Vernberg FJ, Vernberg WB (eds) Pollution and physiology of marine organisms. Academic Press, London New York San Francisco, pp 122–135

Bryan GW (1976a) Heavy metal contamination in the sea. In: Johnston R (ed) Marine pollution. Academic Press, London New York San Francisco, pp 185–302

Bryan GW (1976b) Some aspects of heavy metal tolerance in aquatic organisms. In: Lockwood APM (ed) Effects of pollutants on aquatic organisms. Cambridge Univ Press, Cambridge, pp 7–34

Bull KR, Murton RK, Osborn D, Ward P, Cheng L (1977) High levels of cadmium in Atlantic sea birds and sea skaters. Nature (London) 269:507–509

Burnett M, Patterson CC (1980) Analysis of natural and industrial lead in marine ecosystems. In: Branica M, Konrad Z (eds) Lead in the marine environment. Pergamon Press, Oxford, pp 15–30

Butler JN (1976) Transfer of petroleum residues from sea to air: evaporative weathering. In: Windom HL, Duce RA (eds) Marine pollutant transfer. Lexington Books, Lexington Toronto, pp 201–211

Butler JN, Morris BF, Sleeter TD (1977) The fate of petroleum in the open ocean. Proc Symp Sources, Effects and Sinks of Hydrocarbons in the Aquatic Environment, Washington DC, 9–11 August 1976. Am Inst Biol Sci, pp 287–297

Butler PA (1973) Organochlorine residues in estuarine molluscs, 1965–1972. A report of one segment of the National Pesticide Monitoring Programm. Pestic Monit J 6:238–246

Butler PA, Choldress R, Wilson AJ Jr (1972) The association of DDT residues with losses in marine productivity. In: Ruivo M (ed) Marine pollution and sea life. Fishing News, London, pp 262–266

Cantelmo FR, Rao KR (1978) Effect of pentachlorophenol (PCP) on meiobenthic communities established in an experimental system. Mar Biol 46:17–22

Carey AE, Harvey GR (1978) Metabolism of polychlorinated biphenyls by marine bacteria. Bull Environ Contam Toxicol 20:527–534

Carmody DJ, Pearce JB, Yasso WE (1973) Trace metals in sediments of New York Bay. Mar Pollut Bull 4:132–135

Carr RA, Jones MM, Russ ER (1974) Anomalous mercury in near bottom water of a Mid-Atlantic-Rift valley. Nature (London) 251:489–490

Caspers H (1975) Pollution in coastal waters. An interim report on results of a priority programme of the German Research Society (1966–1974). Boldt, Boppard, 142 pp

Catell FCR, Scott WD (1978) Copper in aerosol particles produced by the ocean. Science 202: 429–430

Chasse C (1978) The ecological impact on and near shores by the Amoco Cadiz oil spill. Mar Pollut Bull 9:298–301

Chester R, Stoner JH (1975) Trace elements in total particulate material from surface sea water. Nature (London) 255:49–51

Chow TJ (1973) Lead pollution records in Southern California coastal sediments. Science 181: 551–552

Clark RB (1978a) Tanker wrecks and the human element. Mar Pollut Bull 9:113

Clark RB (1978b) Oiled seabird rescue and conservation. J Fish Res Board Can 35:675–678

Clark RB (1978c) No refuge in the zoo. Mar Pollut Bull 9:58–59

Clifton AP, Vivian CMG (1975) Retention of mercury from an industrial source in Swansea Bay sediments. Nature (London) 253:621–622

Conover RJ (1971) Some relations between zooplankton and Bunker C oil in Chedabucto Bay following the wreck of the tanker Arrow. J Fish Res Board Can 28:1327–1330

Coulson JC, Deans IR, Potts GR, Robinson J, Crabtree AN (1972) Changes in organochlorine contamination of the marine environment of Eastern Britain monitored by shag eggs. Nature (London) 236:454–456

Cowell EB (1978) Ecological effects of dispersants in the United Kingdom. In: McCarthy LT, Lindblom GP, Walter HF (eds) Chemical dispersants for the control of oil spills. ASTM STP 659, American Society for Testing and Materials, pp 277–292

Davies PH, Spies RB (1980) Infaunal benthos of a natural petroleum seep: Study of community structure. Mar Biol 59:31–41

Delong RL, Gilmartin WG, Simpson JG (1973) Premature births in California sea lions: association with high organochlorine pollutant residue levels. Science 181:1168–1169

Department of the Environment (1972) Out of sight, out of mind, Report of a working party on sludge disposal in Liverpool Bay, vol I. Stationary Office, London, 36 pp

Department of the Environment (1976) Accidental oil pollution of the sea. A report by officials on oil spills and clean-up measures. (Pollution Paper No. 8). Stationary Office, London, pp 1–170

Dethlefsen V (1973) Zur Frage des Fischvorkommens im Dünnsäureverklappungsgebiet nordwestlich Helgolands. Arch Fisch-Wiss 24:65–75

Dethlefsen V (1981) Hamburg beendet Klärschlammverklappung in der Nordsee. Inf Fischwirtsch 28:53–57

Dethlefsen V, Watermann B (1980) Vorkommen von Hauttumoren der Kliesche *(Limanda limanda L.)* im Verbringungsgebiet für Abfälle der Titandioxidproduktion und Vergleichsgebieten. Inf Fischwirtsch 27:57–65

Deutsche Forschungsgemeinschaft (1979) Rückstände in Fischen – Situation und Bewertung. Mitt Kommiss Prüfung von Rückständen in Lebensmitteln 7:1–26

Deutsches Hydrographisches Institut (1978) Jahresbericht 1976/77, p 45

Dickson AG, Riley JP (1976) The distribution of short chain halogenated aliphatic hydrocarbons in some marine organisms. Mar Pollut Bull 7:167–169

Doduchi M (1977) Polychlorinated terphenyls as an environmental pollutant in Japan. Ecotoxicol Environ Safety 1:239–248

Drescher HE, Harms U, Huschenbeth E (1977) Organochlorines and heavy metals in the Harbour Seal *Phoca vitulina* from the German North Sea coast. Mar Biol 41:99–106

Drost R (1966) Über den Seeunfall des Tankers „Anne Mildred Brövig", Folgen und Folgerungen. Ber Dtsch Sekt Int Rat Vogelschutz 6:52–55

Duce RA, Hoffmann GL, Zoller WH (1975) Atmospheric trace metals at remote Northern and Southern Hemisphere sites: pollution or natural? Science 187:59–61

Dunn BP, Fee J (1979) Polycyclic aromatic hydrocarbon carcinogens in commercial sea foods. J Fish Res Board Can 36:1469–1476

Dunn BP, Stich HF (1976) Monitoring procedures for chemical carcinogens in coastal waters. J Fish Res Board Can 33:2040–2046

Dybern BJ, Jensen S (1978) DDT and PCB in fish and mussels in the Kattegat-Skagerrak area. Medd Havsfiskelab Lysekil 232:1–17

Eder G (1976) Polychlorinated biphenyls and compounds of the DDT group in sediments of the Central North Sea and the Norwegian Depression. Chemosphere 5:101–106

Eder G, Schaefer RG, Ernst W, Goerke H (1976) Chlorinated hydrocarbons in animals of the Skagerrak. Veröff Inst Meeresforsch Bremerhaven 16:1–9

Edwards CA (1966) Insecticide residues in soils. Residue Rev 13:83–131

Edwards P (1975) An assessment of possible pollution effects over a centruy on the benthic marine algae of Co. Durham, England. Bot J Linnean Soc 70:269–305

Ehlin U (1974) Kylvatten effekter på miljön. SVN Statens Naturvårdsverk Publ 25:1–104

Ehrhardt M, Derenbach J (1977) Composition and weight per area of pelagic tar collected between Portugal and south of the Canary Islands. Meteor Forschungsergebn A 19:1–9

Enzinger RM, Cooper RC (1976) Role of bacteria and protozoa in the removal of *Escherichia coli* from estuarine waters. Appl Environ Microbiol 31:758–763

Ernährungsbericht (1976) Herausgeber: Deutsche Gesellschaft für Ernährung, Frankfurt, 477 pp

Ernst W (1972) Accumulation and metabolism of DDT-^{14}C (Dichlorodiphenyl-trichloro-ethane) in marine organisms. In: Ruivo M (ed) Marine pollution and sea life. Fishing News, London, pp 260–262

Ernst W (1975a) Pestizide im Meerwasser – Aspekte der Speicherung, Ausscheidung und Umwandlung in marinen Organismen, Schr R Ver Wasser Boden Lufthyg Berlin-Dahlem 46:81–92

Ernst W (1975b) Organochlorverbindungen in Meerestieren. In: European Colloquium, Luxembourg 14–16 May 1974: Problems raised by the contamination of man and his environment by persistent pesticide and organo-halogenated compounds (Commission of the European Communities), pp 67–80

Ernst W (1977) Determination of the bioconcentration potential of marine organisms – a steady state approach. I. Bioconcentration data for seven chlorinated pesticides in mussels *(Mytilus edulis)* and their relation to solubility data. Chemosphere 11:731–740

Ernst W (1980) Effects of pesticides and related organic compounds in the sea. Helgol Meeresunters 33:301–312

Ernst W (1981) Schadstoffe in Meerestieren – aktuelle Belastung und hygienisch-toxikologische Aspekte. In: Noelle H (ed) Nahrung aus dem Meer. Springer, Berlin Heidelberg New York, pp 229–241

Ernst W, Goerke H (1974) Anreicherung, Verteilung, Umwandlung und Ausscheidung von DDT-^{14}C bei *Solea solea* (Pisces: Soleidae). Mar Biol 24:287–304

Ernst W, Weber W (1978) The fate of pentachlorophenol in the Weser Estuary and the German Bight. Veröff Inst Meeresforsch Bremerhaven 17:45–53

Ernst W, Goerke H, Eder G, Schaefer RG (1976) Residues of chlorinated hydrocarbons in marine organisms in relation to size and ecological parameters. I. PCB, DDT, DDE, and DDD in fishes and molluscs from the English Channel. Bull Environ Contam Toxicol 15:55–65

Ernst W, Goerke H, Weber K (1977) Fate of ^{14}C-labelled di-, tri- and pentachlorobiphenyl in the marine annelid *Nereis virens*. II. Degradation and faecal elimination. Chemosphere 6:559–568

Establier R (1972) Concentración de mercurio en los tejidos de algunos peces, moluscos y crustaceos del Golfo de Cadiz y caladeros del norteste Africano. Invest Pesq 36:355–364

Farn RJ (1976) Sinking and dispersing oil. In: Wardley-Smith J (ed) The control of oil pollution on the sea and inland waters. Graham and Trotman, London, pp 159–180

Fimreite N, Bjerk JE, Kveseth N, Brun E (1977) DDE and PCBs in eggs of Norwegian sea birds. Astarte 10:15–20

Finch R (1973) Effects of regulatory guidelines on the intake of mercury from fish – the MECCA project. Fish Bull 71:615–626

Fisher NS, Carpenter EJ, Remsen CC, Wurster CF (1974) Effects of PCB on interspecific competition in natural and gnothobiothic phytoplancton communities in continuous and batch culture. Microb Ecol 1:39–50

Fitzgerald WF, Lyons WB (1975) Mercury concentrations in open-ocean waters: sampling procedure. Limnol Oceanogr 20:468–471

Floodgate GD (1973) A threnody concerning the biodegradation of oil in natural waters. In: Ahearn DG, Meyers SP (eds) The microbial degradation of oil pollutants. Louisiana State Univ, Baton Rouge, pp 17–24

Förstner U, Müller G (1974) Schwermetalle in Flüssen und Seen als Ausdruck der Umweltverschmutzung. Springer, Berlin Heidelberg New York, 225 pp

Förstner U, Reineck H-E (1974) Die Anreicherung von Spurenelementen in den rezenten Sedimenten eines Profilkerns aus der Deutschen Bucht. Senckenberg Marit 6:175–184

Fowler SW (1977) Trace elements in zooplankton particulate products. Nature (London) 269: 51–52

Fowler SW, Benayoun G (1976) Influence of environmental factors on selenium flux in two marine invertebrates. Mar Biol 37:59–68

Fowler SW, Heyraud M, La Rosa J (1978) Factors affecting methyl and inorganic mercury dynamics in mussels and shrimp. Mar Biol 46:267–276

Fries GF (1972) Degradation of chlorinated hydrocarbons under anaerobic conditions. In: Faust SD (ed) Fate of organic pesticides in the aquatic environment. Am Chem Soc, Washington, pp 256–270

Gadow S, Reineck H-E (1969) Ablandiger Sandtransport bei Sturmfluten. Senckenberg Mart 1: 63–78

Gameson ALH, Barnett MJ, Shewbridge JS (1975) The aerobic Thames Estuary. In: Jenkins SH (ed) Advances in water pollution research, 6th Int Conf Jerusalem, June 8–23 1972. Pergamon Press, Oxford, pp 843–852

Garder D (1978) Mercury in fish and waters of the Irish Sea. Nature (London) 272:49–51

Gardner D, Riley JP (1973) Distribution of dissolved mercury in the Irish Sea. Nature (London) 241:526–527

Gerlach SA (1976) Meeresverschmutzung – Diagnose und Therapie. Springer, Berlin Heidelberg New York, 145 pp

Gerlach SA (1978) Quecksilber im Speisefisch und in Seesäugetieren. Öff Gesundheitswes 40: 460–462

Gerlach SA (1980) Land- oder Seebeseitigung von Abfällen? In: Offhaus E (ed) Abfallbeseitigung auf See, Beiheft 17 zu Müll und Abfall. E Schmidt, Berlin, pp 91–94

Gerlach SA (1981) Über die Verschmutzung der Helgoländer Bucht. Naturwiss Rundsch 34:276–823

GESAMP (1976) IMCO/FAO/UNESCO/WMO/IAEA/UN Joint Group of Experts on the Scientific Aspects of Marine Pollution. Review of harmful substances. Rep Stud GESAMP 2: 80 pp

Ghirardelli E (1973) L'iniquinamento del Golfo di Trieste. Atti Mus Civ Stor Nat Trieste 28: 431–450

Giam CS, Chan HS, Neff GS, Atlas EL (1978) Phthalate ester plasticizers: a new class of marine pollutant. Science 199:419

Gibbs CF, Pugh KB, Andrews AR (1975) Quantitative studies on marine biodegradation of oil. II. Effect of temperature. Proc R Soc London Ser B 188:83–94

Gibbs RH Jr, Jarosewich E, Windom HL (1974) Heavy metal concentrations in museum fish specimens: effects of preservation and time. Science 184:475–477

Gienapp H (1973) Strömungen während der Sturmflut vom 2. November 1965 in der Deutschen Bucht und ihre Bedeutung für den Sedimenttransport. Senckenberg Marit 5:135–151

Giere O (1979) The impact of oil pollution on intertidal meiofauna. Field studies after the La Coruña-spill, May 1976. Cahier Biol Mar 20:231–251

Gilbertson M (1974) Seasonal changes in organohaline compounds and mercury in Common Terns of Hamilton Harbour, Ontario. Bull Environ Contam Toxicol 12:726–732

Gillespie DC, Scott DP (1971) Mobilization of mercuric sulfide from sediment into fish under aerobic conditions. J Fish Res Board Can 28:1807–1808

Gocke K (1975) Untersuchungen über den Einfluß des Salzgehaltes auf die Aktivität von Bakterienpopulationen des Süß- und Abwassers. Kieler Meeresforsch 30:99–105

Goerke H, Ernst W (1977) Fate of ^{14}C-labelled di-, tri- and pentachlorobiphenyl in the marine annelid *Neries virens*. I. Accumulation and elimination after oral administration. Chemosphere 6:551–558

Goerke H, Eder G, Weber K, Ernst W (1979) Patterns of organochlorine residues in animals of different trophic levels from the Weser Estuary. Mar Pollut Bull 10:127–133

Goldberg ED (1974) The surprise factor in marine pollution studies. Mar Technol Soc J 8:29–34

Goldberg ED (1975) Synthetic organohalides in the sea. Proc R Soc London Ser B 189:277–289

Goldberg ED (1976) The health of the oceans. UNESCO Press, Paris, 172 pp
Goldberg ED, Bowen VT, Farrington JW, Harvey G, Martin JH, Parker PL, Risebrough RW, Robertson W, Schneider E, Gamble E (1978) The mussel watch. Environ Conserv 5:101–125
Granmo Å, Jørgensen G (1975) Effects on fertilization and development of the common mussel *Mytilus edulis* after long-term exposure to a nonionic surfactant. Mar Biol 33:17–20
Grasshoff K (1974) Chemische Verhältnisse und ihre Veränderlichkeit. In: Magaard L, Rheinheimer F (eds) Meereskunde der Ostsee. Springer, Berlin Heidelberg New York, pp 85–101
Grasshoff K (1975) The hydrochemistry of landlocked basins and fjords. In: Riley JP, Skirrow G (eds) Chemical oceanography, 2nd edn, vol 2. Academic Press, London New York San Francisco, pp 455–597
Greve PA (1971) Chemical wastes in the sea: new forms of marine pollution. Science 173:1021–1022
Grose PL, Mattson JS (1977) The Argo Merchant oil spill. NOAA Special Report. US Dept Commerce, National Oceanic and Atmospheric Administration, Washington, 133 pp
Grossling BF (1977) An estimate of the amounts of oil entering the oceans. In: Sources, effects and sinks of hydrocarbons in the aquatic environment. Symposium Washington DC, 9–11 August 1976. Am Inst Biol Sci, pp 5–36
Guelin A (1974) Sur le pouvoir bactéricide de l'eau de mer. CR Acad Sci Paris (D) 279:871–874
Gundlach ER, Hayes MO, Ruby CH, Ward LG, Blount AE, Fisher IA, Stein RJ (1978) Some guidelines for oil-spill control in coastal environments, based on field studies of four oil spills. ASTM Spec Techn Publ 659:98–118
Hagel P, Tuinstra LGMT (1978) Trends in PCB contamination in Dutch coastal and inland fishery products, 1972–1976. Bull Environ Contam Toxicol 19:671–676
Hagmeier E (1978) Variations in phytoplankton near Helgoland. Rapp PV Reun Cons Int Explor Mer 172:361–363
Hammond AL (1971) Mercury in the environment: natural and human factors. Science 171:788–789
Hammond AL (1977) Oceanography: geochemical tracers offer new insight. Science 195:164–166
Hann RW, Rice L, Trujillo M-C, Young HN (1978) Oil spill cleanup activities. In: Hess WN (ed) The Amoco Cadiz oil spill. NOAA-EPA Special Report, April 1978, pp 229–276
Hapke H-J (1980) Pflanzenschutzmittel-Rückstände in Grundnahrungsmitteln – Situation und Bewertung. Deutsche Forschungsgemeinschaft. Mitt Komiss Pflanzenschutz- Pflanzenbehandlungs- Vorratsschutzmittel 12:72–79
Harada M, Smith AM (1975) Minamata disease: a medical report. In: Smith WE, Smith AM (eds) Minamata, a warning to the world. Chatto & Windus, London, pp 180–192
Harding LW, Phillips JH (1978) Polychlorinated biphenyl (PCB) effects on marine phytoplankton photosynthesis and cell division. Mar Biol 49:93–101
Harms U, Drescher HE, Huschenbeth E (1978) Further data on heavy metals and organochlorines in marine mammals from German coastal waters. Meeresforschung 26:153–161
Harris RC, White DB, MacFarlane RB (1970) Mercury compounds reduce photosynthesis in plankton. Science 170:736–737
Harrison J, Grant P (1976) The Thames transformed. London's river and its waterfowl. A Deutsch, London, 239 pp
Harrison RM, Laxen DPH (1978) Natural source of tetraalkyllead in air. Nature (London) 275:738–740
Harvey GR, Miklas HP, Bowen VT, Steinhauer WG (1974) Observations on the distribution of chlorinated hydrocarbons in Atlantic Ocean organisms. J Mar Res 38:103–118
Helle E, Olsson M, Jensen S (1976) PCB levels correlated with pathological changes in seal uteri. Ambio 5:261–263
Heslinga GA (1976) Effects of copper on the coral-reef echinoid *Echinometra mathaei*. Mar Biol 35:155–160
Hetherington JA (1976) Radioactivity in surface and coastal waters of the British Isles 1974. Fish Radiobiol Labor Techn Rep 11:1–35

Hickel W (1969) Sedimentbeschaffenheit und Bakteriengehalt im Sediment eines zukünftigen Verklappungsgebietes von Industrieabwässern nordwestlich Helgolands. Helgol Wiss Meeres-Unters 19:1−20

Hickel W et al (1980) Phosphat-Eutrophierung der Deutschen Bucht. Jahresber Biol Anst Helgol 1979:29

Higgins IJ, Burns RG (1975) The chemistry and microbiology of pollution. Academic Press, London New York San Francisco, 248 pp

Hodge VF, Folsom TR, Young DR (1973) Retention of fall-out constituents in upper layers of the Pacific Ocean as estimated from studies of a tuna population. In: Radioactive contamination of the marine environment. Proceedings of symposium Seattle, 10−14 July 1972. Int At Energy Ag, Vienna, pp 263−276

Hoerschelmann H, Polzhofer K, Figge K, Ballschmitter K (1979) Organochlorpestizide und polychlorierte Biphenyle in Vogeleiern von den Falklandinseln und aus Norddeutschland. Environm Pollut 13:247−269

Hoffmann W (1979) Phosphor- und Stickstoffzufuhr aus der Landwirtschaft in die Ostsee, insbesondere durch die Schwebstoffe der Gewässer. Wasser Abwasser Forsch Praxis 16:1−106

Holden AV (1970) Source of polychlorinated biphenyl contamination in the marine environment. Nature (London) 228:1220−1221

Hollister CD (1977) The seabed option. Oceanus 20:18−25

Holmes PD (1977) A model for the costing of oil spill clearance operations at sea. Proc 1977 Oil Spill Conf, March 8−10, 1977. Am Petrol Inst, New Orleans Louisiana Washington, pp 39−44

Holmström A (1975) Plastic films on the bottom of the Skagerrak. Nature (London) 255:622−623

Hütter LA (1978) Quecksilbergefährdung durch Leuchtstofflampen? Naturwiss Rundsch 31:17−19

Huschenbeth E (1973) Zur Speicherung von chlorierten Kohlenwasserstoffen in Fisch. Arch Fisch Wiss 24:105−116

Huschenbeth E (1977) Überwachung der Speicherung von chlorierten Kohlenwasserstoffen im Fisch. Arch Fisch Wiss 28:173−186

Hyland JL, Scheider ED (1977) Petroleum hydrocarbons and their effects on marine organisms, populations, communities, and ecosystems. Proc Symp Sources, Effects and Sinks of Hydrocarbons in the Aquatic Environment, Washington DC, 9−11 August 1976. Am Inst Biol Sci, pp 464−506

IAEA (1975) Convention on the prevention of marine pollution by dumping of wastes and other matter. The definition required by Annex I, paragraph 6 to the convention and the recommendations required by Annex II, Sect D. Int At Energy Ag Inf Circ 205:Add 1, 1−21

ICES (1974) Report of the Working Group for the International Study of the Pollution of the North Sea and its Effects on Living Resources and their Exploitation. ICES Cooper Res Rep 39:1−191

International Bird Rescue Research Center (1978) Saving oiled seabirds. Am Petrol Inst, Washington, 35 pp

Jacobs G (1977) Gesamt- und organisch gebundener Quecksilbergehalt in Fischen aus deutschen Fanggründen. Z Lebensmittel Unters Forsch 164:71−76

Jansson BO (1980) Natural systems of the Baltic Sea. Ambio 9:128−136

Jansson B, Jensen S, Olsson M, Renberg L, Sundström G, Vaz R (1975) Identification by GC-MS or phenolic metabolites of PCB and p,p'-DDE isolated from Baltic guillemot and seal. Ambio 4:93−97

Jenkins SH (1981) EEC bathing water standards. Mar Pollut Bull 12: 33−34

Jensen S (1972) The PCB story. Ambio 1:123−131

Jensen S, Lange R, Jernelov A, Palmork KH (1972) Chlorinated byproducts from vinyl chloride production: a new source of marine pollution. In: Ruivo M (ed) Marine pollution and sea life. Fishing News, London, pp 242−244

Jensen S, Kihlström JE, Olsson M, Lundberg C, Örberg J (1977) Effects of PCB and DDT on mink (Mustela vision) during the reproductive season. Ambio 6:239

Jernelöv A (1975) Heavy metals in the marine environment. In: Gameson ALH (ed) Discharge of sewage from sea outfalls. Pergamon Press, Oxford, pp 115–122

Jernelöv A, Rosenberg R, Jensen S (1972) Biological effects and physical properties in the marine environment of aliphatic chlorinated by-products from vinyl chloride production. Water Res 6:1181–1191

Joensen AH, Hansen EB (1977) Oil pollution and sea birds in Denmark 1971–1976. Danish Rev Game Biol 10 (5):1–31

Johannes RE (1975) Pollution and degradation of coral reef communities. In: Ferguson-Wood EJ, Johannes RE (eds) Tropical marine pollution. Elsevier, Amsterdam, pp 13–51

Johnston R (ed) (1976a) Marine pollution. Academic Press, London New York San Francisco, 729 pp

Johnston R (1976b) Mechanisms and problems of marine pollution in relation to commercial fisheries. In: Johnston R (ed) Marine pollution. Academic Press, London New York San Francisco, pp 3–158

Jones GE, Cobet AB (1975) Heavy metal ions as the principal bactericidal agent in Caribbean sea water. In: Gameson ALH (ed) Discharge of sewage from sea outfalls. Pergamon Press, Oxford, pp 199–208

Junge C, Seiler W, Schmidt U, Bock R, Greese KD, Radler F, Rüger H-J (1972) Kohlenmonoxyd- und Wasserstoffproduktion mariner Mikroorganismen im Nährmedium mit synthetischem Seewasser. Naturwissenschaften 59:514–515

Kahn E (1971) Perspective on tuna fish. N Engl J Med 285:49–50

Katzmann W (1974) Regression der Braunalgenbestände im Mittelmeer. Naturwiss Rundsch 27: 480–481

Kautsky H (1972) Zur Beseitigung radioatkiver Abfälle im Meer. Naturwissenschaften 59:19–22

Kautsky H (1973) Radioaktivität im Meer zur Zeit unbedenklich. Umschau 73:527–529

Kautsky H (1977) Strömungen in der Nordsee. Umschau 77:672–673

Keckes S, Miettinen JK (1972) Mercury as a marine pollutant. In: Ruivo M (ed) Marine pollution and sea life. Fishing News, London, pp 276–289

Keiser RK Jr, Amado JA, Murillo R (1974) Pesticide levels in estuarine and marine fish and invertebrates from the Guatemalan Pcific Coast. Bull Mar Sci 23:905–924

Kinne O, Bulnheim H-P (eds) (1980) Protection of life in the sea. 14th European Marine Biology Symposium. Helgol Wiss Meeresunters 33:1–772

Kittredge JS, Takahashi FT, Lindsey J, Lasker R (1973) Chemical signals in the sea: marine allelochemics and evolution. Fish Bull 72:1–11

Koeman JH (1975) The toxicological importance of chemical pollution for marine birds in the Netherlands, Vogelwarte 28:145–150

Koeman JH, Veen J, Brouwer E, Huisman-de Brouwer L, Koolen JL (1968) Residues of chlorinated hydrocarbon insecticides in the North Sea environment. Helgol Wiss Meeresunters 17: 375–380

Koeman JH, Peeters WHM, Koudstaal-Hol CHM, Tjioe PS, Goeij JJM de (1973a) Mercury selenium correlations in marine mammals. Nature (London) 245:385–386

Koeman JH, van Velzen-Blad HCW, de Vries R, Vos JG (1973b) Effects of PCB and DDE in cormorants and evaluation of PCB residues from an experimental study. J Reprod Fertil Suppl 19:353–364

Koons CB, Monaghan PH (1977) Input of hydrocarbons from seeps and recent biogenic sources. In: Proc Symp Sources, Effects and Sinks of Hydrocarbons in the Aquatic Environment. Washington DC, 9–11 August 1976. Am Inst Biol Sci, pp 85–107

Korringa P (1968) Biological consequences of marine pollution with special reference to the North Sea fisheries. Helgol Wiss Meeresunters 17:126–140

Krüger K (1977) Mein Gott, wenn da etwas passiert. Geo-Magazin 6:110–134

Krugmann H, von Hippel F (1977) Radioactive wastes: a comparison of U.S. military and civilian inventories. Science 197:883–884

Kühl H (1977) *Mercierella enigmatica* (Polychaeta: Serpulidae) an der deutschen Nordseeküste. Veröff Inst Meeresforsch Bemerhaven 16:99–104

Kühnhold WW (1977) The effect of mineral oils on the development of eggs and larvae of marine species. A review and comparison of experimental data in regard to possible damage at sea. Rapp PV Reun Cons Int Explor Mer 171:175–183

Kukla GJ, Angell JK, Korshover J, Dronia H, Hoshiai M, Namias J, Rodewald M, Yamamoto R, Iwashima T (1977) New data on climatic trends. Nature (London) 270:573–580

Kurelec B, Britvic S, Rigavec M, Müller WEG, Zahn RK (1977) Benzo(a)pyrene monooxygenase induction in marine fish – molecular response to oil pollution. Mar Biol 44:211–216

Kurtz DA, Kim KC (1976) Residues in fish, wildlife and estuaries: Chlorinated hydrocarbon and PCB residues in tissues and lice of Northern Fur Seals, 1972. Pestic Monit J 10:79–83

Lawrence J, Tosine HM (1977) Polychlorinated biphenyl concentrations in sewage and sludges of some waste treatment plants in Southern Ontario. Bull Environ Contamin Toxicol 17:49–56

Lee RF (1977) Fate of petroleum components in estuarine waters of the Southeastern United States. Proc 1977 Oil Spill Conf, March 8–10, 1977. New Orleans Louisiana (EPA-API-USCG), pp 611–616

Leppäkoski E (1975) Macrobenthic fauna as indicator of oceanization in the Southern Baltic. Havsforskningsinst Skr 239:280–288

Leppäkoski E (1980) Man's impact on the Baltic ecosystem. Ambio 9:174–181

Lindén O (1976) Acute effects of oil and oil/dispersant mixture on larvae of Baltic herring. Ambio 4:130–133

Lindén O (1976) Effects of oil on the reproduction of the amphipod *Gammarus oceanicus*. Ambio 5:36–37

Littler MM, Murray SN (1975) Impact of sewage on the distribution, abundance and community structure of rocky intertidal macro-organisms. Mar Biol 30:277–291

Lock RAC (1975) Uptake of methylmercury by aquatic organisms from water and food. In: Koeman JH, Strik JJTWA (eds) Sublethal effects of toxic chemicals on aquatic animals. Elsevier, Amsterdam Oxford New York, pp 61–79

Longhurst AR, Radford PJ (1975) PCB concentrations in North Atlantic surface water. Nature (London) 256:239

Lovelock JE (1975) Natural halocarbons in the air and in the sea. Nature (London) 256:193–194

Lüneburg H, Schaumann K, Wellershaus S (1975) Physiographie des Weser-Ästuars (Deutsche Bucht). Veröff Inst Meeresforsch Bremerhaven 15:195–226

Lunde G, Gether J, Josefsson B (1975) The sum of chlorinated and of brominated hydrocarbons in water. Bull Environ Contamin Toxicol 13:656–661

MacGregor JS (1974) Changes in the amount and proportions of DDT and its metabolites, DEE and DDD, in the marine environment off Southern California, 1949–1972. Fish Bull 72:275–293

MacGregor JS (1976) DDT and its metabolites in the sediments off Southern California. Fish Bull 74:27–35

MacKay NJ, Kazacos MN, Williams RJ, Leedow MI (1975) Selenium and heavy metals in Black Marlin. Mar Pollut Bull 6:57–61

MacKie PR, Hardy R, Whittle KJ (1978) Preliminary assessment of the presence of oil in the ecosystem at Ekofisk after the blowout, April 22–30, 1977. J Fish Res Board Can 35:544–551

Martin JH, Flegal AR (1975) High copper concentrations in squid livers in association with elevated levels of silver, cadmium and zinc. Mar Biol 30:51–55

Martin JH, Elliott PD, Anderlini VC, Girvin D, Jacobs SA, Risebrough RW, Delong RL, Gilmartin WG (1976) Mercury-selenium-bromine inbalance in premature parturitent California sea lions. Mar Biol 35:91–104

Martoja R, Viale D (1977) Accumulation de granules de séléniure mercurique dans le foie d'Odontocètes (Mammifères, Cétacés); un mécanisme possible de détoxification du méthyl-mercure par le sélénium. CR Acad Sci Paris Ser D 285:109–112

McIntyre AD, Johnston R (1975) Effects of nutrient enrichment from sewage in the sea. In: Gameson ALH (ed) Discharge of sewage from sea outfalls. Pergamon Press, Oxford, pp 131–141

McKellar HN Jr (1977) Metabolism and model of an estuarine bay ecosystem affected by a coastal power plant. Ecol Model 3:85–118

Menzel DW, Anderson J, Radtke R (1970) Marine phytoplankton vary in their response to chlorinated hydrocarbons. Science 167:1724–1726

Mertens EW (1977) The impact of oil on marine life: a summary of field studies. Proc Symp Sources, Effects and Sinks of Hydrocarbons in the Aquatic Environment, Washington DC, 9–11 August 1976. Am Inst Biol Sci, pp 508–514

Meyer V (1972) Zur Situation des Quecksilbergehaltes bei Fischen und ihren Zubereitungen. Arch Fisch Wiss 23 (1):2–20

Miller GE, Grant PM, Kishore R, Steinkruger FJ, Rowland FS, Guinn VP (1973) Mercury concentrations in museum specimens of tuna and swordfish. Science 175:1121–1122

Mitchell R, Chamberlin C (1975) Factors influencing the survival of enteric microorganisms in the sea: an overview. In: Gameson ALH (ed) Discharge of sewage from sea outfalls. Pergamon Press, Oxford, pp 237–251

Miyake Y (1971) Radioactive models. In: Hood DW (ed) Impingement of man on the oceans. Wiley-Interscience, New York, pp 565–588

Moore B (1975) The case against microbial standards for bathing beaches. In: Gameson ALH (ed) Discharge of sewage from sea outfalls. Pergamon Press, Oxford, pp 103–114

Moore B (1977) The EEC bathing water directive. Mar Pollut Bull 8:269–272

Morris BF, Butler JN, Sleeter TD, Cadwallader J (1976) Transfer of particulate hydrocarbon material from the ocean surface to the water column. In: Windom HL, Duce RA (eds) Marine pollutant transfer, Lexington Books, Lexington Toronto, pp 213–234

Mosley JW (1975) Epidemiological aspects of microbial standards for bathing beaches. In: Gameson ALH (ed) Discharge of sewage from sea outfalls. Pergamon Press, Oxford, pp 85–93

Moss JE (1971) Petroleum – the problem. In: Hood DW (ed) Impingement of man on the oceans. Wiley-Interscience, New York, pp 381–419

Müller G (1979) Schwermetalle in den Sedimenten des Rheins – Veränderungen seit 1971. Umschau 79:778–783

Munda J (1974) Changes and succession in the benthic algal associations of slightly polluted habitats. Rev Int Oceanogr Med 34:37–52

Mundt W, Feldt W (1971) Bestimmung von Quecksilber in Thunfisch mit Hilfe der Isotopenverdünnungsanalyse. Arch Fisch Wiss 22:136

Murozumi M, Chow TS, Patterson CC (1969) Chemical concentrations of pollutant aerosols, terrestrial dusts and sea salts in Greenland and Antarctic snow strata. Geochim Cosmochim Acta 33:1247–1294

Murray CN, Kautsky H (1977) Plutonium and americium activities in the North Sea and German coastal regions. Estuarine Coastal Mar Sci 5:319–328

National Academy of Sciences (1975a) Assessing potential ocean pollutants. A report of the study panel on assessing potential ocean pollutants to the ocean affairs board. Washington DC, 438 pp

National Academy of Sciences (1975b) Petroleum in the marine environment. Workshop on inputs, fates, and the effects of petroleum in the marine environment, May 21–25, 1973, Airlie, Virginia. Washington DC, 107 pp

Nimmo DR, Blackman RR, Wilson AJ Jr, Forester J (1971) Toxicity and distribution of Arochlor 1254 in the pink shrimp *Penaeus duorarum*. Mar Biol 11:191–197

Norton MG, Franklin FL (1980) Research into toxicity evaluation and control criteria of oil dispersants. Fish Res Techn Rep MAFF Direct Fish Res 57:1–20

Nuorteva D (1971) Methylquecksilber in den Nahrungsketten der Natur. Naturwiss Rundsch 24:233–243

Nuzzi R (1972) Toxicity of mercury to phytoplankton. Nature (London) 237:38–39

O'Connors HB, Wurster CF, Powers CD, Biggs DC, Rowland RG (1978) Polychlorinated biphenyls may alter marine trophic pathways by reducing phytoplankton size and production. Science 201:737–739

Odum WE, Johannes RE (1975) The response of mangroves to man-induced environmental stress. In: Ferguson-Wood EJ, Johannes RE (eds) Tropical marine pollution. Elsevier, Amsterdam New York, pp 52–62

Offhaus E (ed) (1980) Abfallbeseitigung auf See. Müll Abfall, Beiheft 17:105

Officer CB, Ryther JH (1977) Secondary sewage treatment versus ocean outfalls: an assessment. Science 197:1056–1060

Olivier J-P (1978) Sea-disposal practices for packaged radioactive waste. Proc Int Symp Management of Waste from the LWR Fuel Cycle. Denver, Colorado, USA, 11th of July, 1976, pp 667–677

Olsson M, Johnels AG, Vaz R (1975) DDT and PCB levels in seals from Swedish waters. Proc Symp Seal in the Baltic. Nat Swed Environ Protect Board, SNV PM 591:43–65

Oppenheimer CH, Gunkel W, Gassmann G (1977) Microorganisms and hydrocarbons in the North Sea during July-August 1975. Proc 1977 Oil Spill Conf, March 8–10, 1977. Am Petr Inst, New Orleans Louisiana Washington, p 593

Osterroth C, Smetacek V (1980) Vertical transport of chlorinated hydrocarbons by sedimentation of particulate matter in Kiel Bight. Mar Ecol Prog Ser 2:27–34

O'Sullivan AJ (1978) Red tide on south coast of Ireland. Mar Pollut Bull 9:315–316

Parslow JLF (1971) Oil pollution and sea birds. Rep Colloq Pollution of the Sea by Oil Spills, Brussels 2–6 November 1970, NATO Committee on the Challenges of Modern Society, pp 11.1–11.12

Parslow JLF (1973) Mercury in waters from the Wash. Environ Pollut 5:295–304

Patchineelam SR, Förstner U (1977) Bindungsformen von Schwermetallen in marinen Sedimenten. Untersuchungen an einem Sedimentkern in der Deutschen Bucht. Senckenberg Marit 9: 75–104

Patterson C, Settle D, Schaule B, Burnett M (1976) Transport of pollutant lead to the oceans and within ocean ecosystems. In: Windom HL, Duce RA (eds) Marine pollutant transfer, Lexington Books, Lexington Toronto, pp 23–38

Peakall DB, Miller DS, Kinter WB (1975) Prolonged eggshell thinning caused by DDE in the duck. Nature (London) 254:421

Pearson CR, McConnell G (1975) Chlorinated C_1 and C_2 hydrocarbons in the marine environment. Proc R Soc London Ser B 189:305–332

Pearson EA (1975) Conceptual design of marine waste disposal systems. In: Gameson ALH (ed) Discharge of sewage from sea outfalls. Pergamon Press, Oxford, pp 403–413

Pearson TH, Rosenberg R (1976) A comparative study of the effects on the marine environment of wastes from cellulose industries in Scotland and Sweden. Ambio 5:77–79

Peden JD, Crothers JH, Waterfall CE, Beasley J (1973) Heavy metals in Somerset marine organisms. Mar Pollut Bull 4:7–9

Peirson DH, Cawse PA, Cambray RS (1974) Chemical uniformity of airborne particulate material and a maritime effect. Nature (London) 251:675–679

Pekkari S (1973) Effects of sewage water on benthic vegetation. Oikos Suppl 15:185–188

Pentreath RJ (1978) ^{237}Pu experiments with the thornback ray *Raja clavata*. Mar Biol 48:337–342

Perez JM (1976) Pollution en zones côtières et au large. Voies d'accès – zones de dilution – dispersion. In: Perez JM (ed) La pollution des eaux marines. Gauthier-Villars, Paris, pp 11–29

Peters N (1981) Fischkrankheiten und Gewässerbelastung im Küstenbereich. Verh Dtsch Zool Ges 74 (in press)

Peterson CL, Klawe WL, Sharp GD (1973) Mercury in tunas: a review. Fish Bull 71:603–614

Portmann JE (1977) International marine pollution control. Mar Pollut Bull 8:126–132

Postma H (1978) The nutrient contents of North Sea water: changes in recent years, particularly in the Southern Bight. Rapp PV Reun Cons Int Explor Mer 172:350–357

Pounder B (1974) Wildfowl and pollution in the Tay Estuary. Mar Pollut Bull 5:35–38

Preston A, Fukai R, Volchok HL, Yamagata N (1971) Radioactivity. In: Report of the Seminar on Methods of Detection, Measurement and Monitoring of Pollutants in the Marine Environment, Rome 4–10 December 1970. FAO Fish Rep 99, Suppl 1:87–99

Priebe K (1976) Erfahrungen über die Quecksilberuntersuchung bei Fischen und Fischerzeugnissen des Handels. Fleischwirtschaft 56:1252–1263

Rachor E (1978) Faunenverarmung in einem Schlickgebiet in der Nähe Helgolands. Helgol Wiss Meeresunters 30:633–651

Rachor E (1980) The inner German Bight – an ecologically sensitive area as indicated by the bottom fauna. Helgol Wiss Meeresunters 33:522–530

Rachor E, Gerlach SA (1978) Changes of macrobenthos in a sublittoral sand area of the German Bight, 1967–1975. Rapp PV Reun Cons Int Explor Mer 172:418–431

Rasmussen E (1973) Systematics and ecology of the Isefjord marine fauna (Denmark). Ophelia 11:1–507

Rasmussen RA, Khalil MAK, Dalluge RW (1981) Atmospheric trace gases in Antarctica. Science 211:285–287

Rat von Sachverständigen für Umweltfragen (1980) Umweltprobleme der Nordsee. Sondergutachten Juni 1980. W Kohlhammer, Stuttgart Mainz, 503 pp

Ratasuk S (1972) A simplified method of predicting dissolved oxygen distribution in partially stratified estuaries. Water Res 6:1525–1532

Ravera O (1978) Evaluation of effects of radioactive contamination of the marine biota. Rev Int Oceanogr Med 49:147–161

Reed AW (1975) Ocean waste disposal practices. Noyes Data Corp, Park Ridge New Jersey London, 336 pp

Regnier AP, Park RWA (1972) Faecal pollution of our beaches – how serious is the situation? Nature (London) 239:408–410

Renberg L, Sundström G, Reutergårdh L (1978) Polychlorinated terphenyls (PCT) in Swedish white-tailed eagles and in gray seals. A preliminary study. Chemosphere 6:477–482

Rice SD, Short JW, Karinen JF (1977) Comparative oil toxicity and comparative animal sensitivity. In: Wolfe DA (ed) Fate and effects of petroleum hydrocarbons in marine ecosystems and organisms, Pergamon Press, Oxford, pp 78–94

Rice TR, Wolfe DA (1971) Radioactivity – chemical and biological aspects. In: Hood DW (ed) Impingement of man on the oceans, Wiley-Interscience, New York, pp 325–379

Risebrough RW (1971) Chlorinated hydrocarbons. In: Hood DW (ed) Impingement of man on the oceans. Wiley-Interscience, New York, pp 259–286

Risebrough RW (1977) Transfer of organochlorine pollutants to Antarctica. In: Llano GA (ed) Adaptations within Antarctic ecosystems. Gulf Publ Co, Houston, pp 1203–1210

Risebrough RW, Lappe BW de, Schmidt TT (1976a) Bioaccumulation factors of chlorinated hydrocarbons between mussels and seawater. Mar Pollut Bull 7:225–228

Risebrough RW, de Lappe BW, Walter II W (1976b) Transfer of higher molecular weight chlorinated hydrocarbons to the marine environment. In: Windom HL, Duce RA (eds) Marine pollutant transfer. Lexington Books, Lexington Toronto, pp 261–321

Roberts TM, Heppelston PB, Roberts RD (1976) Distribution of heavy metals in tissues of the common seal. Mar Pollut Bull 7:194–196

Saha JG (1972) Significance of mercury in the environment. Residue Rev 42:103–163

Sawyer TK, Visvesvara GS, Harke BA (1977) Pathogenic amoebas from brackish and ocean sediments, with a description of *Acanthamoeba hatchetti*, n.sp. Science 196:1324–1325

Sayler GS, Thomas R, Colwell RR (1978) Polychlorinated biphenyl (PCB) degrading bacteria and PCB in estuarine and marine environments. Estuarine Coastal Mar Sci 6:553–567

Schaefer RG, Ernst W, Goerke H, Eder G (1976) Residues of chlorinated hydrocarbons in North Sea animals in relation to biological parameters. Ber Dtsch Wiss Komm Meeresforsch 24:225–233

Schaule B, Patterson CC (1980) The occurrence of lead in the Northeast Pacific and the effects of anthropogenic inputs. In: Branica M, Konrad Z (eds) Lead in the marine environment. Pergamon Press, Oxford, pp 31–44

Schulz-Baldes M (1972) Toxizität und Anreicherung von Blei bei der Miesmuschel *Mytilus edulis* im Laborexperiment. Mar Biol 16:226–229

Schulz-Baldes M (1974) Lead uptake from sea water and food, and lead loss in the common mussel, *Mytilus edulis*. Mar Biol 25:177–193

Schulz-Baldes M (1978) Lead transport in the common mussel *Mytilus edulis*. In: McLusky DS, Berry AJ (eds) Physiology and behaviour of marine organisms. Pergamon Press, Oxford New York, pp 211–218

Schulz-Baldes M, Cheng L (1980) Cadmium in *Halobates micans* from the Central and South Atlantic Ocean. Mar Biol 59:163–168

Schulz-Baldes M, Lewin RA (1976) Lead uptake in two marine phytoplankton organisms. Biol Bull (Woods Hole Mass) 150:118–127
Seibold E (1974) Der Meeresboden. Ergebnisse und Probleme der Meeresgeologie. Springer, Berlin Heidelberg New York, 183 pp
Siedler G, Hatje G (1974) Temperatur, Salzgehalt und Dichte. In: Magaard L, Rheinheimer G (eds) Meereskunde der Ostsee. Springer, Berlin Heidelberg New York, pp 43–60
Silva AJ (1977) Physical processes in deep-sea clays. Oceanus 20:31–40
Sindermann CJ (1978) Pollution-associated diseases and abnormalities of fish and shellfish: a review. Fish Bull 76:717–748
Skei JM (1978) Serious mercury contamination of sediments in a Norwegian semi-enclosed bay. Mar Pollut Bull 9:191–193
Smith JE (1968) Torrey Canyon, pollution and marine life. University Press, Cambridge, 192 pp
Smith JE, Smith AM (1975) Minamata: a warning to the world. Chatto & Windus, London, 192 pp
Södergren A (1972) Chlorinated hydrocarbon residues in airborne fallout. Nature (London) 236: 395–397
Somer E (1978) Die Verschmutzung der Nord- und Ostsee: Natürlich bedingt oder vom Menschen verschuldet? Umschau 78:267–270
Southward AJ, Southward EC (1978) Recolonization of rocky shores in Cornwall after use of toxic dispersants to clean up the Torrey Canyon Spill. J Fish Res Board Can 35:682–706
Spangler WJ, Spigarelli JL, Rose JM, Miller HM (1973) Methylmercury: bacterial degradation in lake sediments. Science 180:192–193
Spitzer PR, Risebrough RW, Walter II W, Hernandez R, Poole A, Pulestone D, Nisbet ICT (1978) Productivity of ospreys in Connecticut – Long Island increases as DDE residues decline. Science 202:333–335
Spooner MF (1978) Amoco Cadiz oil spill: Editorial introduction. Mar Pollut Bull 9:281–284
Stebbing ARD (1976) The effects of low metal levels on a colonial hydroid. J Mar Biol Assoc UK 56:977–994
Steimle FW, Sindermann CJ (1978) Review of oxygen depletion and associated mass mortalities of shellfish in the Middle Atlantic Bight in 1976. Mar Fish Rev 40:17–26
Stenersen J, Kvalvag J (1972) Residues of DDT and its degradation products in cod liver from two Norwegian fjords. Bull Environ Contamin Toxicol 8:120–121
Stich HF, Acton AB, Dunn BP, Oishi K, Yamazaki F, Harada T, Peters G, Peters N (1977) Geographic variations in tumor prevalescence among marine fish populations. Int J Cancer 20: 780–791
Stockner JG, Anita NJ (1976) Phytoplankton adaptation to environmental stresses from toxicants, nutrients, and pollutants – a warning. J Fish Res Board Can 33:2089–2096
Stout VF, Beezhold FL, Houle CR (1972) DDT residue levels in some US fishery products and the effectiveness of some treatments in reducing them. In: Ruivo M (ed) Marine pollution and sea life. Fishing News, London, pp 550–553
Straughan D (1976) Sublethal effects of natural chronic exposure to petroleum in the marine environment. Am Petrol Inst Publ 4280:1–119
Stripp K, Gerlach S (1969) Die Bodenfauna im Verklappungsgebiet von Industrieabwässern nordwestlich von Helgoland. Veröff Inst Meeresforsch Bremerhaven 12:149–156
Stukas VJ, Wong CS (1981) Stable lead isotopes as a tracer in coastal waters. Science 211:1424–1427
Su C, Goldberg ED (1976) Environmental concentrations and fluxes of some halocarbons, In: Windom HL, Duce RA (eds) Marine pollutant transfer. Lexington Books, Lexington Toronto, pp 353–374
Sumari P, Partanen T, Hietala S, Heinonen OP (1972) Blood and hair mercury content in fish consumers. A preliminary report. Work Environ Health 9:61–65
Takahashi FT, Kittredge JS (1973) Sublethal effects of the water soluble component of oil: chemical communication in the marine environment. In: Ahearn DG, Meyers SP (eds) Microbial degradation of oil pollutants. Louisiana State Univ, Baton Rouge, pp 259–264
Taruski AG, Olney CE, Winn HE (1975) Chlorinated hydrocarbons in cetaceans. J Fish Res Board Can 32:2205–2209

Teichmann H (1959) Über die Leistung des Geruchssinnes beim Aal (*Anguilla anguilla* L.). Vergl Physiol 42:206–254

Ten Berge WF, Hillebrand H (1974) Organochlorine compounds in several marine organisms from the North Sea and the Dutch Waddensea. Neth J Sea Res 8:361–368

Theede H, Andersson I, Lehnberg W (1979a) Cadmium in *Mytilus edulis* from German coastal waters. Meeresforschung 27:147–155

Theede H, Scholz N, Fischer H (1979b) Temperature and salinity effects on the acute toxicity of cadmium to *Laomedea loveni* (Hydrozoa). Mar Ecol Prog Ser 1:13–19

Thompson S, Eglington G (1978) Composition and sources of pollutant hydrocarbons in the Severn Estuary. Mar Pollut Bull 9:133–136

Thormann D (1975) Über die Wirkung von Cadmium und Blei auf die natürliche heterotrophe Bakterienflora im Brackwasser des Weser-Ästuars. Veröff Inst Meeresforsch Bremerhaven 15: 237–267

Tokuomi H (1969) Medical aspects of Minamata disease. Rev Int Oceanogr Med 13:5–35

Tomita J, Nakamura Y, Yagi Y (1977) Phthalic acid esters in various food stuffs and biological materials. Exotoxicol Environ Safety 1:275–287

Topping G, Davies IM (1980) Loss of mercury from sea water. Mar Pollut Bull 11:83–84

Traxler RW, Bhattacharya LS (1978) Effect of a chemical dispersant on microbial utilization of petroleum hydrocarbons. In: McCarthy LT, Lindblom GP, Walter HF (eds) Chemical dispersants for the control of oil spills. ASTM STP 659, American Society for Testing and Materials, pp 181–187

Trites RW (1972) Gulf of St. Lawrence from a pollution viewpoint. In: Ruivo M (ed) Marine pollution and sea life. Fishing News, London, pp 59–72

Tsytsugina VG, Risik NS, Lazorenko GE (1973) Artificial and natural radionuclides in marine life. Naukova Dumka, Kiev (translated by Israel Program for Scientific Translations, Jerusalem 1975, 115 pp)

Ui J (1969) Minamata disease and water pollution by industrial waste. Rev Int Oceanogr Med 13: 37–44

Ui J (1971) Mercury pollution of sea and fresh water, its accumulation into water biomass. Rev Int Oceanogr Med 22/23:79–128

Underdal B (1971) Mercury in fish and water from a river and a fjord in the Kragerø region, South Norway. Oikos 22:101–105

Vaccaro RF, Grice CD, Rowe GT, Wiebe PH (1972) Acid-iron waste and the summer distribution of standing crops in the New York Bight. Water Res 6:231–256

Vandermeulen JH (1977) The Chedabucto Bay spill – Arrow, 1970. Oceanus 20 (4):31–39

Vandermeulen JH, Gordon DC (1976) Re-entry of 5 year old stranded bunker C fuel oil from a low energy beach into the water, sediments, and biota of Checlabucto Bay, Nova Scotia. J Fish Res Board Can 33:2202–2210

Vandermeulen JH, Ross CW (1977) Assessment of cleanup tests of an oiled salt marsh – the Golden Robin spill in Miguasha, Quebec. I. Residual Bunker C hydrocarbon concentrations and compositions. Fish Environ Can: Environ Impact Assess Rep EPS-8-EC-77-1, pp 1–31

Van Gelder-Ottway S, Knight M (1976) A review of world oil spillages 1960–1975. In: Baker JM (ed) Marine ecology and oil pollution. Appl Sci Publ, Barking, pp 483–520

Vauk G, Lohse H (1978) Biocid-Belastung von Seevögeln sowie einiger Landvögel und Säuger der Insel Helgoland. Veröff Übersee-Mus Bremen (E) 1:3–27

Venrick EL, Backman TW, Bartram WC, Platt CJ, Thornhill MS, Yates RE (1973) Man-made objects on the surface of the Central North Pacific Ocean. Nature (London) 241:271

Vermeer K, Peakall DB (1977) Toxic chemicals in Canadian fish-eating birds. Mar Pollut Bull 8: 205–210

Vosjan HJ, van der Hoek GJ (1972) A continuous culture of *Desulfovibrio* on a medium containing mercury and copper ions. Neth J Sea Res 5:440–444

Wachenfeldt T von (1971) Alg- och fytoplankton-undersögningar i Öresund, ØKV 1965–1970, Report on the investigations of the Swedish-Danish Committee on Pollution of the Sound 1965–1970, Lund, pp 139–172

Wachs B (1972) Größe und Abbau der organischen Substanz im Brack- und Meerwasser. Münchener Beitr Abwasser Fisch Flußbiol 22:38–58

Waldichuk M (1978) Global marine pollution: an overview. UNESCO, Intergovernm Oceanogr Comm Techn Ser 18:1–96

Walker TI (1976) Effects of species, sex, length and locality on the mercury content of School Shark Galeorhinus australis(MacLeay)and Gummy Shark Mustelus antarcticus Guenther from South-East Australian waters. Aust J Mar Freshwater Res 27:603–616

Walz F (1979) Uptake and elimination of antimony in the mussel, *Mytilus edulis*. Veröff Inst Meeresforsch Bremerhaven 18:203–215

Warren Spring Laboratory (1972) Oil pollution of the sea and shore. A study of remedial measures. Her Majesty's Stationary Office, London, 33 pp

Weber K, Ernst W (1978) Occurrence of brominated phenols in the marine polychaete *Lanice conchilega*. Naturwissenschaften 65:262

Weichart G (1972) Neuere Entwicklungen in der Meereschemie. Naturwissenschaften 59:16–19

Weichart G (1975a) Untersuchungen über die Fe-Konzentration im Wasser der Deutschen Bucht im Zusammenhang mit dem Einbringen von Abwässern aus der Titandioxid-Produktion. Dtsch Hydrogr Z 28:49–61

Weichart G (1975b) Untersuchungen über den pH-Wert im Wasser der Deutschen Bucht im Zusammenhang mit dem Einbringen von Abwässern aus der Titandioxid-Produktion. Dtsch Hydrogr Z 28:243–252

Weish P, Gruber E (1979) Radioaktivität und Umwelt, 2. edn. Fischer, Stuttgart, 188 pp

West RH, Hatcher PG (1980) Polychlorinated biphenyls in sewage sludge and sediments of the New York Bight. Mar Pollut Bull 11:126–129

Westernhagen H v, Rosenthal H, Dethlefsen V, Ernst E, Harms U, Hansen PD (1981) Bioaccumulating substances and reproductive success in Baltic flounder *Platichthys flesus*. Aquatic Toxicol (in press)

WHO (1976a) Environmental health criteria 1: Mercury. World Health Organization, Geneva, 131 pp

WHO (1976b) Environmental health criteria 2: Polychlorinated biphenyls and terphenyls. World Health Organization, Geneva, 85 pp

Williston SII (1968) Mercury in the atmosphere. J Geophys Res 73:7051–7055

Windom H, Taylor F, Stickney R (1973) Mercury in North Atlantic plankton. J Cons Int Explor Mer 35:18–21

Wolf P de (1975) Mercury content of mussels from West European coasts. Mar Pollut Bull 6: 61–63

Wong PTS, Chau YK, Luxon PL (1975) Methylation of lead in the environment. Nature (London) 253:263–264

Wood PC (1972) The principles and methods employed for the sanitary control of molluscan shellfish. In: Ruivo M (ed) Marine pollution and sea life. Fishing News, London, pp 560–565

Woodhead DS (1973) Levels of radioactivity in the marine environment and the dose commitment to marine organisms. In: Radioactive contamination of the marine environment. Proc Symp Seattle 10–14 July 1972,. Int At Energy Agency, Vienna, pp 499–525

Woodwell GM, Whittaker RH, Reiners WA, Likens GE, Delwiche CC, Botkin DB (1978) The biota and the world carbon budget. Science 199:141–146

Young DR, Kohnson JN, Soutar A, Isaacs JD (1973) Mercury concentrations in dated varved marine sediments collected off Southern California. Nature (London) 244:273–274

Young DR, Dermott-Ehrlich D, Heesen TC (1977) Sediments as sources of DDT and PCB. Mar Pollut Bull 8:254–257

Young DR, Alexander GV, McDermott-Ehrlich D (1979) Vessel-related contamination of Southern California harbours by copper and other metals. Mar Pollut Bull 10:50–56

Zell M, Ballschmitter K (1980) Baseline studies of the global pollution. II. Global occurrence of hexachlorobenzene and polychlorocamphenes (Toxaphene) (PCC) in biological samples. Fresenius Z Anal Chem 300:387–402

Zieman JC, Ferguson-Wood EJ (1975) Effects of thermal pollution on tropical-type estuaries, with emphasis on Biscayne Bay, Florida. In: Ferguson-Wood EJ, Johannes RE (eds) Tropical marine pollution. Elsevier, Amsterdam, pp 75–98

Zitko V (1976) Levels of chlorinated hydrocarbons in eggs of double-crested cormorants from 1971 to 1975. Bull Environ Contamin Toxicol 16:399–405

Zobell CE (1973) Microbial degradation of oil: present status, problems, and perspectives. In: Ahearn DG, Meyers SP (eds) The microbial degradation of oil pollutants. Louisiana State Univ, Baton Rouge, pp 3–16

Zoller WH, Gladney ES, Duce RA (1974) Atmospheric concentrations and sources of trace metals at the South Pole. Science 183:198–200

Subject Index

2,4,5 T 168

Acanthocardium see cockle
Acartia 125
accumulation 126–136, 194
Acipenser see sturgeon
air see atmosphere
air dust 139–140, 142
air-sea interface 139–141, 160
Alca see razorbill
aldrin 128, 167
aliphatic chlorinated hydrocarbons 53, 164
alkanes 71, 75
ammunition 56
Amoco Cadiz 85–87, 93–97, 101, 103
anaerobic bacteria 7
– conditions 64, 66
– sediment 9, 154–155, 161–162
Anarhichas see catfish
Anguilla see eel
anthracene 71–72, 75, 90
antifouling paint 68, 174, 193
antimony (Sb) 130, 133, 136, 140–142, 157
Arenicola see lugworm
Argo Merchant 89
aromatics 71–72
Arrow 78, 93, 95
arsenic (As) 53, 57, 68, 136, 149, 157, 185, 186, 187
asphaltenes 76
atmosphere 141, 153, 177, 183, 185, 190, 195
atomic see nuclear

background radioactivity see natural radioactivity
bacteria degrading oil 74–76
bacterial decomposition 6–7
bactericidal effect 14
Balanus see barnacle
Baltic Sea 10, 17, 20–23, 25, 156, 159, 166, 170, 176–181, 185–186, 190, 192
Barcelona Convention 186
barnacle 89, 94, 95
bathing waters 14, 16

Benthoctopus see *octopus*
benzene 71–72, 75, 81, 88, 89
benzenehexachloride see HCB
benzo(a)pyrene 75, 79, 80, 89
beryllium (Be) 157, 187
BHC see HCB
biochemical oxygen demand see BOD
bioconcentration see accumulation
biodegradable organic substance 6–8
biogenic hydrocarbons 78–79
biomagnification see accumulation
BOD 6–7, 68
bonito 156
Bonn Convention 189
brominated phenols 165
bromine (Br) 135
Brussels Convention 189
Bucchinum see whelk

Cadmium (Cd) 158–160
– in air dust 142
– in fish 149
–, measures against cadmium pollution 185, 186, 195
– in molluscs 159–160
– in plankton 139, 140
– in sea-birds and mammals 46, 136, 142, 159, 160
– in sea-skater 142–143, 160
– in seawater 139, 141, 157
– in sediment 138, 157
–, sources 63, 66, 69, 157, 160
–, toxicity and accumulation 122, 130, 131, 134, 135
caesium-137 (^{137}Cs) 104, 106, 108, 110, 111, 113–116, 118, 137
Calidris see knot
California 48–51, 69, 154–155, 161–162
Callorhinus see seal
Campanularia see hydroid polyps
cancerogenic compounds 79, 89, 91, 186
carbon-14 (^{14}C) 104, 108, 110, 117
Carcinus 125
Cardium see cockle

213

catfish 148
cellulose plants 9, 123
Cephalopoda 159, 160
Chaetoceros see plankton algae
chelating effect 158
chemical orientation 90, 144
– oxygen demand see COD
Chlamydomonas see plankton algae
chlordan 167, 183
Chlorella see plankton algae
chlorinated dibenzofuranes 168
– hydrocarbons 53, 56, 126, 131, 136, 141, 164–183, 185, 186, 190, 192, 195
chlorine (Cl) 169
chlorine-alkali-plants 40–41
chlorofluoro hydrocarbons 164, 165, 178
chloroform 165
chromium (Cr) 54, 55, 58, 59, 62, 66, 69, 108, 109, 116, 140, 157, 185, 187
Ci see Curie
Clangula 92
Clupea see herring
coalfish 148, 165, 170, 171
cobalt (Co) 110, 116, 126, 130, 157
cockle 94, 129, 172
COD 7, 68
cod 40, 52, 89, 148, 165, 166, 170–173, 179, 183
coli see *Escherichia coli*
concentration factor 126–128, 130, 177
Conger 165
Congeria 125
conservation parks see nature reserves
cooling water 29–33, 165
copper (Cu) 46, 54, 66, 69, 121, 122, 126, 131, 135, 140–142, 149, 157–160, 185, 186, 187
coral reefs 35
cormorant 180, 182, 183
Crangon see shrimp
Crassostrea see oyster
CRISTAL 189
Cristigera 125
Curie 104
cuttlefish 159
cyanides 53
cycloalkanes 71, 72
Cyclotella see plankton algae

dab 59, 62, 116, 172
Daphnia 132
DBP see phthalate esters
DDD 48, 49, 51, 128, 129, 168, 176
DDE 48, 49, 51, 128, 129, 168, 176
DDT 48–52, 166–168

– accumulation 126–129, 177
– in air 183
–, degradation and metabolism 176
– in fish 49–52, 170–173, 179, 183
–, measures against DDT pollution 49, 167 185, 192
–, permissible level in sea-food 169–170, 172
– in sea-birds and mammals 46, 48, 49, 143, 173, 174, 178–183
– in seawater 130, 142, 175, 176, 182
– in sediment 51
–, sources 48, 49, 52, 68, 166–167, 176
–, toxicity 48–49, 179–181, 183
– in various organisms 129, 130, 173, 175, 182
definition of pollution 4
degradation of oil 71–76
DEHP see phthalate esters
detergents 20, 28–29, 192
dibenzofuranes see chlorinated dibenzofuranes
dichlorethane 53, 164
dieldrin 48, 52, 128, 129, 167, 168, 173, 174, 176, 178, 180, 182, 183
diesel oil 72, 89
dioxine see TCDD
diphenylether 166
dispersants 74–76, 84, 86, 97–100
dog fish see shark
domestic effluents 6, 8, 10, 16, 27–28, 33, 79–81, 186, 189, 191, 192, 193
dpm 104
dredge spill 64
dumping 53–54, 66, 187–188, 193

Echinocardium see heart urchin
EDC see dichlorethane
EDC-tar 53, 164, 167
eel 144, 169
eel-grass 36
egg-shell thinning 179–181
eider duck 92, 151
Eirene see hydroid polyps
Ekofisk 73–74, 85–86
Elbe river and estuary 17, 20, 29, 45, 148, 152
elements in seawater 5
elimination 126, 128, 135
emulsification of oil 74, 98
endosulfan 128, 129
endrin 48, 128, 167, 173, 176
Ensis see razor shell
Enteromorpha see sea-weeds
erosion 136, 137
Escherichia coli 14–16
essential trace elements 134–135

estuaries 8, 10, 27, 34, 139, 140
evaporation of oil 73, 75, 88
eutrophication 16–20, 25, 27, 67, 186, 191, 192

fall-out 104–107, 110, 117–118
fecal pellets 140
fish diseases 59–60
flat fish see Pleuronectidae
flounder 183
fluorine hydrocarbons 164, 165
Fratercula see puffin
freon 164
frigene 164
Fucus see sea-weeds
fulmar 151, 174
Fulmaris see fulmar
fur seal see seal

Gadus see cod
Galeorhinus see sharks
Gammarus 90
gannet 173, 174
garbage 69–70, 186, 187, 189, 193
Gavia 174
Glyptocephalus see dab
goosander 151
green cod see coalfish
grey seal se seal
guillemot 92, 151, 155, 156, 174, 176
gull 173, 174

haddock 148, 171
half-life time 106, 128
halibut 150, 171
Halichoerus see seal
Halobates see sea-skater
halogenated hydrocarbons see chlorinated hydrocarbons
– methans 165
harbor seal see seal
HCB 128, 167, 172–174, 183
HCH (α; β; γ-HCH = lindan) 68, 128, 129, 167, 168, 172, 173, 174, 176, 178, 182, 183
heart urchin 94
heavy metals 42, 54, 120, 124, 128, 130, 131, 133, 135–137, 141, 146–163, 187, 193
Helsinki Convention 69, 98, 185–186
heptachlor 167
heptachlorepoxide 128
herbicides 52, 168
herders 99
herring 102, 148, 170–173, 176
hexachlorobenzene see HCB
hexachlorobutadiene 164, 178

hexachlorocyclohexane see HCH
Hippoglossus see halibut
Hippoglossoides see dab
Homarus see lobster
hydrocarbons see oil
hydroid polyps 61, 122

ice 154, 161, 175
IMCO Convention 91, 188–189
incineration 56, 168, 187, 190
industrial effluents 37–52
infections 13
iodine-131 (^{131}J) 108, 110, 116
Irish Sea 109, 116, 148, 152, 159, 165
iron (Fe) 58, 62–63
Isochrysis see plankton algae
Ixtoc 87

Katsuwonus see bonito
Kelp 165
kittiwake 151, 174
knot 47
krill 140
krypton-85 (^{85}Kr) 108, 110

La Hague 109–111
Lamna see sharks
Lanice 126, 127, 165
Laomedea see hydroid polyps
Larus see gull
laverbread see *Porphyra*
laws against pollution 184–190, 195
lead (Pb) 160–163
–, measures against lead.pollution 163, 185, 186, 187, 195
–, metalorganic compounds 120, 141
– in organisms 46, 140, 157
–, permissible amounts in sea-food 160–161
– in seawater, ice, air dust 141–142, 157, 161
– in sediments 54, 138, 140
–, sources 63, 66, 69, 157, 161
–, toxicity and accumulation 121, 123, 131–135, 158, 162
Lepas 77
lignin 9
Limanda see dab
limpet 94, 95, 100, 159
lindan see HCH
Littorina see periwinkle
load-on-top system 188
lobster 79, 125
Loligo see squid
London Convention 186–187
lugworm 94, 129

mackerel 148, 170
macrofauna 9, 10, 22, 35–36, 58, 59, 61, 62, 66, 67, 193
Makaira see marlin
manganese (Mn) 126, 157
mangrove 52, 94, 168
marlin 149
Melanogrammus see haddock
Mercenaria 125
mercury (Hg) 37–47, 146–158
– accumulation 130, 132, 134–136
– accumulation in man and health risk 37–40, 146, 147
– in effluents 37–40, 42, 66
– in fish 39–40, 45, 135, 148–150, 155, 156
– in ice 154
–, maximum permissible amounts in sea-food 147–150
–, measures against mercury pollution 41–42, 185–186, 192, 194
–, metalorganic compounds 120, 124
– in molluscs 39–40, 43
– in sea-birds and mammals 47–57, 135–136, 151, 155, 156
– in seawater 45, 139, 151–152, 156, 157
– in sediments 39–41, 138, 154, 155, 157
–, sources 68, 152–154
–, toxicity 121, 124, 158
– in zooplankton 44, 140
mercuryelenide 135
Mergus see goosander
metallothionin 136, 142
metalorganic compounds 120, 124, 136, 141, 142, 161, 162, 185
methoxychlor 128, 167
methylchloride 165
methyllead 120
methylmercury 38–39, 120, 135, 146
methylnaphthalene 75, 88
Minamata 37–39, 146–147
mink 183
mirex 167
Modiolus see mussel
Monte Urquiola 95
mother's milk 172–174
Munida 172
mussel 28, 43, 69, 95, 116, 127, 128, 130, 132–134, 159, 162, 165, 177
musselwatch 43, 130–131
mustard gas 56
Mustela see mink
Mya 93, 127, 129
Myctophidae 49
Mytilus see mussel

naphthalene 71–72, 75, 88–90
naphthenes see cycloalkanes
natural radioactivity 104, 108, 116–118
nature reserves 35, 192
Nereis 122, 126, 127, 131, 176
New York Bight 54, 59, 64
nickel (Ni) 58, 66, 157, 185, 187
Niigata 38
nitrogen, nitrate (N, NO$_3$) 17–18, 139, 141, 192
Nitzschia see plankton algae
North Sea 17–20, 46–47, 52, 56–62, 66, 84, 91, 92, 95, 102–103, 107, 110, 111, 118, 129, 132, 137, 138, 148–149, 151–152, 170–174, 180, 181, 189
nuclear accidents 109
– bombs 105, 106, 109, 194
– vessels 110, 111
nutrients 17–19, 191

ocean regions 34–35
octopus 159, 172
oil 71–103, 185, 186, 188–190, 193, 194
– accidents 82, 85, 87, 89, 92–98, 131, 141, 186, 189, 193–194
– pollution 71–103
– rig accidents see oil accidents
– seeps, natural 80–81, 87
Oil Spill Committee 189–190
OPOL 189
organic substances 6–8
organophosphorus compounds 185
organosilicium compounds 185, 186
Oslo Convention 186–187
osprey 180
oyster 13, 125, 159, 183
oxygen (O) 7–8, 12, 22–23, 25, 28, 32–33, 191, 192

Pachygrapsus 90
Pandalus see shrimp
Pandion see osprey
paper mills 8, 40, 53
paraffines see alkanes
parathion 47
Paris Convention 185, 187
Patella see limpet
pathogens 13, 191
PCB 166, 168
– accumulation 126, 128, 129, 177–178, 182
– in air 183
– degradation 176
– in fish 171, 172, 173, 179, 183
–, measures against PCB pollution 166, 185
–, permissible amounts in sea-food 170–172

- in sea-birds and mammals 46, 49, 142, 173 174, 175, 178, 180
- in seawater 130, 142, 177, 182
- in sediment 136–137
- , sources 52, 68, 166, 175, 177
- , toxicity 180, 181, 183
- in various organisms 52, 129, 130, 172, 173, 175

PCC see polychlorocamphene
PCP 52, 128, 178, 182
PCT 68, 166
Pecten see scallop
Pelecanus see pelican
pelican 48, 179, 180, 182
penguin 175
pentachlorobenzene 173
pentachlorophenol see PCP
perchlorethylene 164, 165, 178, 182
periwinkle 94–95
persistance 136
pesticide plants 47, 48
petrel 142, 160
petroleum see oil
Petrolisthes 125
Ph$_2$ see diphenylether
Phaeodactylum see plankton algae
Phalacrocorax see cormorant
phenanthrene 72, 88, 89
pheromones 90, 144
Phoca see seal
phosphorus, phosphate (P, PO$_4$) 17, 19–20, 24–27, 139, 141, 185, 191, 192
phthalate esters 169, 178, 182, 183
Phusa see seal
phytane 79
phytoplankton see plankton algae
plaice 107, 116, 171, 172, 173
plankton algae 121, 123, 124, 125, 127, 131, 132, 158, 162, 173, 183
plastic material 185–186
Platichthys see flounder
Pleuronectidae 148, 170
plutonium (Pu) 104, 106–110, 112, 114, 116
PNAH see polycyclic aromatic hydrocarbons
polar bear 173
Pollachius see coalfish
polychlorinated biphenyls see PCB
polychlorinated terphenyls see PCT
polychlorocamphene see toxaphene
polycyclic aromatic hydrocarbons 72, 79, 90, 91
polynuclear aromatic hydrocarbons see polycyclic aromatic hydrocarbons
polyvinylchloride see PVC
population equivalent unit 6

Porphyra 116, 119, 120
potassium-40 (^{40}K) 104, 108, 117, 118
power plants 29–33
ppb, ppm, ppt 4
pristane 72, 79
puffin 92
PVC 53
pyrene 71–72

rad 117
radioactive elements 131, 136
- wastes 104, 112–115
radioactivity 104–119, 186, 190, 194
razorbill 174
razor shell 94
reactors 112
redfish 148, 171
red tide 36, 191
Reinhardtius see halibut
rem 117, 118
reprocessing plants 109–111, 118, 194
residual heat 29–33, 192
Rhine 42, 148, 152, 180
Rissa see kittiwake
rubidium-87 (^{87}Rb) 104, 108
ruthenium-106 (^{106}Ru) 108, 110, 116, 119

Salmo see salmon
salmon 169
salt marsh 94, 95
sardines 170
scallop 159, 165, 172
Scylliorhinus see shark
Scenedesmus see plankton algae
Scomber see mackerell
Scophthalmus 172
sea-birds 47–49, 91–93, 142, 151, 155, 173–175, 179, 181, 182
sea-burial 187
sea-food, permissible levels and concentrations of cadmium 158–159
-, - DDT 169–170, 172
-, - mercury 146–148, 149, 151
-, - PCB 170–172
seal 15, 45–47, 135–136, 151, 158, 161, 166, 173, 174, 176, 178, 180, 181
sea-lion see seal 45, 49
sea-skater 142–143, 160
seawater composition 5, 45
sea-weeds 29, 33–34, 93, 95, 116, 119–120
Sebastes see redfish
sedimentation 136, 138, 155, 194
selenium (Se) 128, 130, 135, 136, 140–142, 149, 157
seston 77

217

sewage see domestic effluents
— sludge 28, 54, 64–65, 68, 193
shag see cormorant
sharks 149–150
shellfish culture 13
shrimp 100, 125, 129, 171, 172, 173
silver (Ag) 66, 131, 142, 157
Skeletonema see plankton algae
skua 151
sole 126–129, 171, 176
Solea see sole
Somateria see eider duck
Somniosus see sharks
sorbents 99
Spartina 93
squid 159, 160, 172
Stercorarius see skua
Sterna see tern
strontium-90 (^{90}Sr) 104, 106–108, 110, 112–118
sturgeon 169
sulfuric acid 57, 63
Sula see gannet
swimming waters 14, 16
swordfish 135, 149, 151
Symplectotheutis see squid

T see 2,4,5 T
taint 72, 90, 185
tanker accidents 80, 84–87
tar 73, 77, 78, 143, 187
TAVALOP 189
TCDD 168
telodrin 48
TDE see DDD
temperature 29–33
tensides 28
tern 48, 179
tetrachlorethylene 128
tetrachlorophenol 178
tetrachlorodibenzodioxine see TCDD
tetraethyllead 141, 161, 162
Thalassiosira see plankton algae
Thames river and estuary 9, 12, 66, 148, 152
tin (Sn) 69, 185
titanium dioxide pigment industry wastes 57, 62, 63
toluene 72, 75, 81, 88, 89
Torrey Canyon 85, 92, 95, 98–99
tope see shark 27–28
toxaphene 167

toxicity 120–125, 194
— of oil 87, 96
— of radioactivity 116–117
trace elements see heavy metals
transuranic elements 106
treatment plants 9–10, 12–13, 15–17, 20, 191
trichlorethane 53, 164, 165, 178
trichlorethylene 164, 165, 178, 182
trichlorobiphenylol 176
trichlorophenol 178
trichlorophenoxyacetic acid see 2,4,5 T
trimethylbenzene 88–89
Trigla 172
tritium (^{3}H) 104–105, 108, 110, 112–113, 116
tuna 149, 150, 155, 156
turbidity cloud 8, 11, 139

Uca 125
Ulva see sea-weeds
units of measurement 4
Uria see guillemot
Ursus see polar bear

vanadium (V) 58, 63, 126, 157, 187
vinylchloride 39, 56, 164
volcano 153, 158
vulnerability index 93–94

Wadden-Sea 35, 103, 138, 159, 181
washing powder 20, 191
waste acids 53, 57, 62
— heat see residual heat
weapons 186
Weser river and estuary 8, 10–11, 15–17, 20, 29, 45, 52, 129, 132, 148, 152, 165
whales 135–136, 174
whelk 165
Windscale 109, 110, 111, 116, 118–119
wolffish see catfish

Xiphias see swordfish
xylene 72

Zalophus see seal
zeolites 20, 191
zinc (Zn) 46, 54–55, 66, 69, 108, 109, 116, 122, 126, 130, 135, 136, 138, 140, 141, 149, 157, 185, 186, 187
zooplankton 44, 173, 175

Now in its 2nd edition – with coverage of all important developments since 1978

U. Förstner, G. T. W. Wittmann

Metal Pollution in the Aquatic Environment

With contributions by F. Prosi, J. H. van Lierde
Foreword by E. D. Goldberg
2nd revised edition. 1981. 102 figures, 94 tables.
XVIII, 486 pages
ISBN 3-540-10724-X

Contents: Introduction. – Toxic Metals. – Metal Concentrations in River, Lake, and Ocean Waters. – Metal Pollution Assessment from Sediment Analysis. – Metal Transfer Between Solid and Aqueous Phases. – Heavy Metals in Aquatic Organisms. – Trace Metals in Water Purification Processes. – Concluding Remarks. – Appendix. – References. – Subject Index.

From the reviews of the first edition:
"...the authors present what is probably the first successful compilation on a worldwide basis, including a critical inventory and evaluation of the prevailing investigations in water, sediments and organisms. ... **Toxic Metals,** by **W. Wittmann,** may be considered as an introductory review which will be gratefully received, both by students and others. **Metal Concentration in River, Lake and Ocean Waters,** by **U. Förstner,** contains a critical evaluation of the recent literature. ...**Metal Pollution Assessment from Sediment Analysis; Metal Transfer between Solid and Aqueous Phase,** by **U. Förstner,** these chapters are the heart of the book, containing an excellent presentation of the actual concepts for assessing metal pollution from sediment analysis. ...In **Heavy Metals in Aquatic Organisms, F. Prosi** discusses the metal enrichment in limnic and marine organisms, and provides a comprehensive reference list in the appendix. ...**Trace Metals in Water Purification Processes,** by **U. Förstner** and **J. H. van Lierde,** of some 30 pages, deals with the most important processes with respect to heavy metal removal in drinking and sewage waters. ...it is an especially useful addition in times when legislation requires more and better treatment of industrial and domestic effluents." *Marine Chemistry*

Springer-Verlag
Berlin
Heidelberg
New York

D. A. Ross

Opportunities and Uses of the Ocean

1980. 144 figures, 48 tables. XI, 320 pages
ISBN 3-540-90448-4

Contents: The Opportunity of the Ocean. – How the Ocean Works. – The Legal Aspects of the Ocean. – Marine Shipping. – The Resources of the Ocean. – Marine Pollution. – Military Uses of the Ocean. – The Coastal Zone. – Innovative Uses of the Ocean. – Index.

This book explores the opportunities that exist for present day and future utilization of the oceans. After a concise introduction to the field of oceanography, the author discusses the legal aspects of the oceans, marine shipping, ocean resources and pollution, military uses and many more topics of crucial concern. A key aim of the author is to focus on those problems that can be solved with available scientific and technical capabilities.

This book will be fascinating and invaluable to a broad range of readers, including all those concerned with our oceans and wise marine policy.

Springer-Verlag
Berlin
Heidelberg
New York